Success Stories in Environmental Science

The Unbelievable Journeys and Shocking Truths That Are Leading the Fight for a Greener Future

Charlotte Fruittree

ISBN: 978-1-77961-269-4
Imprint: Telephasic Workshop
Copyright © 2024 Charlotte Fruittree.
All Rights Reserved.

Contents

Historical Perspectives on Environmental Science 20

Theoretical Frameworks in Environmental Science 30

Methods and Techniques in Environmental Science 39

1.5.3 Remote Sensing and GIS Applications in Environmental Science 46

Challenges and Opportunities in Environmental Science 48

1.6.5 The Importance of Biodiversity 59

Chapter 2: Restoring Ecosystems 101

Chapter 2: Restoring Ecosystems 101

Ecosystem Restoration: Principles and Concepts 106

Case Studies in Ecosystem Restoration 116

Success Stories in Ecosystem Restoration 133

Challenges and Future Directions in Ecosystem Restoration 141

Chapter 3: Conservation Biology and Wildlife Management 157

Chapter 3: Conservation Biology and Wildlife Management 157

Foundations of Conservation Biology 161

Wildlife Management Techniques 171

Case Studies in Conservation Biology 181

Success Stories in Conservation Biology 197

Challenges and Future Directions in Conservation Biology 216

Chapter 4: Sustainable Agriculture and Food Systems 235

Chapter 4: Sustainable Agriculture and Food Systems 235

Principles of Sustainable Agriculture 239

Innovations in Agricultural Technologies 251

Case Studies in Sustainable Agriculture 263

Success Stories in Sustainable Agriculture 278
Challenges and Future Directions in Sustainable Agriculture 293

Chapter 5: Renewable Energy and Green Technologies 313

Chapter 5: Renewable Energy and Green Technologies 313
The Need for Renewable Energy 318
Fossil Fuels and Climate Change 318
Renewable Energy Technologies 328
Case Studies in Renewable Energy 346
Success Stories in Renewable Energy 362
Challenges and Future Directions in Renewable Energy 379

Index 399

CONTENTS

What is Environmental Science?

Environmental science is a multidisciplinary field that examines the interactions between humans and the environment. It encompasses a wide range of topics, including the study of ecosystems, natural resources, pollution, and climate change. By understanding the complex relationships between society and the environment, environmental scientists work towards finding sustainable solutions to environmental problems.

What is Environmental Science?

Environmental science is the study of the Earth's natural processes and how they are impacted by human activities. It involves collecting data, analyzing patterns, and making predictions to better understand and protect the environment. The field incorporates elements of biology, chemistry, physics, geology, and social sciences to tackle the complex challenges facing our planet.

Definition and Scope of Environmental Science

The scope of environmental science is vast, as it covers everything from the smallest microorganisms to global climate patterns. At its core, environmental science seeks to understand the intricate relationships between different components of the environment, such as air, water, soil, plants, and animals. It also investigates the influence of human activities on these components and the subsequent impact on biodiversity, ecosystems, and human well-being.

Interdisciplinary Nature of Environmental Science

One of the defining features of environmental science is its interdisciplinary nature. It draws upon various scientific disciplines to obtain a comprehensive understanding of environmental issues. For example, ecologists study the interactions between organisms and their environment, while chemists analyze the composition of air and water. Geologists examine the Earth's geological processes and history, and social scientists explore how human behavior and policies affect the environment.

This interdisciplinary approach is crucial because environmental problems are complex and require a holistic perspective that considers the interplay between different factors. By integrating knowledge from multiple fields, environmental scientists can develop effective strategies for environmental management and conservation.

Importance of Environmental Science in Today's World

Environmental science plays a vital role in addressing the numerous environmental challenges we face today. It provides the scientific basis for understanding the causes and consequences of pollution, climate change, habitat destruction, and resource depletion. By studying these problems and their impacts, environmental scientists can inform policies and develop sustainable solutions that mitigate the negative effects on ecosystems and human health.

Furthermore, environmental science promotes environmental literacy and awareness among the general public. It helps individuals understand the importance of environmental protection and empowers them to make informed decisions in their daily lives. By fostering environmental stewardship and responsibility, environmental science contributes to the development of a greener and more sustainable future.

Role of Environmental Scientists

Environmental scientists play a crucial role in advancing knowledge and finding solutions to environmental problems. They conduct research, collect data, and analyze information to understand the intricate workings of our environment. This knowledge is used to develop sustainable practices and policies aimed at mitigating the negative impacts of human activities on the environment.

Environmental scientists also collaborate with government agencies, non-profit organizations, and industries to develop and implement environmental regulations and policies. They provide expertise on environmental impact assessments, pollution control measures, and conservation strategies. Ultimately, their goal is to reconcile the needs of society with the protection and preservation of our natural resources.

Historical Perspectives on Environmental Science

Understanding the historical context of environmental science is essential for appreciating the progress made in environmental conservation and recognizing the challenges that remain.

Early Conservation Movements

The conservation movement traces its roots back to the 19th century when concerns about deforestation, wildlife depletion, and land degradation began to emerge. Visionary figures such as John Muir and Theodore Roosevelt played pivotal roles in advocating for the preservation of natural landscapes and the establishment of national parks in the United States.

This early conservation movement marked the beginning of a broader recognition that humans have a responsibility to steward and protect the Earth's resources for future generations. It set the stage for the development of environmental science as a distinct field of study.

Environmentalism in the 20th Century

The 20th century witnessed a significant escalation in environmental problems, driven largely by rapid industrialization and population growth. Rachel Carson's groundbreaking book, "Silent Spring," published in 1962, brought attention to the detrimental effects of pesticide use and sparked the modern environmental movement.

Environmentalism gained momentum in the 1970s with the establishment of Earth Day and the passage of landmark environmental legislation such as the Clean Air Act and the Endangered Species Act in the United States. These developments underscored the need for scientific research and informed decision-making to address the growing environmental challenges.

Milestones in Environmental Science

Over the years, environmental science has achieved significant milestones in understanding and protecting the environment. Some notable milestones include:

1. Discovery of the ozone hole in the 1980s, leading to the implementation of the Montreal Protocol and subsequent phasing out of ozone-depleting substances. 2. The Intergovernmental Panel on Climate Change (IPCC) established in 1988 to provide scientific assessments of climate change and inform international policies. 3. The concept of sustainable development popularized in the 1987 Brundtland Report, emphasizing the need to balance economic development with environmental and social considerations. 4. Advances in remote sensing and GIS technology, enabling scientists to monitor changes in land cover, deforestation, and urban sprawl more effectively. 5. The Rio Earth Summit in 1992, which produced important international agreements such as the United Nations Framework Convention on Climate Change and the Convention on Biological Diversity.

These milestones reflect the ongoing efforts of environmental scientists to deepen our understanding of environmental processes and guide sustainable resource management.

Theoretical Frameworks in Environmental Science

Environmental science relies on various theoretical frameworks that help to explain complex environmental phenomena and guide scientific investigations.

Systems Theory and Complexity

Systems theory provides a framework for understanding the interactions between different components of an ecosystem. It recognizes that ecosystems are composed of interconnected parts and that changes in one component can have far-reaching effects on others. Systems thinking allows environmental scientists to analyze the cascading impacts of human activities on the environment and design holistic strategies for sustainable management.

Complexity theory complements systems thinking by recognizing the dynamic and adaptive nature of ecosystems. It acknowledges that environmental systems are nonlinear and influenced by feedback loops, thresholds, and emergent properties. By considering the complexity of environmental systems, scientists can better anticipate ecological responses and develop more resilient management approaches.

Ecological Principles and Concepts

Ecological principles provide a foundation for understanding the structure and functioning of ecosystems. Some key principles include:

1. The interconnectedness of all living organisms and their environment. 2. The dependence of organisms on their physical and biological surroundings. 3. The flow of energy and matter through ecosystems. 4. The concept of ecological niches and species interactions. 5. The importance of biodiversity in maintaining ecosystem resilience.

These principles guide environmental scientists in studying and preserving the intricate web of life on Earth.

1 4.3 Environmental Ethics

Environmental ethics examines the moral and philosophical dimensions of human-environment relationships. It considers questions of value, rights, and responsibilities in relation to the natural world. Environmental ethics provides a framework for addressing ethical dilemmas in environmental decision-making and advocating for the intrinsic value of nature.

Principles such as environmental justice, intergenerational equity, and the precautionary principle are prominent in the field of environmental ethics. These concepts emphasize the need to consider the well-being of all communities, both human and non-human, and to take proactive measures to prevent harm to the environment.

Methods and Techniques in Environmental Science

Environmental science employs a variety of methods and techniques to collect and analyze data, investigate environmental processes, and monitor changes in the environment.

Sampling and Data Collection

Sampling is a fundamental technique used in environmental science to gather data on various environmental parameters. Random sampling allows scientists to collect representative samples from a larger population, enabling accurate and unbiased measurements. Data collection methods range from field surveys and sample collection to remote sensing and satellite imagery analysis.

Laboratory Techniques in Environmental Analysis

Laboratory techniques play a crucial role in analyzing samples collected from the environment. These techniques include chemical analysis, DNA sequencing, microscopy, and spectroscopy. They allow scientists to determine the composition of air, water, soil, and biological samples and assess the presence of pollutants or pathogens.

Remote Sensing and GIS Applications in Environmental Science

Remote sensing involves the use of satellite or aerial imagery to monitor and study the Earth's surface. It provides valuable information about land use, vegetation cover, pollution patterns, and other environmental indicators. Geographic Information Systems (GIS) help in the integration and analysis of spatial data, enabling scientists to map and model environmental processes effectively.

These methods and techniques facilitate the collection of accurate and comprehensive data, which form the basis for informed decision-making and effective environmental management.

Challenges and Opportunities in Environmental Science

Environmental science is confronted with numerous challenges in addressing contemporary environmental issues. However, it also presents opportunities for innovative solutions and positive change.

Global Climate Change

Global climate change, primarily driven by human activities, poses one of the most significant challenges to environmental science. Rising greenhouse gas emissions are leading to increased temperatures, changing weather patterns, and sea-level rise. Environmental scientists are tasked with understanding the impacts of climate change on ecosystems and human communities and developing strategies for mitigation and adaptation.

Causes and Impacts of Climate Change

Understanding the causes and impacts of climate change is crucial for formulating effective responses. Environmental scientists study the emission sources and processes that contribute to climate change, such as burning fossil fuels and deforestation. They also investigate the consequences of climate change on ecosystems, including habitat loss, species extinction, and altered ecological dynamics.

Mitigation and Adaptation Strategies

Mitigation strategies aim to reduce greenhouse gas emissions and slow down the pace of climate change. These strategies include transitioning to renewable energy sources, improving energy efficiency, and implementing sustainable land-use practices. Adaptation strategies, on the other hand, focus on minimizing the impacts of climate change on vulnerable communities and ecosystems, such as building resilient infrastructure and implementing early warning systems.

Biodiversity Loss

Biodiversity loss is another pressing environmental challenge that environmental science seeks to address. Human activities, such as habitat destruction, pollution, and climate change, are causing the extinction of numerous species and the degradation of ecosystems. Environmental scientists work towards understanding the drivers of biodiversity loss and developing conservation strategies to protect endangered species and preserve important habitats.

The Importance of Biodiversity

Biodiversity is essential for the functioning of ecosystems and provides numerous ecosystem services, including pollination, nutrient cycling, and climate regulation. Environmental scientists explore the intricate relationships between species and ecosystems, recognizing the value of biodiversity for maintaining ecological balance and human well-being.

Threats to Biodiversity

Environmental science identifies and investigates the threats to biodiversity, such as habitat loss, invasive species, pollution, and overexploitation. By understanding these threats, scientists can develop effective conservation measures and management strategies to protect and restore biodiversity.

Conservation Efforts and Success Stories

Conservation efforts around the world have yielded remarkable success stories in protecting biodiversity and restoring ecosystems. Environmental scientists play a crucial role in these efforts, collaborating with local communities, governments, and non-profit organizations.

Examples of conservation success stories include the recovery of the bald eagle population in the United States, the restoration of the degraded Loess Plateau in China, and the establishment of protected areas for endangered species conservation, such as the Great Barrier Reef Marine Park in Australia.

Pollution and Environmental Health

Pollution poses significant risks to both the environment and human health. Environmental science investigates various types and sources of pollution, including air pollution, water contamination, and soil degradation. It examines the impacts of pollution on ecosystems, wildlife, and human populations, as well as the development of pollution prevention and control measures.

Types and Sources of Pollution

Environmental scientists study the different types and sources of pollution, such as industrial emissions, agricultural runoff, and the improper disposal of waste. They analyze the chemical composition of pollutants, their transport and fate in the environment, and their potential impacts on environmental and human health.

Human Health Impacts

Understanding the link between environmental pollution and human health is essential for developing effective public health strategies. Environmental scientists investigate the health impacts of pollution, such as respiratory diseases, waterborne illnesses, and the long-term effects of exposure to pollutants. This knowledge helps in designing policies and practices that protect human health and promote environmental justice.

Pollution Prevention and Control Measures

To mitigate pollution, environmental scientists work on developing pollution prevention and control measures. These include the implementation of clean technologies, the adoption of sustainable practices, and the establishment of regulatory frameworks. The aim is to reduce the release of pollutants into the environment and minimize their adverse effects on ecosystems and human health.

Sustainable Development

Sustainable development aims to meet the needs of the present generation without compromising the ability of future generations to meet their own needs. Environmental science plays a crucial role in understanding the principles and practices of sustainable development.

Principles of Sustainable Development

Environmental scientists explore the principles of sustainable development, such as the integration of economic, social, and environmental considerations, the promotion of equity and justice, and the wise use of natural resources. These principles guide decision-making processes to ensure that development is ecologically sound, economically viable, and socially responsible.

Sustainable Resource Management

Sustainable resource management is a critical component of sustainable development. Environmental science provides tools and approaches for managing natural resources in a way that balances economic development with environmental conservation. This includes sustainable forestry practices, responsible fisheries management, and the promotion of renewable energy sources.

Green Technologies and Innovations

Environmental science explores and promotes green technologies and innovations that help reduce environmental impacts and contribute to sustainable development. Examples include renewable energy technologies, energy-efficient buildings, waste recycling and reuse, and sustainable transportation solutions.

These innovations hold great promise for transitioning towards a greener future and reducing our reliance on non-renewable resources.

In summary, environmental science is a multidisciplinary field that seeks to understand the complex interactions between humans and the environment. It incorporates principles from various scientific disciplines and employs a range of methods and techniques to study environmental processes, address environmental challenges, and develop sustainable solutions. By embracing an interdisciplinary approach, environmental science plays a crucial role in shaping a greener and more sustainable future for generations to come.

1.1 What is Environmental Science?

Environmental science is a multidisciplinary field that explores the interactions between humans and the natural environment. It involves the study of the physical, chemical, and biological processes that occur in the environment, as well as the impact of human activities on these processes. Environmental science combines elements of various scientific disciplines, such as biology, chemistry, physics, geology, and ecology, to understand and address environmental issues.

Definition and Scope of Environmental Science

At its core, environmental science is the study of how organisms interact with each other and with their surroundings. It focuses on the complex web of relationships that exist in ecosystems and the impact of human actions on these systems. This includes examining natural phenomena, such as climate change, biodiversity loss, pollution, and resource depletion, and understanding how these processes affect the health and well-being of both humans and the planet.

The scope of environmental science is broad and encompasses a diverse range of topics. It includes the study of natural resources, such as water, air, and soil, and the ways in which they are affected by pollution and degradation. It also involves analyzing the impacts of human activities, such as deforestation, urbanization, industrialization, and agriculture, on ecosystems and biodiversity. Additionally, environmental science explores strategies for sustainable development, conservation, and the mitigation of environmental risks.

Interdisciplinary Nature of Environmental Science

One of the defining characteristics of environmental science is its interdisciplinary nature. It draws upon knowledge and techniques from a variety of scientific disciplines to study and solve environmental problems. For example, understanding the impact of pollution on aquatic ecosystems requires knowledge of chemistry, biology, and ecology. Similarly, studying climate change involves concepts from physics, atmospheric science, and geology.

The interdisciplinary nature of environmental science allows for a holistic approach to environmental problem-solving. By considering multiple perspectives and utilizing a wide range of tools and techniques, environmental scientists can develop more comprehensive and effective solutions.

Importance of Environmental Science in Today's World

In today's world, environmental science plays a crucial role in addressing pressing environmental challenges. As human activities continue to exert unprecedented pressures on the planet, understanding and mitigating their environmental impact is of paramount importance. Environmental science provides the knowledge and tools necessary to inform policy decisions, develop sustainable practices, and create a greener future.

By studying the complexity of environmental systems, environmental scientists can identify and evaluate potential environmental risks. This information can then be used to develop strategies for minimizing and managing these risks, such as pollution prevention measures and climate change adaptation strategies.

Furthermore, environmental science helps us to understand the delicate balance of ecosystems and the critical services they provide, such as clean air, water, and food production. By recognizing the value of these ecosystem services, we can better appreciate and conserve the natural environment.

Role of Environmental Scientists

Environmental scientists play a crucial role in advancing our understanding of the environment and developing sustainable solutions. They are involved in various activities, such as conducting research, collecting and analyzing data, as well as advising policymakers and industries on environmental issues.

Research is a fundamental aspect of environmental science, as it allows scientists to expand knowledge, explore new theories, and develop innovative solutions. Through fieldwork, laboratory experiments, and data analysis, researchers contribute to our understanding of environmental processes and the impacts of human activities.

Environmental scientists also play a vital role in policy-making. They provide valuable insights and scientific evidence to inform decision-making processes at local, national, and international levels. By communicating their findings effectively, they can influence policy and contribute to the development of sustainable practices and regulations.

Additionally, environmental scientists work closely with industries and other stakeholders to promote sustainable practices and reduce environmental impact. They can provide expertise and guidance on issues such as waste management, pollution control, and resource conservation, helping to shape more environmentally-friendly practices.

Overall, environmental scientists are at the forefront of the fight for a greener future, working to ensure that human activities are sustainable and that the Earth's ecosystems are protected for future generations.

Summary

In summary, environmental science is a multidisciplinary field that investigates the interactions between humans and the natural environment. It encompasses the study of environmental processes, human impacts on the environment, and sustainable solutions to environmental challenges. Environmental science is characterized by its interdisciplinary nature, drawing upon knowledge from various scientific fields. It is of critical importance in today's world, as it provides the tools and knowledge needed to address environmental issues and pave the way for a greener and more sustainable future. Environmental scientists play a vital role in conducting research, informing policy decisions, and working with industries to develop sustainable practices.

Interdisciplinary Nature of Environmental Science

Environmental science is a highly interdisciplinary field that draws upon knowledge and tools from various scientific disciplines to study and understand the environment and its interactions with human activities. It encompasses a wide range of scientific areas, including biology, chemistry, geology, physics, ecology, sociology, economics, and political science. The interdisciplinary nature of environmental science allows for a holistic approach to studying and solving environmental problems.

1. **Biology:** Biology provides a foundation for understanding the living components of the environment, including plants, animals, and microorganisms. It helps environmental scientists study the structure and function of ecosystems, biodiversity, and the impacts of human activities on natural populations.

2. **Chemistry:** Chemistry is essential in understanding chemical reactions and processes in the environment. Environmental scientists use chemistry to analyze pollutants, study air and water quality, assess soil composition, and investigate chemical reactions occurring in ecosystems.

3. **Geology:** Geology plays a crucial role in understanding Earth's processes, including the formation of rocks, minerals, and landforms. Environmental scientists rely on geology to study the distribution and availability of natural resources, assess the impacts of geological hazards, such as earthquakes and volcanic eruptions, and analyze geological factors influencing ecosystems.

4. **Physics:** Physics provides the fundamental principles to understand physical processes occurring in the environment. Environmental scientists use physics to study the flow of water, air, and energy in ecosystems, analyze climate patterns, and assess the impacts of natural disasters like hurricanes and floods.

5. **Ecology:** Ecology is the study of the interactions between organisms and their environment. It provides a framework for understanding the complex relationships between living organisms and their surroundings. Environmental scientists use ecological principles to study the structure and function of ecosystems, assess the impacts of human activities on biodiversity, and develop conservation strategies.

6. **Sociology:** Sociology helps environmental scientists understand the social and cultural factors that influence human interactions with the environment. It examines how human attitudes, beliefs, and behaviors shape environmental problems and solutions. Environmental scientists work with sociologists to study the social impacts of environmental policies, assess community perceptions of environmental issues, and promote sustainable behaviors.

7. **Economics:** Economics provides insights into the allocation and management of scarce environmental resources. Environmental scientists use economic principles to analyze the costs and benefits of environmental policies, assess the economic value of ecosystem services, and develop strategies for sustainable resource management.

8. **Political Science:** Political science explores the political processes and institutions that shape environmental decision-making. Environmental scientists collaborate with political scientists to study environmental policies, analyze the role of governance in environmental management, and advocate for sustainable development.

The integration of these disciplines is essential for addressing complex environmental challenges. For example, understanding the impacts of climate change requires knowledge of atmospheric science (physics and chemistry), ecological systems (biology and ecology), and human behavior (sociology and economics). By combining expertise from multiple disciplines, environmental scientists can develop comprehensive solutions to environmental problems.

Key Concepts

1. **Holistic Approach:** The interdisciplinary nature of environmental science allows for a holistic approach to environmental problem-solving. This approach recognizes that environmental issues are interconnected and require a comprehensive understanding of their scientific, social, and economic dimensions.

2. **Systems Thinking:** Environmental science employs systems thinking, which involves analyzing the interactions and feedback loops between different components of the environment. By viewing the environment as a complex system, scientists can identify the underlying causes of environmental problems and develop effective solutions.

3. **Sustainability:** Sustainability is a central concept in environmental science. It refers to the responsible use and management of natural resources to meet the needs of the present generation without compromising the ability of future generations to meet their own needs. Environmental scientists work towards achieving sustainable development by balancing environmental, social, and economic considerations.

4. **Interdisciplinary Research:** Environmental science encourages interdisciplinary research collaborations to address complex environmental challenges. By combining expertise from different disciplines, scientists can gain new insights and develop innovative approaches to environmental problem-solving.

Case Study: Climate Change

Climate change is a prime example of the interdisciplinary nature of environmental science. It involves the study of physical processes (physics), the analysis of greenhouse gases and their impacts on the atmosphere (chemistry), the examination of ecosystems' responses to changing climate patterns (ecology), the assessment of social and economic impacts (sociology and economics), and the development of policy measures (political science).

To understand climate change, environmental scientists use physics to study the greenhouse effect and the role of atmospheric gases in trapping heat. Chemistry plays a crucial role in analyzing the composition of the atmosphere and monitoring greenhouse gas emissions. Ecologists study the impacts of changing climate on ecosystems, including shifts in species distributions and altered ecological interactions. Sociologists and economists assess the social and economic impacts of climate change, such as the displacement of communities and the costs of adaptation and mitigation measures. Political scientists study international agreements and policies aimed at addressing climate change, such as the United Nations Framework Convention on Climate Change (UNFCCC) and the Paris Agreement.

By combining knowledge and tools from these interdisciplinary fields, environmental scientists can provide a comprehensive understanding of climate change and its impacts. This interdisciplinary approach is essential for developing effective strategies to mitigate climate change and adapt to its consequences.

Exercises

1. Research and summarize a case study that demonstrates the interdisciplinary nature of environmental science. Present your findings in a short presentation highlighting the contributions of different disciplines to address the environmental issue.

2. Discuss the benefits and challenges of interdisciplinary research in environmental science. Provide examples of successful interdisciplinary collaborations in the field and explain how they have contributed to solving environmental problems.

3. Imagine you are part of an interdisciplinary team working on a project to address a specific environmental challenge. Identify the disciplines that would be involved in your team and explain the role of each discipline in solving the problem.

4. Investigate a current environmental issue and analyze how it could be approached from an interdisciplinary perspective. Identify the relevant disciplines and explain how their contributions could lead to a more comprehensive understanding and effective solutions to the problem.

5. Explore the concept of sustainability and its application in different sectors (e.g., energy, agriculture, urban planning). Discuss how interdisciplinary collaboration is critical for achieving sustainability goals and overcoming sector-specific challenges.

Resources

1. "Environmental Science: A Global Concern" by William P. Cunningham and Mary Ann Cunningham 2. "Environmental Science: Earth as a Living Planet" by

CONTENTS

Daniel B. Botkin and Edward A. Keller 3. "Environmental Science for Dummies" by Alecia M. Spooner 4. "Introducing Environmental Science and Sustainability: A Global Perspective" by Neil Stutzman and Andrew Stutzman 5. "Ecology: Concepts and Applications" by Manuel C. Molles Jr.

Importance of Environmental Science in Today's World

Environmental science plays a critical role in understanding and addressing the complex environmental challenges we face in the modern world. It is an interdisciplinary field that combines knowledge from various scientific disciplines such as biology, chemistry, geology, physics, and social sciences to study the environment and find sustainable solutions to environmental problems. In this section, we will explore the importance of environmental science in today's world, focusing on its relevance to global challenges such as climate change, biodiversity loss, pollution, and sustainable development.

Understanding Climate Change

One of the most pressing environmental issues today is climate change. Environmental science provides invaluable insights into the causes, impacts, and mitigation strategies of climate change. Through the study of atmospheric processes, ocean dynamics, and interactions between living organisms and their environment, environmental scientists can assess the impact of greenhouse gas emissions on the Earth's climate system.

By analyzing temperature records, ice cores, and carbon dioxide levels in the atmosphere over time, environmental scientists have been able to demonstrate the link between human activities and the rise in global temperatures. This knowledge is crucial for policymakers and governments to develop effective climate change mitigation and adaptation strategies.

Preserving Biodiversity

Biodiversity, the variety of life on Earth, is essential for maintaining the stability and resilience of ecosystems. Environmental science plays a crucial role in understanding the drivers of biodiversity loss and developing conservation strategies to protect endangered species and fragile ecosystems.

Through field research and ecological modeling, environmental scientists can assess the impact of habitat destruction, invasive species, pollution, and climate change on biodiversity. This knowledge helps in identifying key areas for conservation, establishing protected areas, and implementing measures to restore degraded habitats.

For example, environmental scientists have worked closely with conservation organizations to protect endangered species such as the giant panda in China, sea turtles in coastal regions, and orangutans in Borneo and Sumatra. Their research and conservation efforts have contributed significantly to the preservation of these species and their habitats.

Addressing Pollution and Environmental Health

Pollution, both of air and water, poses significant threats to human health and the environment. Environmental science provides the necessary tools and knowledge to monitor, analyze, and mitigate different types of pollution.

By studying the sources and impacts of pollution, environmental scientists can develop strategies to minimize pollution and its effects on ecosystems and human health. They conduct field studies, collect and analyze samples, and use advanced laboratory techniques to detect and quantify pollutants in air, water, and soil.

Furthermore, environmental scientists play a crucial role in assessing the health risks associated with pollution and providing recommendations for pollution prevention and control measures. For instance, their research on the health impacts of air pollution has played a pivotal role in shaping regulations on emissions from industrial sources and vehicles.

Promoting Sustainable Development

Sustainable development aims to meet the needs of the present generation without compromising the ability of future generations to meet their own needs. Environmental science provides the necessary knowledge and tools to achieve this balance between economic development, social equity, and environmental protection.

Environmental scientists study the impact of human activities on natural resources such as water, land, and forests. They develop models and tools to assess the environmental implications of various development projects, such as renewable energy installations, urban planning, and agricultural practices.

By analyzing the trade-offs and potential synergies between economic development and environmental conservation, environmental scientists help policymakers make informed decisions that promote sustainable development. They contribute to the development of sustainable resource management strategies, green technologies, and innovations that minimize environmental impacts while promoting economic growth.

Concluding Remarks

The importance of environmental science in today's world cannot be overstated. It provides the knowledge, tools, and solutions needed to address pressing environmental challenges such as climate change, biodiversity loss, pollution, and sustainable development. Through interdisciplinary research and collaboration, environmental scientists are contributing to a greener and more sustainable future for all.

Exercises

1. Research and discuss a recent case study where environmental science played a crucial role in addressing a specific environmental challenge. Highlight the key findings and recommendations of the study.

2. Conduct a local environmental study in your area, focusing on a specific environmental issue such as air pollution, water contamination, or habitat degradation. Collect data, analyze the findings, and propose recommendations for mitigating the issue.

3. With the knowledge gained from this section, develop a proposal for a sustainable development project in your community. Outline the project goals, strategies, and expected outcomes in terms of environmental, social, and economic impacts.

Additional Resources

1. "Environmental Science: A Global Concern" by William P. Cunningham and Mary Ann Cunningham - This comprehensive textbook provides an in-depth understanding of environmental science, covering key concepts, case studies, and current research.

2. National Geographic Environment - This website features articles, videos, and interactive resources on a wide range of environmental topics, including climate change, biodiversity, and sustainable development.

3. United Nations Environment Programme (UNEP) - UNEP's website offers access to reports, publications, and resources on global environmental issues, conservation initiatives, and sustainable development goals.

Role of Environmental Scientists

Environmental scientists play a crucial role in our society by studying and understanding the interactions between the natural world and human activities. They work to identify and solve environmental problems, develop strategies for sustainable resource management, and provide guidance for policymakers and stakeholders. In this section, we will explore the various responsibilities and roles that environmental scientists undertake in their work.

Assessing Environmental Impact

One of the primary roles of environmental scientists is to assess the impact of human activities on the environment. They conduct comprehensive studies to evaluate the potential environmental consequences of proposed projects, such as construction of new infrastructures or industrial activities. Environmental scientists use various tools and techniques to collect data, monitor air and water quality, analyze soil samples, and assess biodiversity in the project area.

To measure the environmental impact, environmental scientists employ mathematical models and statistical analysis to quantify the potential risks and predict possible outcomes. They also consider factors like climate change, habitat loss, and pollution to develop a comprehensive understanding of the project's ecological consequences. By identifying and analyzing potential environmental risks, scientists can provide recommendations and guidelines to mitigate the harm and ensure sustainable development.

Developing Conservation Strategies

Another crucial role of environmental scientists is to develop conservation strategies to protect and preserve our natural resources. They study ecosystems, species, and biodiversity patterns to identify areas of high conservation value. By conducting field surveys and research, environmental scientists gather data on endangered species, vulnerable habitats, and critical ecosystems.

With this information, environmental scientists can propose conservation plans and management strategies to safeguard these valuable resources. They work closely with government agencies, non-profit organizations, and local communities to implement conservation measures and establish protected areas. By advocating for conservation practices and raising awareness about the importance of biodiversity, environmental scientists play a vital role in safeguarding our planet's natural heritage.

CONTENTS   19

Policy Development and Advocacy

Environmental scientists also contribute to policy development and advocacy efforts. They provide scientific expertise and evidence-based recommendations to policymakers and stakeholders. Environmental scientists analyze the potential impacts of existing policies and propose new regulations to address emerging environmental issues.

They play a vital role in shaping environmental policies related to climate change, pollution control, and natural resource management. By conducting research and generating scientific evidence, environmental scientists provide policymakers with the necessary information to make informed decisions for sustainable development.

Furthermore, environmental scientists serve as advocates for the environment, working to raise awareness and promote sustainable practices. They engage in public outreach programs, educate communities about environmental issues, and empower individuals to take action to protect the environment.

Collaboration and Communication

Environmental scientists must have strong collaboration and communication skills as they often work in multidisciplinary teams. They collaborate with other scientists, engineers, policymakers, and stakeholders from diverse backgrounds to address complex environmental challenges.

Effective communication is crucial for environmental scientists to convey their findings, recommendations, and research outcomes to different audiences. They must be able to present their work in a clear and understandable manner to policymakers, community members, and the general public. Environmental scientists also publish their research in scientific journals and participate in conferences to share their findings and contribute to the advancement of knowledge in the field.

Ethics and Responsibility

Environmental scientists have a responsibility to uphold ethical standards in their work. They must prioritize accuracy, integrity, and transparency in their research and ensure that their findings are not influenced by external factors. By adhering to ethical principles, environmental scientists contribute to the credibility and trustworthiness of their research.

Furthermore, environmental scientists have an ethical responsibility to advocate for sustainable practices and conservation efforts. They must consider the social, economic, and ecological impacts of their recommendations and strive for solutions that promote long-term sustainability.

Conclusion

The role of environmental scientists is paramount in addressing the environmental challenges we face today. Through their assessments of environmental impact, development of conservation strategies, policy development and advocacy, collaboration and communication, and adherence to ethical standards, environmental scientists play a crucial role in safeguarding our planet's resources for future generations. Their work contributes to the development of sustainable practices, policies, and technologies, paving the way for a greener and more resilient future.

Historical Perspectives on Environmental Science

Early Conservation Movements

The early conservation movements in the field of environmental science laid the foundation for the modern understanding and practice of environmental conservation. These early efforts were driven by the recognition of the negative impacts of human activities on the natural environment and the need for sustainable resource management. In this section, we will explore some of the key movements and individuals that shaped the early conservation movement.

The Romantic Movement and the Birth of Conservation

The Romantic Movement of the late 18th and early 19th centuries played a significant role in shaping the early conservation movements. This artistic and intellectual movement emphasized the importance of nature and its preservation. Romantic thinkers, such as Henry David Thoreau and John Muir, celebrated the beauty and spiritual connection with nature, inspiring a sense of awe and reverence for the natural world.

Henry David Thoreau, an American writer and naturalist, is known for his book "Walden," in which he documented his two-year experience living in harmony with nature at Walden Pond. Thoreau's writings emphasized the importance of self-reliance and simplicity in living, urging individuals to live in harmony with the natural world.

HISTORICAL PERSPECTIVES ON ENVIRONMENTAL SCIENCE 21

John Muir, often referred to as the "Father of the National Parks," was a Scottish-American naturalist and advocate for wilderness preservation. Muir's activism and writings played a crucial role in the establishment of national parks, including Yosemite National Park. His efforts helped shape the conservation movement and influenced future generations of environmentalists.

The Conservation Movement and the Founding of National Parks

The conservation movement gained momentum in the late 19th century with the establishment of the world's first national park, Yellowstone National Park, in 1872. This marked a significant milestone in the history of conservation, demonstrating the recognition of the need to protect and preserve natural areas for future generations.

Theodore Roosevelt, the 26th President of the United States, was a passionate conservationist and played a pivotal role in expanding the national park system. During his presidency, he established five national parks, including Grand Canyon National Park, and signed the Antiquities Act of 1906, which allowed for the preservation of historic landmarks and natural areas as national monuments.

The conservation movement in the United States was further strengthened by the establishment of organizations such as the Sierra Club and the Audubon Society. These organizations advocated for the protection of wilderness areas, wildlife conservation, and the promotion of sustainable land and resource management practices.

Early Conservation Efforts in Europe

While the United States led the way in the establishment of national parks, the conservation movement also gained traction in Europe. In 1889, the Swiss government established the world's first national park, the Swiss National Park, recognizing the value of preserving natural areas for scientific research and public enjoyment.

In the United Kingdom, the creation of national parks began with the establishment of Peak District National Park in 1951. Since then, additional national parks have been designated in England, Scotland, and Wales, preserving and protecting areas of outstanding natural beauty and ecological importance.

Furthermore, the early conservation movements in Europe also focused on protecting biodiversity and natural habitats. Efforts were made to establish nature reserves, such as the Wicken Fen Nature Reserve in England, to safeguard vulnerable species and restore degraded ecosystems.

Legacy and Impact of Early Conservation Movements

The early conservation movements laid the groundwork for our modern understanding of environmental conservation. They emphasized the importance of protecting and preserving natural areas, the need for sustainable resource management, and the recognition of the intrinsic value of nature.

These movements also highlighted the role of individuals in advocating for conservation and the power of public support in influencing policy decisions. The legacies of key figures, such as Henry David Thoreau, John Muir, and Theodore Roosevelt, continue to inspire environmental activists and conservationists today.

Additionally, the establishment of national parks and protected areas sparked a global movement towards nature conservation, creating models for other countries to follow. The success of early conservation efforts led to the development of international agreements and organizations dedicated to the conservation of biodiversity, such as the Convention on Biological Diversity and the International Union for Conservation of Nature.

Despite the achievements of early conservation movements, significant challenges remain. The increasing pressures of human population growth, climate change, habitat destruction, and pollution require continued efforts and innovative approaches to ensure the long-term sustainability and conservation of our natural resources.

Overall, the early conservation movements marked a turning point in the way we perceive and interact with the natural world. They set the stage for ongoing conservation efforts and serve as a reminder of the collective responsibility we have in protecting and preserving the environment for future generations.

Exercises

1. Research and write a short essay on the contributions of one of the key figures mentioned in the early conservation movements, such as Henry David Thoreau, John Muir, or Theodore Roosevelt. Highlight their significant achievements and lasting impacts on the conservation movement.

2. Identify a national park or nature reserve in your country or region. Research its history, biodiversity, and conservation efforts. Write a report outlining the challenges it faces and potential solutions for its long-term sustainability.

3. Choose a contemporary environmental issue, such as deforestation or water pollution, and explore the historical context and evolution of efforts to address this

HISTORICAL PERSPECTIVES ON ENVIRONMENTAL SCIENCE   23

issue. Discuss the successes and failures of conservation strategies implemented and propose innovative solutions for future conservation efforts.

4. Interview a local conservationist or environmentalist and learn about their work and motivations. Write a profile piece highlighting their contributions to the field of conservation and their perspectives on current environmental challenges.

5. Organize a debate or panel discussion on the topic of conservation ethics and the balance between human development and environmental preservation. Encourage participants to present different viewpoints and propose strategies for sustainable resource management.

Additional Resources

1. "Walden" by Henry David Thoreau - This book offers insights into Thoreau's philosophy on nature and sustainable living.

2. "My First Summer in the Sierra" by John Muir - Muir's firsthand account of his experiences in the Sierra Nevada Mountains and his deep connection with nature.

3. "The Wilderness Warrior: Theodore Roosevelt and the Crusade for America" by Douglas Brinkley - This biography explores Theodore Roosevelt's conservation legacy and his efforts to protect America's natural resources.

4. "Silent Spring" by Rachel Carson - A groundbreaking book that raised public awareness about the environmental impacts of pesticides and the need for conservation.

5. "A Sand County Almanac" by Aldo Leopold - This collection of essays by renowned conservationist Aldo Leopold reflects on his experiences in nature and promotes a land ethic for the preservation of the environment.

6. International Union for Conservation of Nature (IUCN) - An international organization that works towards nature conservation and the sustainable use of natural resources. Their website offers valuable resources and updates on global conservation efforts.

Environmentalism in the 20th Century

In the 20th century, environmentalism emerged as a powerful movement advocating for the protection and preservation of the natural world. As industrialization and technology advanced, so did the awareness of the negative impacts of human activities on the environment. This section will explore key milestones and influential figures that shaped the environmentalism movement in the 20th century.

Rachel Carson and the Birth of the Modern Environmental Movement

One of the most significant events in the history of environmentalism was the publication of Rachel Carson's book, "Silent Spring," in 1962. Carson, a biologist and writer, focused on the detrimental effects of pesticides, particularly DDT, on wildlife and the environment. Her book raised public awareness about the dangers of chemical pollution and is often credited with sparking the modern environmental movement.

Carson's work led to increased public concern and a push for stricter regulations on pesticide use. It also laid the groundwork for the establishment of the Environmental Protection Agency (EPA) in 1970, a governmental agency dedicated to protecting human health and the environment.

Earth Day and the Rise of Environmental Activism

In 1970, Earth Day was celebrated for the first time, marking a significant milestone in the environmentalism movement. On April 22nd, millions of people gathered to protest against environmental degradation and advocate for a more sustainable future. The event served as a catalyst for raising public awareness and mobilizing grassroots efforts to address environmental issues.

Earth Day not only brought environmental concerns to the forefront of public consciousness but also played a key role in shaping environmental policy. The movement resulted in the passage of several important environmental laws, including the Clean Air Act, the Clean Water Act, and the Endangered Species Act, among others.

The Chernobyl Disaster and Nuclear Awareness

The Chernobyl nuclear disaster in 1986 was a turning point in public perception of nuclear energy and its potential risks. The explosion and subsequent release of radioactive materials into the environment highlighted the devastating consequences of nuclear accidents.

The incident ignited a global debate on the safety and viability of nuclear power. It led to increased scrutiny of nuclear facilities, improved safety regulations, and a decline in public support for nuclear energy in many countries. The Chernobyl disaster served as a wake-up call for the world, underscoring the need for environmental accountability and the potential dangers associated with certain industrial practices.

HISTORICAL PERSPECTIVES ON ENVIRONMENTAL SCIENCE

Climate Change Awareness and the Intergovernmental Panel on Climate Change (IPCC)

Throughout the latter half of the 20th century, scientists began to recognize the potential impacts of human-induced climate change. The establishment of the Intergovernmental Panel on Climate Change (IPCC) in 1988 marked a significant development in global efforts to understand and address climate change.

The IPCC brings together scientists and policymakers from around the world to assess the latest scientific research on climate change and provide guidance for decision-making. Its reports have been instrumental in shaping international climate agreements, such as the United Nations Framework Convention on Climate Change (UNFCCC) and the Paris Agreement.

The recognition and understanding of climate change have propelled environmentalism into the 21st century, with a focus on mitigating greenhouse gas emissions, promoting renewable energy, and implementing strategies for adaptation to a changing climate.

Key Concepts and Principles

Precautionary Principle

The precautionary principle is a guiding principle in environmental decision-making that advocates for taking preventive action in the face of uncertain risks. It suggests that in situations where the potential for harm exists but the scientific evidence is inconclusive, action should still be taken to prevent or minimize potential damage to the environment or human health.

Applying the precautionary principle often involves prioritizing the protection of the environment and human well-being over economic interests. It emphasizes the need for responsible decision-making in the face of uncertainty, helping to avoid irreversible harm and promoting sustainable practices.

Sustainability

Sustainability is a fundamental concept in environmentalism, focusing on meeting the needs of the present without compromising the ability of future generations to meet their own needs. It involves finding a balance between economic development, environmental protection, and social equity.

Sustainable practices aim to minimize negative environmental impacts, promote the efficient use of resources, and ensure the long-term viability of ecosystems. This

Case Study: The Love Canal Disaster

One of the most prominent environmental disasters in the 20th century was the Love Canal disaster in Niagara Falls, New York. The Love Canal area, originally intended to be a neighborhood, became a toxic waste dump in the 1940s and 1950s.

In the 1970s, residents began experiencing health issues and discovered that their homes were built on top of buried chemical waste. Chemicals, including carcinogens and other hazardous substances, were leaking into the soil and groundwater, causing severe health problems among the community.

The Love Canal disaster served as a wakeup call, highlighting the importance of proper waste management and the potential long-term consequences of environmental pollution. It led to the passage of the Comprehensive Environmental Response, Compensation, and Liability Act (CERCLA), also known as Superfund, which provides funds for the cleanup of hazardous waste sites.

Conclusion

Environmentalism in the 20th century witnessed significant advancements in public awareness and the formulation of environmental policies. From Rachel Carson's seminal work to Earth Day and the global recognition of climate change, the environmentalism movement has shaped attitudes, policies, and actions to protect the environment.

Key principles such as the precautionary principle and sustainability have guided environmental decision-making, emphasizing the importance of responsible stewardship of the planet. These concepts, along with case studies like the Love Canal disaster, illustrate the ongoing struggle to strike a balance between human activities and the health of ecosystems.

Looking forward, the challenges of the 21st century, including climate change, biodiversity loss, and resource depletion, call for continued environmental activism and innovative solutions to shape a greener future. Environmentalism remains a vital force in promoting a sustainable and harmonious relationship between humans and the natural world.

Resources for Further Reading

+ Carson, R. (1962). *Silent Spring*. Houghton Mifflin.

HISTORICAL PERSPECTIVES ON ENVIRONMENTAL SCIENCE

- Orr, D. W. (2004). *Earth in Mind: On Education, Environment, and the Human Prospect.* Island Press.

- Steffen, W., et al. (2018). *The Anthropocene: A New Era for the Earth System.* Science, 361(6408), eaav2361.

- Intergovernmental Panel on Climate Change. (2014). *Climate Change 2014: Mitigation of Climate Change.*

- United Nations Framework Convention on Climate Change. (1992). *Text of the Convention.* Retrieved from: www.unfccc.int/resource/docs/convkp/conveng.pdf

Exercises

1. Research and discuss the legacy of Rachel Carson and the impact of "Silent Spring" on the environmental movement.

2. Investigate a local environmental issue in your community and analyze it from the perspective of the precautionary principle. What actions would you recommend based on the principle?

3. Choose a case study on a successful environmental restoration project and prepare a presentation highlighting the key strategies and outcomes.

4. Conduct research on a renewable energy technology and create a persuasive argument for its implementation on a large scale.

Milestones in Environmental Science

Environmental science has undergone significant advancements and milestones throughout history. These milestones have shaped our understanding of the environment, helped identify key environmental issues, and paved the way for innovative solutions. In this section, we will explore some of the key milestones in environmental science that have had a profound impact on our understanding of the world around us.

The Silent Spring and the Birth of the Environmental Movement

One of the most significant milestones in environmental science was the publication of Rachel Carson's book, "Silent Spring," in 1962. This groundbreaking work shed light on the detrimental effects of pesticides, particularly DDT, on wildlife and the environment. Carson's book sparked public concern and outrage, leading to a wider recognition of the need for environmental protection.

The impact of "Silent Spring" was far-reaching. It not only fueled the formation of various environmental organizations and advocacy groups but also prompted legislative actions, such as the ban on DDT in the United States. The book's influence on public opinion and policy paved the way for the birth of the modern environmental movement.

Establishment of Environmental Regulations

As the environmental movement gained momentum, governments around the world recognized the need to enact laws and regulations to protect the environment. One significant milestone in this regard was the establishment of the U.S. Environmental Protection Agency (EPA) in 1970. The EPA has since played a crucial role in developing and enforcing environmental regulations in the United States.

Similarly, other countries established their own environmental protection agencies and regulatory frameworks. These regulations addressed various environmental issues, including air and water pollution, waste management, and the preservation of natural resources. The establishment of environmental regulations marked a turning point in environmental science, as it highlighted the importance of governmental intervention in protecting the environment.

Montreal Protocol and the Ozone Layer

The discovery of the ozone hole in the 1980s raised alarm bells worldwide. Scientists determined that chlorofluorocarbons (CFCs), commonly used in aerosol cans and refrigeration, were responsible for depleting the ozone layer. The depletion of the ozone layer posed serious threats to human health and ecosystems.

In response to this environmental crisis, the international community came together to address the issue. The Montreal Protocol on Substances that Deplete the Ozone Layer, signed in 1987, aimed to phase out the production and consumption of ozone-depleting substances. The successful implementation of the Montreal Protocol led to a gradual recovery of the ozone layer and demonstrated the effectiveness of international cooperation in solving global environmental problems.

Intergovernmental Panel on Climate Change (IPCC)

The recognition of climate change as a significant environmental issue led to the formation of the Intergovernmental Panel on Climate Change (IPCC). Established in 1988 by the United Nations Environment Programme (UNEP) and the World Meteorological Organization (WMO), the IPCC is responsible for assessing scientific information related to climate change.

The IPCC plays a crucial role in synthesizing scientific research, providing policymakers with up-to-date and comprehensive assessments on climate change and its impacts. Through its reports, the IPCC has helped raise awareness about the urgency of mitigating climate change and provided the scientific basis for international climate negotiations, such as the United Nations Framework Convention on Climate Change (UNFCCC).

Paris Agreement and Global Climate Action

The Paris Agreement, adopted in 2015, marked a significant milestone in global climate action. It brought together 195 countries to address the threat of climate change and pursue efforts to limit global temperature rise to well below 2 degrees Celsius above pre-industrial levels.

The agreement emphasizes the importance of reducing greenhouse gas emissions, adapting to the impacts of climate change, and mobilizing finance and technology to support climate action. The Paris Agreement demonstrates global recognition of the need to address climate change on a collective and cooperative basis, highlighting the importance of international cooperation in tackling environmental challenges.

Conclusion

Milestones in environmental science have played a crucial role in shaping our understanding of the environment and guiding environmental action. From the publication of "Silent Spring" to the establishment of environmental agencies and the signing of international agreements, these milestones have propelled us toward a more sustainable future.

It is essential to recognize these milestones and build upon them as we face new challenges in the field of environmental science. By learning from the successes and failures of the past, we can continue to evolve our understanding and develop innovative solutions to protect and preserve our planet for future generations.

Theoretical Frameworks in Environmental Science

Systems Theory and Complexity

In environmental science, understanding the interconnectedness and complexity of natural systems is crucial for effective management and decision-making. Systems theory provides a framework for studying and analyzing these complex systems. This section will introduce the basic principles of systems theory and its application in environmental science.

Principles of Systems Theory

Systems theory is based on the concept that a system is composed of interconnected and interdependent parts that work together to achieve a common goal. It recognizes that changes in one part of the system can have ripple effects on other parts, creating feedback loops and emergent properties.

There are several key principles of systems theory:

1. **Holism:** Systems thinking emphasizes the holistic approach, considering the entire system as a whole rather than focusing solely on individual components. This approach recognizes that the behavior of a system cannot be fully understood by examining its individual parts in isolation.

2. **Emergence:** Systems can exhibit emergent properties that arise from the interactions between the components of the system. These properties are not present in the individual components themselves and can only be observed at the system level. For example, the collective behavior of a flock of birds or a school of fish emerges from the interactions between individual birds or fish.

3. **Feedback Loops:** Feedback loops are a fundamental concept in systems theory. There are two types of feedback loops: positive feedback loops, where a change in one part of the system amplifies or reinforces the change, and negative feedback loops, where a change in one part of the system counteracts or stabilizes the change. Feedback loops can have significant impacts on the behavior and stability of systems.

4. **Hierarchy:** Systems can often be organized into hierarchies, with smaller subsystems nested within larger systems. Each level of the hierarchy has its own properties and interactions, but is also influenced by the larger system

THEORETICAL FRAMEWORKS IN ENVIRONMENTAL SCIENCE   31

in which it is embedded. Understanding these hierarchical relationships is essential for understanding the behavior of complex systems.

Application in Environmental Science

Systems theory provides a valuable framework for studying and understanding complex environmental systems. By applying systems thinking, scientists can gain insights into the interactions between different components of the environment and how changes in one part can impact the entire system.

One key application of systems theory in environmental science is the study of ecosystems. Ecosystems are complex systems composed of biotic (living) and abiotic (non-living) components that interact with each other. Understanding these interactions is crucial for effective ecosystem management and conservation.

For example, applying systems theory to a forest ecosystem involves studying the interactions between plants, animals, microorganisms, soil, water, and other environmental factors. By analyzing the feedback loops and emergent properties within the ecosystem, scientists can gain a deeper understanding of how different disturbances, such as wildfires or climate change, can affect the overall health and resilience of the system.

Systems theory is also relevant in understanding the dynamics of human-environment systems. These systems involve the interactions between human societies and their natural environments. By applying systems thinking, researchers can examine the feedback loops between human activities, resource use, ecosystem services, and societal well-being.

For instance, consider the issue of deforestation in a developing country. Systems theory would consider not only the direct impacts of deforestation, such as loss of biodiversity and carbon emissions but also the indirect effects on local communities, economy, and regional climate patterns. This holistic approach helps policymakers and stakeholders develop more effective strategies for sustainable land management and conservation.

Challenges and Future Directions

While systems theory provides valuable insights into the complexity of environmental systems, there are challenges in its application. One challenge is the availability of data and the difficulty of quantifying the interactions within a system. Collecting comprehensive and accurate data on all components and their interactions can be time-consuming and resource-intensive.

Another challenge is the scale at which systems are studied. Environmental systems can span different spatial and temporal scales, and understanding the interactions and dynamics at each level can be complex. Researchers need to carefully consider the appropriate scale for their study and the potential impacts of scaling up or scaling down their analysis.

Moreover, systems theory requires interdisciplinary collaboration and integration of different scientific disciplines. To truly understand the complexity of environmental systems, researchers from various fields, such as ecology, hydrology, sociology, and economics, need to work together and share their expertise.

In the future, advancements in technology and computational modeling can help address some of these challenges. Technological innovations, such as remote sensing and advanced data analysis techniques, can provide more detailed and comprehensive data on environmental systems. Computational models can also simulate the behavior of complex systems and help identify potential interventions or management strategies.

Overall, systems theory offers a powerful framework for understanding the complexity of environmental systems. By embracing a holistic and interdisciplinary approach, environmental scientists can gain a deeper understanding of the interconnectedness of natural systems and develop more effective strategies for environmental management and conservation.

Ecological Principles and Concepts

Ecological principles and concepts form the foundation of understanding the complex interactions between living organisms and their environment. In this section, we will explore the fundamental principles that govern ecological systems and the key concepts used to study and analyze them.

Ecosystem Structure and Function

Ecological systems are composed of living organisms, such as plants, animals, and microorganisms, as well as their physical environment, including air, water, soil, and climate. Ecosystems can range in scale from small ponds to vast forests, and they exhibit a hierarchical structure with interacting components.

One key principle is the concept of trophic levels, which refers to the feeding relationships within an ecosystem. Organisms can be classified into different trophic levels based on their source of energy and nutrient acquisition. For example, primary producers, such as plants, harness energy from sunlight through photosynthesis and convert it into organic matter. Herbivores, which consume

THEORETICAL FRAMEWORKS IN ENVIRONMENTAL SCIENCE   33

plants, occupy the next trophic level. Carnivores, which feed on herbivores or other carnivores, occupy higher trophic levels.

The flow of energy through an ecosystem is governed by the second law of thermodynamics, which states that energy is not recycled, but rather flows through a system and gradually dissipates as heat. This principle helps us understand why ecological systems have a pyramid-like structure, with fewer individuals and less energy available at higher trophic levels.

Population Dynamics

Population dynamics refers to the changes in the size and composition of populations over time. Understanding population dynamics is crucial for managing wildlife populations, conserving endangered species, and predicting the spread of invasive species.

The growth rate of a population is influenced by birth rates, death rates, immigration, and emigration. The exponential growth model describes the idealized population growth in an unlimited environment with abundant resources. However, real populations are often constrained by limited resources, predation, disease, and competition. The logistic growth model incorporates these constraints and predicts a sigmoidal growth curve, with population growth slowing down as it approaches the carrying capacity of the environment.

In addition to growth, population dynamics also involves fluctuations and cycles. For example, predator-prey interactions can lead to cyclical changes in populations as predator and prey populations oscillate in response to each other's abundance.

Biodiversity and Community Dynamics

Biodiversity is a fundamental concept in ecology and refers to the variety and abundance of living organisms in an ecosystem. It encompasses three main components: species diversity, genetic diversity, and ecosystem diversity.

Species diversity is a measure of the number of species in a given area and includes both species richness (the total number of species) and species evenness (the relative abundance of each species). High species diversity is often associated with stable ecosystem functioning and resilience to disturbances.

Community dynamics focuses on the interactions between species within a community. The concept of niche is central to understanding community dynamics. A niche refers to the specific role that a species plays within its environment, including its habitat requirements, food sources, and interactions with other species. The competitive exclusion principle states that two species with identical niches cannot coexist in the same ecosystem, as one species will outcompete the other.

Communities can also exhibit symbiotic relationships, such as mutualism (both species benefit), commensalism (one species benefits, while the other is unaffected), and parasitism (one species benefits at the expense of the other). These interactions shape community structure and function.

Ecosystem Services and Resilience

Ecosystem services are the benefits that humans derive from ecosystems. They can be categorized into four main types: provisioning services (e.g., food, water, and timber), regulating services (e.g., climate regulation and water purification), cultural services (e.g., recreational activities and spiritual values), and supporting services (e.g., nutrient cycling and soil formation). Understanding the value of ecosystem services is essential for making informed decisions about land use, conservation, and sustainable development.

Ecosystem resilience refers to the ability of an ecosystem to withstand disturbances and recover its original state. Resilient ecosystems can adapt to changes and maintain their structure and function over time. However, human activities, such as habitat destruction, pollution, and climate change, can reduce ecosystem resilience and increase the risk of irreversible changes.

Conservation and Sustainability

Conservation biology is a multidisciplinary field that aims to protect and restore biodiversity and ecosystems. Conservation efforts involve both in-situ conservation (protecting species and ecosystems in their natural habitats) and ex-situ conservation (preserving species in captivity or in protected areas).

Sustainable development seeks to meet the needs of the present generation without compromising the ability of future generations to meet their own needs. It involves integrating environmental, social, and economic considerations in decision-making processes. Sustainable agriculture, renewable energy, and green technologies are examples of sustainable practices that promote long-term environmental sustainability.

In conclusion, ecological principles and concepts provide the framework for understanding and managing complex ecological systems. By applying these principles, we can address pressing environmental issues, conserve biodiversity, and build a sustainable future for generations to come.

Environmental Ethics

Environmental ethics is a branch of philosophy that deals with ethical questions related to the environment and our responsibilities towards it. It explores the moral principles and values that guide our interactions with the natural world and seeks to provide ethical guidelines for making decisions that affect the environment. In this section, we will delve into the theory and concepts of environmental ethics, discussing its importance and relevance in today's world.

Foundations of Environmental Ethics

The foundations of environmental ethics can be traced back to different philosophical traditions and perspectives. One such perspective is anthropocentrism, which places human beings at the center of the moral framework and values the environment primarily for its instrumental worth to humans. On the contrary, ecocentrism considers all living organisms and ecosystems as intrinsically valuable, independent of their usefulness to humans. Another perspective is biocentrism, which asserts that all living organisms have inherent worth and should be respected and protected. Lastly, there is deep ecology, which views humans as equal members of the Earth's living community and emphasizes the interconnectedness and interdependence of all beings.

Ethical Principles in Environmental Ethics

Various ethical principles are central to environmental ethics, guiding our actions and decisions concerning the environment. One such principle is the principle of intrinsic value, which asserts that the environment and its components have inherent worth and deserve moral consideration. This principle is closely linked to the concepts of ecocentrism and biocentrism.

The principle of intergenerational equity emphasizes our responsibility to future generations, advocating for sustainable practices that ensure the well-being of future inhabitants of the Earth. It calls for the preservation of natural resources and ecosystems for the benefit of all humanity across time.

The precautionary principle is another ethical principle that plays a significant role in environmental ethics. It suggests that when there is a potential risk of harm to the environment or human health, the burden of proof falls on those undertaking the activity to demonstrate that it is safe and will not cause significant harm. This principle highlights the need for prudence and careful consideration of the potential impacts of our actions on the environment.

Applications of Environmental Ethics

Environmental ethics has practical applications in various areas of environmental management and decision-making. One such application is in the field of conservation biology, where ethical considerations guide the establishment of protected areas, the management of endangered species, and the evaluation of trade-offs between conservation and human development. Ethical frameworks help scientists and policymakers navigate complex decisions by considering not only ecological and economic factors but also the ethical implications of their choices.

Another application of environmental ethics is in the realm of environmental policy and law. Ethical perspectives inform the development of environmental regulations and legislation, ensuring that human activities are guided by a sense of responsibility towards the environment and future generations.

Challenges and Debates in Environmental Ethics

The field of environmental ethics is not without its challenges and debates. One ongoing debate is between environmental preservation and economic development. Critics argue that strict environmental regulations may hinder economic growth, while proponents of environmental ethics advocate for a balance between environmental protection and sustainable development.

Another challenging question is the ethical consideration of non-human animals. How should we treat animals in our interactions with them and in our use of them for food, research, or entertainment? The debate between animal rights and animal welfare poses complex ethical questions and requires thoughtful consideration.

Furthermore, the issue of environmental justice raises concerns about the equitable distribution of environmental benefits and burdens among different social groups. Ethical frameworks are needed to address disparities in access to clean air, clean water, and healthy environments.

Case Study: Environmental Ethics and Climate Change

Climate change poses significant ethical dilemmas and challenges that require the application of environmental ethics. The burning of fossil fuels, deforestation, and industrial activities contribute to greenhouse gas emissions, causing global warming and climate instability. The impacts of climate change disproportionately affect vulnerable communities and future generations, highlighting the need for ethical action.

THEORETICAL FRAMEWORKS IN ENVIRONMENTAL SCIENCE

An ethical perspective on climate change acknowledges our responsibility to limit our carbon footprint, transition to renewable energy sources, and support adaptation measures for those affected by climate change. It prompts us to pursue climate justice by advocating for the rights of marginalized communities and working towards a sustainable future.

Conclusion

Environmental ethics provides a moral framework for addressing environmental issues and making decisions that consider the well-being of the environment, future generations, and all living beings. It emphasizes the interconnectedness and interdependence of all components of the natural world and challenges us to reflect on our role and responsibilities as stewards of the Earth. By integrating ethical considerations into environmental decision-making, we can work towards a more sustainable and equitable future.

Key Concepts

- Anthropocentrism: A perspective that values the environment primarily for its instrumental worth to humans.

- Ecocentrism: A perspective that considers all living organisms and ecosystems as intrinsically valuable, independent of their usefulness to humans.

- Biocentrism: A perspective that asserts that all living organisms have inherent worth and should be respected and protected.

- Deep ecology: A perspective that views humans as equal members of the Earth's living community and emphasizes the interconnectedness and interdependence of all beings.

- Intrinsic value: The inherent worth of the environment and its components, independent of their usefulness to humans.

- Intergenerational equity: The principle that emphasizes our responsibility to future generations and calls for sustainable practices.

- Precautionary principle: The principle that places the burden of proof on those undertaking potentially harmful activities to demonstrate their safety.

- Environmental preservation: The protection and conservation of natural resources and ecosystems.

- Environmental justice: The equitable distribution of environmental benefits and burdens among different social groups.

Explore More

- *A Sand County Almanac* by Aldo Leopold: A classic book in environmental ethics that explores the author's perspective on land ethics and the importance of ecological thinking.

- *Environmental Ethics: An Introduction to Environmental Philosophy* by Joseph R. DesJardins and John J. McMillan: A comprehensive introduction to environmental ethics and its applications.

- *The Ethics of What We Eat* by Peter Singer and Jim Mason: A thought-provoking exploration of the ethical implications of our food choices and the environmental impact of our diets.

Exercises

1. Consider a scenario where a company is planning to build a factory in a pristine natural area. Apply the principles of environmental ethics to evaluate the ethical implications of this decision.

2. Research and discuss a case study that exemplifies the application of environmental ethics in environmental management or policy.

3. Reflect on your own environmental values and beliefs. How do they align with different ethical perspectives discussed in this section? Discuss any potential conflicts or areas of growth.

Remember to ponder these ethical questions and engage in open and respectful discussions with others to deepen your understanding of environmental ethics and its practical applications.

Methods and Techniques in Environmental Science

Sampling and Data Collection

Sampling and data collection are essential steps in conducting environmental science research. They involve the careful selection and collection of representative samples from a larger population or ecosystem. This section will discuss the importance of sampling and data collection, various sampling techniques, and methods for collecting accurate and reliable data.

The Importance of Sampling

In environmental science, it is often impractical or impossible to study an entire population or ecosystem. Sampling allows scientists to collect data from a smaller portion of the population or ecosystem and make inferences about the entire population. By selecting representative samples, researchers can obtain reliable information about the characteristics, distribution, and behavior of the larger population.

Sampling is crucial for several reasons. Firstly, it saves time and resources. It would be impractical to study every individual organism or every square meter of a large ecosystem. Sampling enables researchers to collect enough data to draw meaningful conclusions in a more efficient way.

Secondly, sampling minimizes the impact on the environment. In some cases, collecting data from an entire population might cause harm or disrupt the ecosystem. By selecting representative samples, researchers can reduce their ecological footprint while still obtaining valuable information.

Lastly, sampling provides a manageable amount of data for analysis. Collecting data from an entire population can result in an overwhelming amount of information. By focusing on representative samples, researchers can analyze the data more effectively and uncover patterns and trends.

Sampling Techniques

There are several sampling techniques commonly used in environmental science research. The choice of sampling technique depends on the research objectives, the characteristics of the population or ecosystem, and practical considerations.

1. **Random Sampling:** This technique involves randomly selecting samples from the population or ecosystem. Random sampling ensures each individual or area has an equal chance of being selected, minimizing bias and increasing the representativeness of the sample.

2. **Stratified Sampling:** This technique divides the population or ecosystem into homogeneous subgroups or strata based on certain characteristics (e.g., age groups, geographic locations, vegetation types). Samples are then randomly selected from each stratum, ensuring that each subgroup is adequately represented in the sample.

3. **Systematic Sampling:** Systematic sampling involves selecting samples at regular intervals or based on a predetermined pattern. For example, a researcher may sample every 10th tree along a transect line. This technique is useful when studying patterns or changes over a defined spatial or temporal scale.

4. **Cluster Sampling:** Cluster sampling involves dividing the population or ecosystem into clusters (e.g., geographical units or habitat patches) and randomly selecting a subset of clusters for sampling. This technique is useful when it is impractical to sample each individual or when the clusters themselves exhibit similar characteristics.

5. **Convenience Sampling:** Convenience sampling involves selecting samples based on their accessibility or convenience. While this technique is easy to implement, it may introduce bias and limit the representativeness of the sample. Convenience sampling should be used with caution and only in situations where other methods are not feasible.

It is important to choose the most appropriate sampling technique based on the research objectives and constraints. A thorough understanding of the population or ecosystem being studied is essential for selecting an appropriate sampling technique.

Data Collection Methods

Once the samples have been selected, data collection begins. The choice of data collection methods depends on the research questions, the characteristics of the samples, and the available resources. Here are some common data collection methods used in environmental science:

1. **Direct Observation:** Direct observation involves visually or audibly recording information about the samples in their natural environment. This method is useful for studying behaviors, population densities, or distribution patterns of organisms. Direct observation can be done through field surveys, aerial surveys, or camera traps.

2. **Measurement and Monitoring:** Measurement and monitoring involve collecting quantitative data about specific variables of interest, such as temperature, pH, pollutant concentrations, or biomass. Measurement techniques may include using sensors, meters, or laboratory equipment. It is essential to calibrate and validate instruments to ensure accurate and reliable measurements.

METHODS AND TECHNIQUES IN ENVIRONMENTAL SCIENCE   41

3. **Sampling Techniques:** Various sampling techniques are used to collect physical samples for further analysis in the laboratory. Examples include water sampling for chemical analysis, soil sampling for nutrient content, or plant sampling for species identification. Sampling methods should follow standardized protocols to ensure consistency and comparability of results.

4. **Remote Sensing:** Remote sensing involves using aerial or satellite-based sensors to collect information about the Earth's surface or atmosphere. Remote sensing data can provide valuable insights into land cover, vegetation health, climate patterns, and other environmental parameters. It is a useful tool for large-scale studies and monitoring.

5. **Interviews and Surveys:** Interviews and surveys are commonly used to collect data from human populations, stakeholders, or local communities. They can provide valuable insights into human perceptions, attitudes, and behaviors related to the environment. Care should be taken to design unbiased and well-structured questionnaires to ensure the reliability of the data.

Ensuring Data Quality

To ensure the quality and reliability of the collected data, several steps should be followed:

1. **Standardization:** Standardized protocols and methodologies should be followed during sampling and data collection to ensure consistency and comparability of results. This includes using calibrated instruments, following established guidelines, and conducting quality control checks.

2. **Data Validation:** Collected data should undergo a thorough validation process to identify and correct errors, anomalies, or outliers. Validation may involve cross-checking with existing data, replicating measurements, or performing statistical checks for consistency.

3. **Data Accuracy and Precision:** Accuracy refers to how close the measured values are to the true values, while precision refers to the level of reproducibility or consistency of the measurements. Both accuracy and precision should be considered when evaluating data quality. Proper instrument calibration, replication of measurements, and statistical analysis can help ensure accurate and precise data.

4. **Data Management:** Proper data management is essential for organizing, storing, and analyzing the collected data. This includes following standardized formats, maintaining backup copies, establishing data security protocols, and documenting metadata (e.g., sampling locations, dates, and methodologies). Good data management practices facilitate data sharing, reproducibility, and long-term usability.

Case Study: Monitoring Water Quality in a River

To illustrate the principles and techniques of sampling and data collection, let's consider a case study on monitoring water quality in a river. The objective is to assess the concentrations of various pollutants at different locations along the river.

1. **Sampling Design:** The river is divided into several segments based on land use and potential pollution sources. Stratified sampling is used, where samples are collected from each segment to ensure representation. Random sampling points are selected within each segment.

2. **Sample Collection:** Water samples are collected using standardized protocols, such as submerging a sampling bottle at a specific depth. The samples are stored in clean, labeled containers and kept at appropriate temperatures to preserve their integrity during transportation to the laboratory.

3. **Laboratory Analysis:** In the laboratory, samples are analyzed using appropriate techniques for each pollutant of interest. For example, spectrophotometry may be used to measure nutrient concentrations, while gas chromatography may be used to measure pesticide concentrations. Calibration curves and quality control checks are performed to ensure accurate and precise measurements.

4. **Data Analysis:** The collected data, including pollutant concentrations and corresponding sampling locations, are compiled and analyzed. Statistical tests and spatial mapping techniques may be used to identify pollution hotspots, assess the overall water quality, and detect temporal trends.

5. **Interpretation and Reporting:** The results are interpreted in the context of water quality standards, regulations, or guidelines. The findings are typically summarized in a report or scientific publication, highlighting key findings, implications, and recommendations for management or remedial actions.

The case study demonstrates the importance of using appropriate sampling techniques, following standardized protocols, and ensuring accurate laboratory analysis to obtain reliable water quality data.

Conclusion

Sampling and data collection are fundamental steps in environmental science research. The choice of sampling techniques and data collection methods depends on research objectives, the characteristics of the populations or ecosystems, and practical considerations. To ensure data quality, standardization, validation, accuracy, precision, and good data management practices are essential. Through

METHODS AND TECHNIQUES IN ENVIRONMENTAL SCIENCE

proper sampling and data collection, scientists can gather valuable information to understand and address environmental challenges effectively.

Laboratory Techniques in Environmental Analysis

In the field of environmental science, laboratory techniques play a crucial role in providing accurate and reliable data for analysis. These techniques allow scientists to measure and quantify various parameters related to the environment, such as water quality, air pollution, soil composition, and chemical contaminants. In this section, we will explore some of the common laboratory techniques used in environmental analysis and their applications.

Sample Preparation

Before any analysis can be conducted, it is essential to properly prepare the samples. Sample preparation involves collecting representative samples from the environment and treating them to remove impurities or concentrate the analytes of interest. The choice of sample preparation method depends on the nature of the sample and the analytes to be measured.

One commonly used technique is filtration, which involves passing a liquid sample through a filter to remove suspended solids. This technique is used in water analysis to remove particles that may interfere with subsequent analysis. Another technique is solid-phase extraction (SPE), where analytes are extracted from a liquid sample onto a solid sorbent material and then eluted for analysis. SPE is often used to concentrate trace contaminants from water or soil samples.

Chemical Analysis

Chemical analysis is the core of environmental laboratory techniques. It involves the detection and measurement of various chemical constituents present in environmental samples. Here are a few commonly used techniques in chemical analysis:

1. Spectroscopy Spectroscopy is the study of the interaction between matter and electromagnetic radiation. In environmental analysis, spectroscopic techniques such as UV-Visible spectroscopy, infrared spectroscopy (IR), and atomic absorption spectroscopy (AAS) are widely used. UV-Visible spectroscopy is used to determine the concentration of certain compounds that absorb light in the ultraviolet and visible regions. IR spectroscopy is used to identify functional groups in organic compounds. AAS is used for the determination of metal ions in samples.

2. Chromatography Chromatography is a separation technique based on the differential partitioning of analytes between a stationary phase and a mobile phase. It is extensively used in environmental analysis for the separation and quantification of complex mixtures. Common chromatographic techniques include gas chromatography (GC), high-performance liquid chromatography (HPLC), and ion chromatography (IC).

GC is used for the separation and quantification of volatile organic compounds (VOCs) in air and water samples. HPLC is used for the separation of non-volatile compounds, such as pesticides and pharmaceuticals, in water and soil samples. IC is used for the analysis of inorganic ions in water samples.

3. Mass Spectrometry Mass spectrometry (MS) is a technique that measures the mass-to-charge ratio of ions to identify and quantify compounds. It is commonly coupled with chromatographic techniques to enhance the separation and detection capabilities. Environmental applications of MS include the analysis of organic pollutants, pesticides, and metabolites in samples.

Quality Assurance and Quality Control

To ensure the accuracy and reliability of laboratory results, quality assurance (QA) and quality control (QC) measures are implemented throughout the analysis process. QA refers to the overall management system that includes standard operating procedures, instrument calibration, and staff training. QC involves the use of internal and external quality control samples to monitor the performance of instruments and analytical methods.

One common QC measure is the use of certified reference materials (CRMs), which are standardized materials with known concentrations of analytes. CRMs are used as a reference to validate the accuracy of the measurement. Additionally, laboratory proficiency testing programs are often conducted to assess the laboratory's performance against other accredited laboratories.

Emerging Techniques

In recent years, technological advancements have led to the development of new laboratory techniques for environmental analysis. These emerging techniques aim

METHODS AND TECHNIQUES IN ENVIRONMENTAL SCIENCE

to improve detection limits, reduce analysis time, and enhance the accuracy of measurements. Here are a few examples:

1. Microplastics Analysis With the increasing concern about microplastic pollution, analytical techniques for the detection and quantification of microplastics have been developed. These techniques include Fourier-transform infrared spectroscopy (FTIR), Raman microscopy, and microscopy coupled with image analysis.

2. Molecular Techniques Molecular techniques, such as polymerase chain reaction (PCR) and DNA sequencing, are being used in environmental analysis to identify and quantify microorganisms, pathogens, and genetically modified organisms (GMOs) in samples. These techniques provide valuable information about biodiversity and ecological interactions.

3. Sensor Technologies Advancements in sensor technologies have enabled the development of portable and real-time monitoring devices for environmental analysis. These sensors can measure parameters such as pH, dissolved oxygen, temperature, and pollutant concentrations in situ. They are particularly useful for field studies and continuous monitoring applications.

Conclusion

Laboratory techniques in environmental analysis are essential for understanding and monitoring the health of our environment. From sample preparation to chemical analysis, these techniques provide crucial data for decision-making and policy development. With the continuous advancements in technology, we can expect to see more innovative laboratory techniques that will further enhance our ability to assess and protect the environment.

1.5.3 Remote Sensing and GIS Applications in Environmental Science

Introduction

Environmental science is a multidisciplinary field that utilizes various techniques and tools to gather data about the environment and understand its processes. One such technique is remote sensing, which involves the acquisition of information about the Earth's surface without direct physical contact. Geographic Information Systems (GIS) provide a framework for storing, analyzing, and visualizing this spatial data. In this section, we will explore the applications of remote sensing and GIS in the field of environmental science.

Remote Sensing

Remote sensing relies on sensors that capture electromagnetic radiation reflected or emitted by objects on the Earth's surface. These sensors, placed on satellites or aircraft, record data in various wavelength regions, ranging from visible light to thermal infrared. The acquired data can be used to assess the health of ecosystems, monitor land cover changes, map natural resources, and study the impacts of climate change. Let's explore some common applications of remote sensing in environmental science.

Vegetation Monitoring

One of the key applications of remote sensing is monitoring vegetation dynamics. By analyzing the spectral signatures of plants, remote sensing can provide valuable information about vegetation health, biomass, and growth rates. This data helps researchers study the effects of climate change, identify areas at risk of deforestation, and assess the success of reforestation programs.

For example, NASA's Moderate Resolution Imaging Spectroradiometer (MODIS) satellite captures data in multiple spectral bands, allowing scientists to monitor global vegetation dynamics. By tracking the greenness index, which indicates the density and vitality of vegetation, researchers can identify areas experiencing drought stress or changes in land use.

Land Cover Classification

Remote sensing data can be used to classify and map different land cover types, such as forests, wetlands, agricultural areas, and urban zones. By analyzing the reflected

1.5.3 REMOTE SENSING AND GIS APPLICATIONS IN ENVIRONMENTAL SCIENCE

or emitted radiation from the Earth's surface, algorithms can distinguish between different land cover classes based on their unique spectral signatures.

These land cover maps are crucial for managing land resources, planning urban development, and assessing the impact of land use changes. For example, the European Space Agency's Sentinel-2 satellite provides high-resolution multispectral imagery, enabling accurate land cover classification and monitoring of land cover changes over time.

GIS Applications

Geographic Information Systems (GIS) combine spatial data with attribute data to create visual representations of the Earth's surface. GIS applications in environmental science are vast, ranging from disaster management and environmental planning to biodiversity conservation and natural resource management. Let's explore some common applications of GIS in environmental science.

Habitat Modeling and Conservation

GIS allows scientists to model and map habitats for endangered species and identify areas of high biodiversity. By integrating spatial data on species distribution, land cover, and environmental variables, researchers can identify critical habitats and design effective conservation strategies.

For example, GIS has been used to identify and protect critical nesting areas for sea turtles. By mapping coastal areas with suitable characteristics, such as sandy beaches and low light pollution, conservationists can prioritize efforts to protect these vital habitats.

Climate Change Impact Assessment

GIS plays a crucial role in assessing the impacts of climate change on the environment. By integrating climate models with spatial data on ecosystems, land cover, and socio-economic factors, researchers can predict the vulnerability of different regions to climate change and develop adaptation strategies.

For instance, using GIS, scientists can analyze the potential effects of sea-level rise on coastal areas, identify areas at risk of flooding, and develop coastal management plans to mitigate these risks.

Conclusion

Remote sensing and GIS are invaluable tools for environmental scientists. Remote sensing provides a means to gather data about the Earth's surface from a distance, while GIS allows for the storage, analysis, and visualization of spatial data. Together, these technologies enable researchers to monitor vegetation, classify land cover, model habitats, assess climate change impacts, and implement effective conservation and management strategies. As technology continues to advance, the applications of remote sensing and GIS in environmental science will only continue to grow, contributing to a greener and more sustainable future.

Challenges and Opportunities in Environmental Science

Global Climate Change

Climate change is one of the most pressing global issues of our time. It refers to long-term shifts in weather patterns and average temperatures across the Earth's surface due to human activities and natural processes. The main driver of climate change in recent years is the increased concentration of greenhouse gases in the atmosphere, primarily carbon dioxide (CO_2) from burning fossil fuels.

Causes and Impacts of Climate Change

The burning of fossil fuels for energy and transportation releases significant amounts of greenhouse gases, primarily CO_2, into the atmosphere. These gases trap heat from the sun, causing a phenomenon known as the greenhouse effect and resulting in a gradual increase in global temperatures.

The impacts of climate change are far-reaching and diverse. Rising global temperatures lead to more frequent and severe heatwaves, droughts, and extreme weather events such as hurricanes and heavy rainfall. Melting ice caps and glaciers contribute to rising sea levels, threatening coastal communities and ecosystems. Changes in rainfall patterns can lead to water scarcity and agricultural disruptions, affecting food security. The loss of biodiversity and increased risks to human health are additional consequences of climate change.

CHALLENGES AND OPPORTUNITIES IN ENVIRONMENTAL SCIENCE

Mitigation and Adaptation Strategies

Addressing climate change requires both mitigation and adaptation strategies. Mitigation focuses on reducing greenhouse gas emissions to slow down the rate of global warming. This can be achieved through actions such as transitioning to renewable energy sources, improving energy efficiency, and establishing policies to promote sustainable practices.

Adaptation involves adjusting to the impacts of climate change to minimize vulnerability and enhance resilience. This includes measures such as building infrastructure to withstand extreme weather events, implementing water management strategies, and incorporating climate considerations into urban planning.

International Efforts

The United Nations Framework Convention on Climate Change (UNFCCC) was established in 1992 as a global response to climate change. Its goal is to stabilize greenhouse gas concentrations in the atmosphere at a level that would prevent dangerous interference with the climate system.

The UNFCCC has led to several international agreements, including the Kyoto Protocol and the Paris Agreement. The Kyoto Protocol, adopted in 1997, set binding emissions reduction targets for developed countries. The Paris Agreement, reached in 2015, aims to keep global temperature rise well below 2 degrees Celsius above pre-industrial levels and pursue efforts to limit the temperature increase to 1.5 degrees Celsius.

Case Studies and Success Stories

There have been notable success stories in addressing climate change. One such example is Costa Rica's renewable energy revolution. Through significant investments in hydropower, geothermal energy, and wind power, Costa Rica now generates more than 99% of its electricity from renewable sources.

Another success story is the city of Copenhagen in Denmark. By prioritizing cycling infrastructure, improving public transportation, and implementing energy-efficient buildings, Copenhagen has reduced its carbon emissions by 50% since 1990 while experiencing economic growth.

Challenges and Future Directions

Despite progress, there are several challenges in addressing climate change. The transition to renewable energy faces barriers such as upfront costs and the need for infrastructure development. Political will and cooperation at the international level are crucial for effective action. Additionally, the involvement and engagement of various stakeholders, including governments, businesses, and individuals, are necessary for widespread adoption of sustainable practices.

Future directions in addressing climate change include advancing green technologies, increasing research and development in renewable energy, and promoting sustainable agriculture and land use. Educating the public about the impacts of climate change and the importance of individual actions is also vital for collective action.

In conclusion, global climate change poses significant challenges to the planet and human well-being. However, through a combination of mitigation and adaptation strategies, international cooperation, and individual actions, we can work towards a greener and more sustainable future. It is crucial to act now and take responsibility for the preservation of our planet for future generations.

Causes and Impacts of Climate Change

Climate change is one of the most pressing global challenges of our time, with far-reaching impacts on the environment, human societies, and the economy. In this section, we will explore the causes of climate change and its various impacts on different aspects of our lives.

Causes of Climate Change

Climate change is primarily caused by the increase in greenhouse gas concentrations in the Earth's atmosphere. Greenhouse gases, such as carbon dioxide (CO_2), methane (CH_4), and nitrous oxide (N_2O), trap heat from the sun and prevent it from escaping back into space, leading to a gradual increase in global temperatures. The main sources of these greenhouse gases include:

- **Burning of Fossil Fuels:** The combustion of coal, oil, and natural gas for energy production and transportation is the largest contributor to CO_2 emissions. These fossil fuels release vast amounts of CO_2 into the atmosphere, accelerating the greenhouse effect.

- **Deforestation:** The removal of forests, especially tropical rainforests, contributes significantly to climate change. Trees absorb CO_2 from the

CHALLENGES AND OPPORTUNITIES IN ENVIRONMENTAL SCIENCE

atmosphere through photosynthesis and release oxygen, acting as a natural carbon sink. Deforestation disrupts this balance, leading to increased CO_2 levels.

- **Industrial Processes:** Certain industrial activities, such as cement production, steel manufacturing, and chemical production, release large amounts of CO_2 and other greenhouse gases as byproducts. These emissions contribute to climate change.

- **Agriculture:** Agricultural activities, including livestock farming and the use of synthetic fertilizers, produce considerable amounts of CH_4 and N_2O. These gases have a much higher warming potential than CO_2 and contribute to the greenhouse effect.

- **Land-use Changes:** Converting forests and grasslands into agricultural or urban areas alters the natural balance of greenhouse gases. Changes in land use can lead to an increase in CO_2 emissions and a decrease in natural carbon sinks.

- **Waste Generation:** Decomposing organic waste in landfills produces CH_4, a potent greenhouse gas. Inefficient waste management practices contribute to the accumulation of methane in the atmosphere.

The release of greenhouse gases from these activities has led to a significant increase in atmospheric CO_2 concentrations since the beginning of the Industrial Revolution. This increase is closely correlated with rising global temperatures and climate change.

Impacts of Climate Change

Climate change has a wide range of impacts on our planet and all its inhabitants. These impacts can be grouped into various categories:

1. Environmental Impacts:

- *Rising Temperatures:* Global warming leads to higher temperatures, which disrupt the balance of ecosystems. Many species struggle to adapt to these changes, which can result in shifts in wildlife habitats, population declines, and increased extinction rates.

- *Extreme Weather Events:* Climate change intensifies extreme weather events such as hurricanes, droughts, heatwaves, and heavy rainfall. These events have significant consequences for both human populations and natural ecosystems, including property damage, loss of life, and crop failure.

- *Melting Ice and Rising Sea Levels:* As temperatures rise, glaciers and ice caps melt, contributing to rising sea levels. This poses a threat to coastal cities, low-lying islands, and vulnerable ecosystems. Increased coastal erosion, saltwater intrusion into freshwater sources, and loss of critical habitats are some of the consequences.

2. **Economic Impacts:**

- *Decreased Agricultural Productivity:* Climate change affects agricultural productivity due to changing rainfall patterns, increased frequency of extreme weather events, and loss of suitable farming areas. Crop yields decline, leading to food shortages, increased food prices, and reduced income for farmers.

- *Increased Health Risks:* Climate change brings health risks in the form of heat-related illnesses, exacerbated air pollution, expanding disease vectors, and increased waterborne diseases. These factors put additional pressure on healthcare systems and can lead to economic losses due to decreased workforce productivity.

- *Loss of Ecosystem Services:* Ecosystems provide essential services such as water purification, pollination, and carbon sequestration. Climate change disrupts these services, resulting in additional costs for providing alternative solutions or dealing with the consequences, such as water treatment costs or crop loss due to pollinator decline.

3. **Social Impacts:**

- *Displacement of Communities:* Rising sea levels, extreme weather events, and changing environmental conditions can displace vulnerable communities, particularly in low-lying coastal areas or regions prone to desertification. This displacement creates social unrest, economic burdens, and challenges surrounding migration and resettlement.

- *Water Scarcity and Conflict:* Climate change exacerbates water scarcity in many regions, leading to competition for resources, conflicts between different user groups, and challenges in managing transboundary water sources. This can strain diplomatic relationships and social cohesion.

- *Indigenous and Traditional Knowledge Loss:* Climate change threatens the cultural heritage and traditional knowledge of indigenous peoples, who are often the most affected by its impacts. Loss of traditional practices and knowledge can lead to the erosion of cultural diversity and the disconnection of communities from their ancestral lands.

Addressing the causes and impacts of climate change is crucial for building a sustainable and resilient future. By reducing greenhouse gas emissions, transitioning to renewable energy sources, implementing sustainable land-use practices, and promoting adaptation measures, we can mitigate the severity of climate change and protect our planet for future generations.

Summary

In this section, we explored the causes and impacts of climate change. We discussed how human activities, such as the burning of fossil fuels, deforestation, and industrial processes, contribute to the increase in greenhouse gas concentrations in the atmosphere. This leads to rising global temperatures and various environmental, economic, and social impacts.

The environmental impacts of climate change include rising temperatures, extreme weather events, and melting ice, leading to habitat loss, ecosystem disruption, and rising sea levels. Economically, climate change affects agricultural productivity, increases health risks, and disrupts ecosystem services. Socially, climate change results in the displacement of communities, water scarcity, and loss of indigenous knowledge.

To address climate change, it is essential to reduce greenhouse gas emissions, promote sustainable land-use practices, and invest in renewable energy sources. By taking action, we can mitigate the impacts of climate change and create a more sustainable and resilient future.

Mitigation and Adaptation Strategies

Mitigation and adaptation are two key strategies in addressing the challenges posed by global climate change. Mitigation involves efforts to reduce or prevent the emission of greenhouse gases (GHGs) into the atmosphere, thereby reducing the extent and severity of climate change. Adaptation, on the other hand, focuses on adjusting societal and natural systems to better cope with the impacts of climate change that are already occurring or are expected to occur in the future.

Mitigation Strategies

Mitigation strategies aim to reduce the amount of GHGs emitted into the atmosphere and slow down the rate of climate change. These strategies can be implemented at various levels - individual, community, national, and global. Some key mitigation strategies include:

1. **Renewable Energy Transition:** Shifting from fossil fuel-based energy production to renewable sources such as solar, wind, hydroelectric, geothermal, and biomass energy can significantly reduce GHG emissions. This transition involves investments in renewable energy infrastructure, research and development of new technologies, and policy support.

2. **Energy Efficiency:** Improving energy efficiency in buildings, industries, and transportation can lead to significant reductions in GHG emissions. This involves

using energy-efficient appliances, improving insulation, implementing efficient industrial processes, and promoting public transportation and electric vehicles.

3. **Land Use Change:** Protecting and restoring forests, grasslands, and other ecosystems can help mitigate climate change by sequestering carbon dioxide (CO2) through photosynthesis. Additionally, reducing deforestation and promoting afforestation and reforestation efforts can help offset GHG emissions.

4. **Carbon Capture and Storage (CCS):** CCS involves capturing CO2 emissions from power plants and industrial processes and storing it underground or utilizing it for industrial purposes. This technology can help reduce GHG emissions from large point sources.

5. **Transition to Low-carbon Agriculture:** Implementing sustainable agricultural practices such as organic farming, agroforestry, and regenerative agriculture can help reduce GHG emissions from the agricultural sector. These practices promote soil health, enhance carbon sequestration, and minimize the use of synthetic fertilizers and pesticides.

6. **Policy Instruments:** Governments can implement various policy instruments to encourage mitigation actions, including carbon pricing mechanisms such as carbon taxes or cap-and-trade systems. These measures create economic incentives for reducing GHG emissions and promote investment in low-carbon technologies.

Adaptation Strategies

Adaptation strategies aim to reduce vulnerability and enhance resilience in the face of climate change impacts. These strategies involve adjusting social and natural systems to cope with changes in temperature, precipitation patterns, sea-level rise, and extreme weather events. Some key adaptation strategies include:

1. **Infrastructure Resilience:** Building resilient infrastructure that can withstand the impacts of climate change, such as designing buildings to withstand extreme weather events, improving drainage systems to deal with increased rainfall, and constructing seawalls to protect coastal areas from sea-level rise.

2. **Natural Ecosystem Conservation:** Protecting and restoring natural ecosystems such as wetlands, mangroves, and coral reefs can provide natural buffers against extreme weather events, reduce coastal erosion, and maintain biodiversity. These ecosystems also have the potential to sequester carbon and mitigate climate change.

3. **Water Management:** Developing effective water management strategies is crucial for adapting to changing precipitation patterns and ensuring water availability in the face of droughts or increased flooding. This can include

CHALLENGES AND OPPORTUNITIES IN ENVIRONMENTAL SCIENCE

implementing water conservation measures, improving irrigation efficiency, and developing water storage and distribution systems.

4. **Climate-resilient Agriculture:** Implementing climate-smart agricultural practices that are adapted to local conditions, such as crop diversification, agroforestry, and improved water management, can help farmers cope with changing climate conditions and ensure food security.

5. **Health and Disaster Preparedness:** Strengthening healthcare systems and disaster preparedness is essential in the face of climate change impacts. This includes developing early warning systems, improving emergency response mechanisms, and enhancing public health infrastructure to prevent and manage climate-related health risks.

6. **Community Engagement and Education:** Engaging local communities and providing education and training on climate change impacts and adaptation strategies can empower individuals and communities to take proactive measures. This can involve raising awareness, facilitating knowledge sharing, and building community resilience networks.

Overall, a combination of mitigation and adaptation strategies is needed to effectively address the challenges of climate change. While mitigation efforts aim to reduce the root causes of climate change by reducing GHG emissions, adaptation strategies are crucial for managing the unavoidable impacts that have already been set in motion. The implementation of these strategies requires collaboration and coordination at all levels, from individuals and communities to national and international organizations. By taking collective action, we can work towards a more sustainable and climate-resilient future.

Case Study: The Great Barrier Reef

The Great Barrier Reef, located off the coast of Australia, is one of the world's most iconic and diverse ecosystems. However, it is facing numerous threats, including climate change, pollution, and overfishing. As a case study, let's explore the mitigation and adaptation strategies being employed to protect this valuable ecosystem.

Mitigation Strategies: - Renewable Energy Transition: Australia has made significant investments in renewable energy sources such as solar and wind power to reduce its carbon emissions. - Carbon Capture and Storage: Research is underway to explore the feasibility of capturing and storing carbon dioxide emissions from industrial sources. - Water Quality Improvement: Efforts are being made to reduce pollution runoff from agricultural and urban areas to improve water quality and reduce stress on the reef.

Adaptation Strategies: - Reef Restoration: Restoration programs are in place to assist the recovery of damaged and degraded areas of the reef, including the planting of coral fragments and the removal of invasive species. - Climate Resilience Planning: The Australian government is developing plans to enhance the resilience of the Great Barrier Reef by addressing key climate change impacts such as rising sea temperatures and ocean acidification. - Partnerships and Collaborations: Stakeholders including government agencies, local communities, and scientific organizations are working together to monitor and assess the reef's health, identify threats, and implement adaptive management strategies.

Despite these efforts, the Great Barrier Reef continues to face significant challenges. The pace of climate change remains a major concern, with rising sea temperatures and increased frequency of coral bleaching events threatening the survival of coral communities. Continued monitoring, research, and adaptation measures are essential to ensure the long-term resilience and survival of this unique ecosystem.

In conclusion, mitigation and adaptation strategies play a crucial role in addressing the challenges posed by climate change. These strategies, when implemented effectively, can help reduce greenhouse gas emissions, protect vulnerable ecosystems, build resilience in communities, and ensure a sustainable future for generations to come. However, a collective effort is required at all levels to mobilize resources, support research and innovation, enact policies, and implement measures that can mitigate climate change and enhance resilience. By adopting a proactive approach, we can work towards a greener, more sustainable future.

Biodiversity Loss

Biodiversity loss is a pressing issue in today's world, with significant consequences for the environment and human well-being. Biodiversity refers to the variety of life on Earth, including the genetic, species, and ecosystem diversity. It is an essential component of the planet's natural capital and plays a vital role in maintaining ecological balance and offering numerous benefits to humanity.

Causes of Biodiversity Loss

The loss of biodiversity is primarily caused by human activities, which have resulted in extensive habitat destruction, overexploitation of natural resources, pollution, climate change, and invasive species. These factors interact and contribute to the decline of species and ecosystems worldwide.

CHALLENGES AND OPPORTUNITIES IN ENVIRONMENTAL SCIENCE

Habitat destruction is one of the leading causes of biodiversity loss. As human populations grow, there is an increasing demand for land for agriculture, urbanization, and infrastructure development. Large-scale deforestation and land conversion for agriculture, such as for livestock grazing and monoculture crop production, have resulted in the destruction of natural habitats, leading to the displacement and extinction of many species.

Overexploitation of natural resources, including hunting, fishing, and logging, has also played a significant role in biodiversity loss. Unsustainable fishing practices, such as overfishing and destructive fishing methods, have led to the depletion of fish stocks and the disruption of marine ecosystems. Similarly, illegal logging and unsustainable timber harvesting practices have resulted in the destruction of forests and the loss of habitat for countless species.

Pollution, particularly from industrial and agricultural activities, poses a severe threat to biodiversity. Water and air pollution from chemical contaminants, such as pesticides, fertilizers, and industrial waste, can have detrimental effects on aquatic and terrestrial ecosystems. Pollution can disrupt the balance of ecosystems, harm biodiversity, and jeopardize the health of species.

Climate change exacerbates the loss of biodiversity by altering ecosystems, disrupting natural processes, and changing the distribution and abundance of species. Rising temperatures, changing rainfall patterns, and extreme weather events can lead to the decline of vulnerable species, such as polar bears and coral reefs. Climate change also affects ecological interactions, such as pollination and species interactions, further impacting biodiversity.

The introduction of invasive species is another significant contributor to biodiversity loss. Invasive species can outcompete native species for resources, disrupt ecosystems, and cause the extinction of native plants and animals. The spread of invasive species is often facilitated by human activities, such as international trade and transport.

Impacts of Biodiversity Loss

The loss of biodiversity has far-reaching consequences for both the environment and human societies. Biodiversity loss disrupts ecosystem functioning and reduces the resilience of ecosystems to environmental changes. This can lead to a cascade of negative effects, including decreased soil fertility, reduced water quality, decreased carbon sequestration, and increased vulnerability to natural disasters.

Biodiversity loss also affects food security and the availability of natural resources. Many species play a crucial role in pollination, seed dispersal, and nutrient cycling, which are essential for agriculture and the functioning of ecosystems. Declines in pollinators, such as bees and butterflies, can have significant impacts on crop production and food availability.

Furthermore, biodiversity loss has implications for human health. Natural ecosystems provide numerous ecosystem services that contribute to human well-being, including the provision of clean air and water, regulation of climate, and the development of medicines. The loss of biodiversity reduces the availability of these services, increasing the vulnerability of communities to diseases, pollution, and natural disasters.

Conservation Efforts and Success Stories

Efforts to conserve biodiversity are crucial to mitigating further loss and preserving the planet's unique ecosystems and species. Conservation strategies involve a combination of measures, including protected areas, habitat restoration, species conservation, and sustainable management of natural resources.

Protected areas, such as national parks and reserves, play a vital role in preserving biodiversity. These areas provide safe havens for species and ecosystems, allowing them to thrive and recover. Many success stories, such as the establishment of the Galapagos Islands National Park and the Great Barrier Reef Marine Park, demonstrate how protected areas can contribute to the conservation of biodiversity.

Habitat restoration initiatives are essential for recovering degraded ecosystems and creating suitable habitats for endangered species. Projects like the restoration of wetlands, reforestation efforts, and urban ecological restoration have shown positive results in reversing biodiversity loss and creating sustainable habitats for plants and animals.

Species conservation programs aim to protect and recover endangered species. These initiatives involve captive breeding programs, habitat protection, and reintroduction efforts. Success stories include the recovery of species such as the black-footed ferret and the California condor through concerted conservation efforts.

Sustainable management of natural resources is another key aspect of biodiversity conservation. By promoting sustainable practices, such as sustainable forestry and fisheries management, we can reduce the impact on ecosystems and ensure the long-term viability of natural resources. Good governance, community involvement, and international cooperation are vital for the success of these initiatives.

Challenges and Future Directions

Despite numerous conservation efforts, biodiversity loss remains an ongoing challenge. Addressing this challenge requires a multi-faceted approach that includes policy changes, increased awareness, and sustainable practices.

One of the main challenges is the lack of political will and commitment to biodiversity conservation. Strong policies and regulations are needed to protect critical habitats, control deforestation, and enforce sustainable resource management practices. Additionally, international collaboration and agreements, such as the Convention on Biological Diversity, are crucial for promoting global biodiversity conservation.

Another challenge is the need to integrate indigenous knowledge and perspectives into biodiversity conservation efforts. Indigenous peoples have profound traditional knowledge and sustainable practices that can contribute to effective biodiversity conservation. Recognizing and respecting their rights and involvement in decision-making are essential for achieving successful conservation outcomes.

Climate change poses an additional hurdle to biodiversity conservation. Adapting to the impacts of climate change and reducing greenhouse gas emissions are crucial for preserving ecosystems and species. Conservation efforts need to factor in climate change resilience and incorporate adaptive management strategies.

Education and public awareness play a vital role in biodiversity conservation. Promoting environmental education, sustainable lifestyles, and raising awareness about the value of biodiversity can foster a sense of responsibility and encourage conservation action at individual and collective levels.

In conclusion, biodiversity loss is a significant environmental concern with severe consequences for ecosystems and human well-being. Understanding the causes, impacts, and conservation strategies is crucial for addressing this global challenge. By implementing effective conservation measures, promoting sustainable practices, and integrating diverse perspectives, we can work towards preserving biodiversity and ensuring a greener future.

1.6.5 The Importance of Biodiversity

Introduction

Biodiversity is a term used to describe the variety of life on Earth, including the diversity of species, ecosystems, and genetic variations within species. It is crucial for the functioning of ecosystems and provides numerous benefits to human society. In this section, we will explore the importance of biodiversity and its role in maintaining ecological stability, supporting human livelihoods, and contributing to the overall health of the planet.

Ecosystem Functioning

Biodiversity is essential for the proper functioning of ecosystems. Each species within an ecosystem plays a unique role, and the interactions between species contribute to the stability and resilience of the ecosystem as a whole. The loss of even a single species can have cascading effects on the entire ecosystem.

One of the key functions of biodiversity is the regulation of ecological processes. For example, a diverse community of plant species supports the cycling of nutrients, water purification, and soil formation. In forests, different tree species interact to enhance carbon sequestration and regulate the climate. Biodiversity also plays a critical role in the pollination of plants, which is vital for the reproduction of many crops and wildflowers.

Economic Importance

Biodiversity provides a wide range of economic benefits to human society. It forms the basis for numerous industries such as agriculture, forestry, fisheries, and pharmaceuticals. Genetic diversity within species ensures that crops and livestock have the ability to adapt to changing environmental conditions and disease outbreaks, which is essential for food security.

Many pharmaceutical drugs are derived from natural compounds found in plants, animals, and microorganisms. For example, the anti-cancer drug Taxol is derived from the Pacific yew tree. Biodiversity also contributes to the development of new technologies and materials. The study of natural forms, structures, and processes has inspired innovations in engineering, architecture, and design.

Cultural and Aesthetic Value

Biodiversity is deeply intertwined with human culture and has significant aesthetic value. Many societies have rich cultural traditions and rituals that revolve around local ecosystems and species. Indigenous communities, in particular, have traditional knowledge systems that rely on a deep understanding of local biodiversity for survival and well-being.

The natural beauty and diversity of species also provide important aesthetic and recreational value. People are drawn to nature for its tranquility, inspiration, and

1.6.5 THE IMPORTANCE OF BIODIVERSITY

connection to the larger web of life. Protected areas, such as national parks and wildlife reserves, are established to preserve and appreciate the beauty and ecological significance of natural habitats.

Ecological Resilience

Biodiversity plays a crucial role in maintaining the resilience of ecosystems. A diverse ecosystem is better able to withstand and recover from disturbances, such as natural disasters, climate change, and invasive species. It provides a buffer against environmental changes and ensures the continued functioning of essential ecosystem services.

Species diversity increases the stability and productivity of ecosystems by reducing the vulnerability to diseases and pests. It allows for efficient resource utilization and prevents the dominance of particular species, which can lead to imbalances and ecosystem degradation. In essence, biodiversity acts as a safety net, ensuring the long-term health and sustainability of ecosystems.

Addressing Biodiversity Loss

Despite its critical importance, biodiversity is facing unprecedented loss due to human activities. Habitat destruction, pollution, climate change, invasive species, overexploitation, and other drivers of biodiversity loss are putting ecosystems and species at risk. Addressing this issue requires a multifaceted approach, encompassing conservation efforts, sustainable resource management, and policy interventions.

Conservation efforts play a vital role in protecting and restoring biodiversity. Strategies include the establishment of protected areas, habitat restoration projects, endangered species protection, and the reduction of wildlife trade. Moreover, incorporating local communities and indigenous peoples in conservation initiatives is crucial for the sustainable management of natural resources.

Sustainable resource management practices, such as sustainable agriculture and fisheries, can minimize the negative impacts on biodiversity. By adopting practices that promote biodiversity conservation, such as organic farming and responsible fishing methods, we can ensure the long-term viability of ecosystems and the species within them.

Policy interventions at local, national, and international levels are needed to address the drivers of biodiversity loss. These include regulations to protect habitats, enforce sustainable practices, and promote environmental education and awareness. Collaboration between governments, organizations, scientists, and communities is key to achieving effective biodiversity conservation.

Conclusion

Biodiversity is of paramount importance for the health and well-being of both ecosystems and human society. It provides critical ecosystem services, supports livelihoods, contributes to cultural identity, and enhances ecological resilience. Protecting and restoring biodiversity is a shared responsibility, requiring the concerted efforts of individuals, governments, and organizations. By recognizing the value of biodiversity and taking action to conserve it, we can ensure a greener and more sustainable future for generations to come.

Key Terms

- Biodiversity: The variety of life on Earth, including the diversity of species, ecosystems, and genetic variations within species.

- Ecosystem functioning: The processes and interactions that occur within an ecosystem, such as nutrient cycling, energy flow, and species interactions.

- Ecological resilience: The ability of an ecosystem to withstand and recover from disturbances, maintaining its essential functions and services.

- Protected areas: Designated regions that are set aside and managed to protect biodiversity and natural habitats.

- Sustainable resource management: Practices that aim to use natural resources in a way that maintains their availability for future generations, without degrading ecosystems.

- Conservation: The protection, management, and restoration of biodiversity and natural resources.

Further Reading

1. Noss, R. F. (1990). Indicators for monitoring biodiversity: A hierarchical approach. Conservation Biology, 4(4), 355-364.

2. Cardinale, B. J., et al. (2012). Biodiversity loss and its impact on humanity. Nature, 486(7401), 59-67.

1.6.5 THE IMPORTANCE OF BIODIVERSITY

3. Díaz, S., et al. (2019). Pervasive human-driven decline of life on Earth points to the need for transformative change. Science, 366(6471), eaax3100.

Discussion Questions

1. Why is biodiversity important for the proper functioning of ecosystems?

2. Give an example of how biodiversity contributes to the economic value of natural resources.

3. How can communities and individuals contribute to biodiversity conservation?

4. What are the challenges faced in addressing biodiversity loss, and what can be done to overcome them?

Threats to Biodiversity

Biodiversity refers to the variety of life on Earth, encompassing the diverse range of species, ecosystems, and genetic diversity within species. It is essential for maintaining the stability and functioning of ecosystems, providing valuable ecosystem services such as food production, water purification, and climate regulation. However, biodiversity is currently facing numerous threats, many of which are caused by human activities. In this section, we will explore some of the major threats to biodiversity and examine the implications for our planet's ecosystems.

Habitat Loss and Fragmentation

One of the leading causes of biodiversity loss is habitat loss and fragmentation. Human activities, such as deforestation, urbanization, and conversion of natural habitats for agriculture, result in the destruction of habitats that support many species. As habitats are cleared or altered, species may lose their homes, become isolated, and face difficulties in finding resources and mates. Fragmentation disrupts ecological processes, reduces genetic diversity, and increases the vulnerability of populations to extinction.

Example: The Amazon rainforest, one of the most biodiverse regions on Earth, is rapidly losing its habitats due to deforestation for agricultural purposes. This ongoing destruction threatens countless species, including the iconic jaguar, giant otter, and harpy eagle, as their habitats shrink and become fragmented.

Solution: Conservation efforts should focus on protecting and restoring habitats, minimizing the conversion of natural areas, and promoting sustainable land-use practices. This includes establishing protected areas, implementing reforestation and habitat restoration projects, and implementing land-use planning that considers biodiversity conservation.

Climate Change

Climate change poses a significant threat to biodiversity worldwide. Rising temperatures, altered precipitation patterns, and extreme weather events can disrupt ecosystems and threaten the survival of many species. Climate change also affects species distribution, as some species may be unable to adapt quickly enough to changing climatic conditions. Additionally, climate change can lead to the loss of important habitats, such as coral reefs and polar ice caps, which support unique and highly specialized species.

Example: The Great Barrier Reef, the world's largest coral reef system, is under immense stress due to climate change. Rising ocean temperatures, combined with ocean acidification, have caused widespread coral bleaching, leading to the loss of biodiversity in this fragile ecosystem.

Solution: Mitigating climate change by reducing greenhouse gas emissions is crucial for protecting biodiversity. This includes transitioning to renewable energy sources, implementing energy-efficient practices, and promoting sustainable transportation. Conservation strategies should also focus on enhancing the resilience of ecosystems to climate change through measures like ecosystem restoration and conservation planning that accounts for climate change impacts.

Invasive Species

Invasive species are non-native species that are introduced intentionally or unintentionally into new environments. They can outcompete and displace native species, disrupt ecological processes, and alter ecosystem dynamics. Invasive species often have no natural predators or diseases in their new habitats, allowing them to thrive and rapidly reproduce. Their presence can lead to a decline in native species populations and negatively impact biodiversity.

Example: The brown tree snake in Guam is an invasive species that has caused the extinction or decline of many bird species on the island. This snake, which is not native to Guam, preys on native birds and their eggs, leading to dramatic declines in bird populations.

1.6.5 THE IMPORTANCE OF BIODIVERSITY

Solution: Preventing the introduction and spread of invasive species is crucial for protecting biodiversity. Strategies include strengthening biosecurity measures, conducting risk assessments for introduced species, and implementing early detection and rapid response programs. Additionally, ecosystem restoration and habitat management can help reduce the vulnerability of native species to invasive species.

Overexploitation and Illegal Wildlife Trade

Overexploitation of natural resources, including hunting, fishing, and poaching, poses a significant threat to biodiversity. Unsustainable harvesting of plants and animals for food, medicine, or commercial purposes can lead to population declines and even extinction of species. Illegal wildlife trade, fueled by demand for exotic pets, traditional medicines, and luxury goods made from endangered species, further exacerbates the problem.

Example: The illegal trade of ivory, derived from poached elephants, has led to a significant decline in elephant populations across Africa. Poaching and the demand for elephant tusks have driven elephants to the brink of extinction in some regions.

Solution: Conservation efforts should focus on implementing and enforcing regulations to prevent overexploitation and illegal wildlife trade. This includes strengthening law enforcement, raising awareness about the impacts of the trade, and promoting sustainable alternatives to the use of endangered species. Community-based conservation approaches that involve local communities in the management of natural resources can also contribute to sustainable resource use.

Pollution

Pollution, including air and water pollution, poses a threat to biodiversity and ecosystems. Industrial activities, agricultural runoff, and improper waste management introduce pollutants into the environment, affecting the health and survival of many species. Pollution can contaminate habitats, disrupt ecological processes, and negatively impact reproductive capabilities, leading to population declines and loss of biodiversity.

Example: Pesticide use in agriculture has been linked to declines in pollinator populations, such as bees and butterflies. The accumulation of pesticide residues in their habitats and food sources impairs their reproduction, behavior, and immune systems, ultimately threatening their populations.

Solution: To address pollution threats to biodiversity, it is essential to adopt cleaner production practices, promote the use of environmentally friendly alternatives, and improve waste management systems. Strict regulation and enforcement of pollution control measures are crucial to minimizing the impacts of pollution on biodiversity.

Resource Extraction and Habitat Degradation

The extraction of natural resources, such as minerals, oil, and gas, often involves the destruction or degradation of habitats. Mining, logging, and infrastructure development can result in the loss of critical ecosystems and the displacement of species. Habitat degradation through activities like soil erosion, stream channelization, and land degradation also poses a threat to biodiversity.

Example: The expansion of palm oil plantations in Southeast Asia has resulted in extensive deforestation, leading to the destruction of habitats for numerous plant and animal species, including orangutans, tigers, and rhinos.

Solution: Sustainable resource extraction practices that prioritize environmental conservation and restoration are crucial for minimizing the impacts on biodiversity. This includes adopting responsible mining techniques, promoting sustainable forestry practices, and implementing measures to avoid and mitigate habitat degradation. It is also essential to consider the social and cultural impacts of resource extraction on local communities.

Conclusion

Biodiversity is facing unprecedented threats, largely due to human activities. The loss of biodiversity not only compromises the ecological integrity of ecosystems but also puts human well-being at risk. It is crucial to recognize the importance of biodiversity conservation and take immediate action to address the threats it faces. By implementing effective conservation strategies, embracing sustainable practices, and promoting awareness and education, we can protect biodiversity and contribute to a more sustainable and greener future.

Conservation Efforts and Success Stories

Conservation efforts play a crucial role in protecting and preserving our natural environment. They aim to prevent the loss of biodiversity, protect ecosystems, and promote sustainable use of resources. In this section, we will explore some inspiring success stories in conservation and highlight the importance of these efforts in safeguarding our planet for future generations.

1.6.5 THE IMPORTANCE OF BIODIVERSITY 67

Conservation Efforts

Conservation efforts encompass a wide range of activities and strategies aimed at preserving the Earth's biodiversity and ecosystems. These efforts can be categorized into several key areas:

Protected Areas and National Parks: One of the most effective ways to conserve natural habitats and biodiversity is through the establishment of protected areas and national parks. These areas provide a sanctuary for endangered species and allow for the conservation of unique ecosystems. For example, the Galapagos Islands in Ecuador have been designated as a national park and a World Heritage site, protecting its rich biodiversity and contributing to scientific research and ecotourism.

Species Conservation: Many organizations and institutions work tirelessly to conserve endangered species and prevent their extinction. These efforts often involve habitat restoration, captive breeding programs, and monitoring of populations. The recovery of the California condor, one of the world's most endangered birds, is a remarkable conservation success story. Through a combination of captive breeding and reintroduction efforts, the population has increased from a mere 27 individuals in the 1980s to over 400 birds today.

Community-led Conservation: Engaging local communities in conservation efforts is essential for long-term success. By involving local residents in decision-making processes and providing them with economic incentives, conservation initiatives become more sustainable. A prime example is the indigenous-led conservation initiatives in the Amazon rainforest. Indigenous people, who have deep cultural and spiritual connections to the land, play a vital role in protecting the forest and its biodiversity from illegal logging and deforestation.

Marine Conservation: The world's oceans face numerous threats, including overfishing, pollution, and habitat destruction. Effective marine conservation efforts involve the establishment of marine protected areas, sustainable fishing practices, and education and awareness campaigns. The establishment of marine protected areas in the Great Barrier Reef, such as the Great Barrier Reef Marine Park, has contributed to the conservation of this iconic ecosystem and its abundant marine life.

Conservation through Technology: Advances in technology have revolutionized conservation efforts. Remote sensing and satellite imagery, for example, allow scientists to monitor changes in land cover and assess the health of ecosystems from a global scale. Drones and camera traps help in the monitoring of wildlife populations and provide valuable data for conservation planning. Technology also enables the development of innovative solutions, such as using artificial intelligence algorithms to detect and combat illegal wildlife trade online.

Success Stories in Conservation

Conservation efforts have yielded significant success stories, demonstrating that positive change is possible. Here, we highlight a few notable examples:

The Recovery of the Bald Eagle: The bald eagle, the national bird of the United States, was once on the brink of extinction due to habitat destruction and the widespread use of the pesticide DDT. However, thanks to conservation efforts and the banning of DDT, the population of bald eagles has made a remarkable recovery. It is now considered a conservation success story and a symbol of successful wildlife management.

The Revival of the Iberian Lynx: The Iberian lynx, the world's most endangered cat species, faced imminent extinction due to habitat loss and a decline in its main prey, the European rabbit. Conservation efforts, including habitat restoration and captive breeding programs, have led to a significant increase in the lynx population. This success story showcases the importance of targeted conservation measures for endangered species.

The Conservation of the Giant Panda: The giant panda, a beloved symbol of wildlife conservation, was once on the verge of extinction. Through rigorous conservation efforts, including habitat protection and captive breeding programs, the population of giant pandas has seen a steady increase. This success story demonstrates the positive impact of conservation initiatives on charismatic and emblematic species.

The Protection of the Amazon Rainforest: The Amazon rainforest, often referred to as the "lungs of the Earth," is a biodiverse hotspot facing severe threats from deforestation. Conservation organizations, local communities, and governments have collaborated to establish protected areas, implement sustainable

1.6.5 THE IMPORTANCE OF BIODIVERSITY

land-use practices, and combat illegal logging. These efforts have contributed to the preservation of the Amazon rainforest, one of the world's most valuable ecosystems.

The Restoration of the Wolf Population in Yellowstone National Park: The reintroduction of gray wolves into Yellowstone National Park in the United States has had a profound impact on the ecosystem. The presence of wolves has led to a cascade of ecological effects, including the control of herbivore populations and the regeneration of riparian vegetation. This success story demonstrates the essential role of top predators in ecosystem functioning and the potential for ecological restoration.

Challenges and Future Directions in Conservation

While there have been notable achievements in conservation, many challenges persist. Addressing these challenges is crucial for the long-term success of conservation efforts. Here are some of the key challenges and future directions in conservation:

Human-Wildlife Conflicts: As human populations expand and encroach upon natural habitats, conflicts between humans and wildlife become more common. Finding effective ways to mitigate these conflicts while ensuring the welfare of both humans and animals is a pressing challenge in conservation biology. This often involves developing sustainable land-use practices, implementing predator-prey management strategies, and fostering community-based conservation initiatives.

Illegal Wildlife Trade: The illegal trade of wildlife and wildlife products, including ivory, rhino horn, and exotic pets, poses a severe threat to many species. Addressing this issue requires international cooperation, strengthened law enforcement, and education and awareness campaigns to reduce demand. The use of innovative approaches, such as DNA barcoding to trace the origin of illegally traded wildlife products, can aid in combating this illegal trade.

Habitat Fragmentation and Loss: Habitat fragmentation, caused by human activities such as agriculture and urbanization, disrupts ecosystems and isolates populations, leading to a loss of biodiversity. Conservation efforts must focus on preserving and restoring habitat connectivity to allow for the movement of species and the exchange of genetic material. This often involves the establishment of ecological corridors and the restoration of degraded habitats.

Climate Change and Species Conservation: Climate change poses a significant threat to biodiversity, affecting species distribution, phenology, and ecosystem dynamics. Conservation strategies must incorporate climate change adaptation and mitigation measures. This may include protecting climate refugia, promoting ecosystem resilience, and reducing greenhouse gas emissions through the use of renewable energy sources.

Engaging Local Communities in Conservation Efforts: Meaningful engagement with local communities is essential for the success of conservation initiatives. This includes recognizing and respecting traditional knowledge and cultural values, involving local residents in decision-making processes, and providing economic incentives for sustainable resource use. Collaborative approaches that empower communities and promote sustainable livelihoods can lead to more effective and sustainable conservation outcomes.

In conclusion, conservation efforts and success stories demonstrate the vital importance of protecting and preserving our natural environment. Through targeted conservation measures, engaging local communities, and harnessing technological advancements, we can safeguard biodiversity, restore ecosystems, and create a more sustainable future. However, challenges such as human-wildlife conflicts, illegal wildlife trade, habitat fragmentation, and climate change require ongoing attention and creative solutions. By working together, we can ensure the long-term survival and well-being of both current and future generations.

Pollution and Environmental Health

Pollution is the introduction of harmful substances or contaminants into the environment, resulting in adverse effects on living organisms and their surroundings. It is a major concern in environmental science as it poses serious threats to human health and the ecosystem. In this section, we will explore the different types and sources of pollution, the impacts on human health, and the measures taken to prevent and control pollution.

Types and Sources of Pollution

There are various types of pollution that we encounter in our daily lives. Understanding these types and their sources is crucial for designing effective pollution control strategies.

1.6.5 THE IMPORTANCE OF BIODIVERSITY

Air Pollution   Air pollution refers to the release of harmful substances into the atmosphere, resulting in the deterioration of air quality. The main sources of air pollution include industrial emissions, vehicular exhaust, burning of fossil fuels, and wildfires. Pollutants such as carbon monoxide, sulfur dioxide, nitrogen oxides, and particulate matter can have serious health effects, including respiratory issues, cardiovascular problems, and increased susceptibility to lung infections.

Water Pollution   Water pollution occurs when harmful substances contaminate water bodies, such as rivers, lakes, and oceans. Sources of water pollution include industrial discharge, agricultural runoff, improper waste disposal, and sewage leakage. Pollutants such as heavy metals, pesticides, fertilizers, and pathogens can pose a risk to human health and aquatic ecosystems. Water pollution can lead to waterborne diseases, ecological imbalances, and reduced biodiversity.

Soil Pollution   Soil pollution refers to the contamination of soil with toxic substances, making it unsuitable for plant growth and harming soil organisms. Industrial activities, mining operations, use of agrochemicals, and improper waste disposal contribute to soil pollution. Contaminants like heavy metals, pesticides, and petroleum hydrocarbons can persist in the soil for a long time, affecting food safety and ecosystem health.

Noise Pollution   Noise pollution is the excessive or disturbing noise that disrupts normal human activities and causes stress and annoyance. Sources of noise pollution include transportation systems, industrial machinery, construction sites, and loud recreational activities. Prolonged exposure to high levels of noise can lead to hearing loss, sleep disturbances, and increased risk of cardiovascular diseases.

Light Pollution   Light pollution refers to the excessive or misdirected artificial light that obscures the visibility of stars and disrupts natural light-dark cycles. It is mainly caused by streetlights, commercial buildings, and advertising signs. Light pollution not only affects astronomical observations but also disrupts the behavior and habitats of nocturnal animals. It can also have detrimental effects on human health, including sleep disorders and disturbance of circadian rhythms.

Human Health Impacts

Pollution has significant impacts on human health, ranging from acute to long-term effects. The severity of health impacts depends on the type and level of pollutants, as well as the duration and frequency of exposure.

Respiratory Diseases Exposure to air pollution, particularly fine particulate matter (PM2.5) and pollutants like ozone and nitrogen dioxide, can lead to respiratory issues such as asthma, bronchitis, and chronic obstructive pulmonary disease (COPD). These pollutants can irritate the respiratory system, cause inflammation, and compromise lung function.

Cardiovascular Diseases Pollution, especially air pollution, is associated with an increased risk of cardiovascular diseases such as heart attacks, strokes, and high blood pressure. Fine particulate matter, hazardous air pollutants, and traffic-related pollutants can enter the bloodstream, causing systemic inflammation, oxidative stress, and dysfunction of blood vessels.

Cancer Certain pollutants, such as benzene, asbestos, and polycyclic aromatic hydrocarbons (PAHs), are recognized as carcinogens and can increase the risk of various types of cancer. Industrial processes, vehicle exhaust, and tobacco smoke are common sources of these carcinogenic pollutants.

Neurological Disorders Exposure to pollutants, including lead, mercury, and pesticides, is associated with an increased risk of neurodevelopmental disorders in children and neurodegenerative diseases in adults. These pollutants can impair brain development, affect cognitive function, and contribute to conditions like autism, attention deficit hyperactivity disorder (ADHD), and Alzheimer's disease.

Reproductive and Maternal Health Pollution can have adverse effects on reproductive health, causing infertility, hormonal imbalances, and complications during pregnancy. Contaminants like endocrine-disrupting chemicals, heavy metals, and air pollutants can interfere with reproductive processes and fetal development.

Pollution Prevention and Control Measures

To address the risks and impacts of pollution, various prevention and control measures are implemented at different levels, including individuals, industries, and governments.

Reducing Emissions One of the primary strategies to control pollution is to reduce emissions of harmful pollutants. This can be achieved through the adoption of cleaner technologies and alternative energy sources, as well as the

1.6.5 THE IMPORTANCE OF BIODIVERSITY

implementation of stringent emission standards for industries and vehicles. Encouraging energy-efficient practices, such as promoting public transportation and adopting renewable energy, can contribute to reducing air pollution.

Waste Management Proper waste management is crucial in preventing pollution. Implementing recycling programs, promoting the use of biodegradable materials, and ensuring safe disposal of hazardous waste can minimize the release of pollutants into the environment. Effective waste management also involves reducing plastic consumption and implementing measures to tackle plastic pollution in water bodies.

Environmental Regulations Governments play a vital role in regulating pollution through the formulation and enforcement of environmental policies and regulations. These regulations set emission standards, prescribe limits on pollutant releases, and impose penalties for non-compliance. By monitoring and enforcing these regulations, governments can ensure industries and individuals adhere to environmentally responsible practices.

Public Awareness and Education Raising public awareness about the impacts of pollution and promoting responsible environmental behavior is essential. Educating individuals about ways to reduce pollution, conserve resources, and adopt sustainable practices can empower them to make informed choices that contribute to a greener and healthier future.

International Collaboration Pollution is a global issue that requires international collaboration to address its causes and impacts. Cooperation among nations in sharing knowledge, technologies, and best practices can lead to innovative solutions for pollution prevention and control. International agreements and conventions, such as the Stockholm Convention on Persistent Organic Pollutants and the Paris Agreement on Climate Change, provide frameworks for global action against pollution.

Case Study: Air Pollution in Delhi, India

To illustrate the real-world impact of pollution on environmental health, let's examine the case of air pollution in Delhi, India. Delhi has been grappling with severe air pollution levels for several years, particularly during the winter months. The main contributors to air pollution in Delhi include vehicular emissions, industrial pollution, construction activities, and agricultural burning.

The high levels of air pollution in Delhi have serious implications for public health. The exposure to fine particulate matter (PM2.5) and other pollutants has led to a significant increase in respiratory diseases, cardiovascular problems, and overall reduced quality of life for the residents of the city. Children, the elderly, and individuals with pre-existing respiratory conditions are particularly vulnerable to the health impacts of air pollution.

To tackle the air pollution crisis, the government of Delhi has implemented several measures. These include the introduction of stricter emission norms for vehicles, promoting the use of cleaner fuels like compressed natural gas (CNG), and restricting the operation of polluting industries. The government has also launched initiatives to curb crop residue burning, which is a major contributor to air pollution in the region.

Public awareness campaigns and the involvement of citizens in monitoring air quality have played a crucial role in addressing the issue. The government has set up real-time air quality monitoring stations across the city, providing people with access to up-to-date information on air pollution levels. This has enabled individuals to take necessary precautions and make informed decisions to protect their health.

While significant progress has been made, air pollution in Delhi remains a complex and challenging problem. It requires sustained efforts from all stakeholders, including the government, industries, and individuals, to ensure long-term improvements in air quality and safeguard the health of the population.

Key Takeaways

- Pollution is the introduction of harmful substances or contaminants into the environment, leading to adverse effects on human health and ecosystems.

- Different types of pollution include air, water, soil, noise, and light pollution, each with unique sources and impacts.

- Pollution can cause a range of health issues, including respiratory diseases, cardiovascular problems, cancer, neurological disorders, and reproductive complications.

- Pollution prevention and control measures involve reducing emissions, proper waste management, implementing environmental regulations, raising public awareness, and fostering international collaboration.

- Case studies, such as air pollution in Delhi, highlight the real-world consequences of pollution and the need for effective interventions.

It is crucial to address pollution and prioritize environmental health to create a greener and sustainable future for all. By understanding the sources, impacts, and control measures of pollution, we can make informed decisions and contribute to the

1.6.5 THE IMPORTANCE OF BIODIVERSITY

preservation of our planet. Together, we can pave the way for a healthier and more sustainable world.

Types and Sources of Pollution

Pollution is the introduction of harmful substances or contaminants into the environment, causing adverse effects on living organisms and ecosystems. It can occur in various forms and arise from different sources. Understanding the types and sources of pollution is crucial in developing effective solutions to mitigate its impacts. In this section, we will explore the major types and sources of pollution and their implications for environmental science and sustainability.

Air Pollution

Air pollution refers to the presence of harmful substances or pollutants in the earth's atmosphere. It can be caused by both human activities and natural processes. Some common sources of air pollution include:

- **Industrial Emissions:** The combustion of fossil fuels in industrial processes releases pollutants such as sulfur dioxide (SO_2), nitrogen oxides (NO_x), and particulate matter. These emissions contribute to respiratory diseases, acid rain, and smog formation.

- **Vehicle Emissions:** The burning of gasoline and diesel fuels in vehicles releases pollutants, including carbon monoxide (CO), nitrogen oxides, volatile organic compounds (VOCs), and particulate matter. Exposure to vehicle emissions can lead to respiratory problems and cardiovascular diseases.

- **Power Generation:** Power plants that burn coal, oil, or natural gas release pollutants, including sulfur dioxide, nitrogen oxides, and mercury. These emissions contribute to regional and global air pollution, as well as climate change.

- **Agriculture:** Agricultural activities such as crop burning, livestock production, and the use of fertilizers and pesticides release pollutants, including ammonia, methane, and volatile organic compounds. These emissions contribute to air pollution and climate change.

- **Waste Management:** Improper disposal of solid waste, including open burning and landfills, releases pollutants such as methane, volatile organic compounds, and hazardous chemicals. These emissions pose health risks and contribute to greenhouse gas emissions.

+ **Natural Sources:** Volcanic eruptions, wildfires, and dust storms are natural sources of air pollution. While their impact is relatively localized and temporary, they can cause significant air quality issues in affected areas.

Water Pollution

Water pollution refers to the contamination of water bodies, such as rivers, lakes, oceans, and groundwater, by harmful substances or pollutants. The sources of water pollution can be categorized as follows:

+ **Industrial Discharges:** Industrial processes release a wide range of pollutants into water bodies, including heavy metals, toxic chemicals, and organic compounds. These discharges can significantly degrade water quality and harm aquatic ecosystems.

+ **Agricultural Runoff:** The use of fertilizers, pesticides, and animal waste in agriculture leads to the contamination of water bodies through runoff. Excessive nutrients, such as nitrogen and phosphorus, can cause eutrophication, leading to oxygen depletion and impaired aquatic life.

+ **Municipal Wastewater:** Improperly treated or untreated sewage from residential and commercial areas can contain pathogens, nutrients, and toxic substances. Discharging untreated wastewater directly into water bodies can spread diseases and degrade water quality.

+ **Oil Spills:** Accidental or intentional releases of oil into water bodies, usually from transportation and offshore drilling activities, can have severe impacts on aquatic ecosystems. Oil spills can suffocate marine organisms, contaminate their habitats, and disrupt the food chain.

+ **Plastic Pollution:** Improper disposal and inadequate waste management of plastic products result in their accumulation in water bodies. Plastics pose a significant threat to marine life through entanglement, ingestion, and the leaching of toxic chemicals.

+ **Acid Rain:** Emissions of sulfur dioxide and nitrogen oxides from industrial activities and fossil fuel combustion can combine with atmospheric moisture, forming acidic compounds that eventually fall as acid rain. Acid rain can acidify water bodies, harming aquatic organisms and ecosystems.

Soil Pollution

Soil pollution refers to the contamination of the soil by substances that are toxic to plants, animals, or humans. The sources of soil pollution include:

+ **Industrial Activities:** Improper disposal of industrial waste and the release of pollutants into the soil can contaminate large areas of land. Heavy metals, solvents, pesticides, and petroleum hydrocarbons are common contaminants.

+ **Agricultural Practices:** The use of pesticides, fertilizers, and animal manure in agriculture can lead to the accumulation of harmful substances in the soil. Overuse of pesticides can kill beneficial organisms and degrade soil quality.

+ **Mining and Resource Extraction:** Mining operations can release heavy metals and toxic chemicals into the soil, rendering it unsuitable for plant growth and disrupting the ecological balance. Oil and gas extraction can also lead to soil contamination through spills and leakage.

+ **Improper Waste Disposal:** Dumping of solid waste, including household trash, construction debris, and electronic waste, inappropriately can contaminate the soil. Hazardous substances from improperly disposed waste can leach into the soil and pose risks to human health and the environment.

+ **Accidental Spills:** Accidental spills of chemicals, fuels, or hazardous materials can contaminate the surrounding soil. These spills may occur during transportation, storage, or industrial accidents, causing localized soil pollution.

+ **Deforestation:** Clearing of forests and vegetation can lead to soil erosion and degradation. Without vegetation cover, topsoil is exposed to erosion by wind and rainfall, resulting in the loss of essential nutrients and fertile soil.

Noise Pollution

Noise pollution refers to excessive or unwanted noise that disrupts the environment and negatively affects human health and well-being. Sources of noise pollution include:

+ **Transportation:** Road traffic, aircraft, and trains are major contributors to noise pollution in urban areas. The constant exposure to high levels of traffic noise can lead to stress, hearing loss, and sleep disturbances.

- **Industrial Activities:** Industrial machinery, construction sites, and power plants generate high levels of noise. Prolonged exposure to industrial noise can cause hearing impairment and other health problems among workers.

- **Recreational Activities:** Recreational activities such as concerts, sporting events, and leisure venues can produce loud noises. The excessive noise generated by these activities can disrupt nearby residential areas and impact the quality of life.

- **Household Appliances:** Household appliances such as air conditioners, vacuum cleaners, and blenders can contribute to noise pollution, especially in densely populated areas or poorly soundproofed buildings.

- **Construction and Demolition:** Construction and demolition activities generate significant noise levels, especially during excavation, drilling, and blasting. Uncontrolled construction noise can cause annoyance and affect the well-being of nearby residents.

Light Pollution

Light pollution refers to excessive or misdirected artificial light that affects the natural environment and interferes with astronomical observations. The main sources of light pollution include:

- **Street Lighting:** Poorly designed or high-intensity street lighting can contribute to light pollution. Excessive artificial lighting can disrupt natural light-dark cycles and affect the behavior and habitats of nocturnal animals.

- **Outdoor Advertising:** Bright and poorly directed lights used in billboards, neon signs, and commercial displays can contribute to light pollution. These lights often point upward or outward, wasting energy and creating unnecessary glare.

- **Urbanization:** The expansion of cities and urban areas leads to increased lighting needs. Urban sky glow, caused by the scattering and reflection of artificial lights in the atmosphere, obscures celestial views and hinders astronomical observations.

- **Sports Facilities:** Outdoor sports facilities, stadiums, and arenas require powerful lighting for night events. The excessive light spill from these venues leads to light pollution, affecting nearby residential areas and ecosystems.

1.6.5 THE IMPORTANCE OF BIODIVERSITY

* **Residential Lighting:** Improperly designed or over-illuminated residential lighting can contribute to light pollution. The use of inefficient lighting fixtures and excessive outdoor lighting can waste energy and disturb the natural night environment.

Thermal Pollution

Thermal pollution refers to the degradation of water quality due to changes in water temperature caused by human activities. The main sources of thermal pollution include:

* **Power Plants:** Power plants that use freshwater or seawater for cooling purposes release heated water back into rivers, lakes, or oceans. The increased water temperature can disrupt aquatic ecosystems, harm fish and other organisms, and reduce dissolved oxygen levels.

* **Industrial Processes:** Industrial activities that involve the use of heated water and subsequent discharge into water bodies can contribute to thermal pollution. The sudden increase in water temperature can stress or kill aquatic plants and animals.

* **Urbanization:** The extensive use of concrete and asphalt in urban areas leads to the urban heat island effect, where the temperature is significantly higher than surrounding rural areas. The increased air and water temperatures can have adverse effects on human and ecosystem health.

* **Deforestation:** Clearing of forests reduces shading, leading to increased solar radiation absorption and higher water temperatures in streams and rivers. These elevated temperatures can negatively impact aquatic organisms adapted to specific temperature ranges.

Radioactive Pollution

Radioactive pollution refers to the contamination of the environment with radioactive substances, usually resulting from nuclear activities and accidents. The main sources of radioactive pollution include:

* **Nuclear Power Plants:** Accidents, leaks, or improper disposal of radioactive waste from nuclear power plants can result in the release of radioactive substances into the environment. These substances can have long-term harmful effects on living organisms and ecosystems.

- **Nuclear Testing and Weapons:** Nuclear weapon testing and the production of nuclear weapons can release radioactive materials into the atmosphere, soil, and water. The fallout from these activities can contaminate large areas and pose a threat to human health.

- **Medical Applications:** The use of radioactive materials in medical diagnosis and treatment can result in the production of radioactive waste. Improper management and disposal of these materials can lead to radioactive pollution.

- **Mining and Extraction:** Mining of radioactive materials, such as uranium and thorium, can release radioactive substances into the environment. These substances can contaminate soil, water, and air, posing risks to workers and nearby communities.

- **Accidents and Incidents:** Accidental releases of radioactive materials, such as the Chernobyl and Fukushima disasters, can cause widespread contamination of the environment and long-term health effects for exposed populations.

Understanding the types and sources of pollution is essential for devising effective strategies and policies to prevent or mitigate their negative environmental impacts. Addressing pollution requires interdisciplinary approaches, involving environmental scientists, policymakers, industries, and communities. By identifying the causes and sources of pollution, we can implement sustainable practices and technologies to reduce pollution levels and create a healthier and more sustainable environment for future generations.

Case Study: Industrial Air Pollution Regulation

To illustrate the importance of addressing air pollution, let's consider a case study on industrial air pollution regulation. Industrial emissions are a significant source of air pollution, contributing to respiratory diseases, climate change, and other adverse effects. To tackle this issue, governments and environmental agencies establish regulations and standards to control and reduce industrial air pollution levels.

One successful example is the Clean Air Act implemented in the United States. The Clean Air Act, initially enacted in 1970 and amended in subsequent years, sets limits on pollutants released by industrial facilities and requires them to adopt pollution control technologies. Standards are set for various pollutants, including sulfur dioxide, nitrogen oxides, particulate matter, and volatile organic compounds.

1.6.5 THE IMPORTANCE OF BIODIVERSITY 81

Under the Clean Air Act, industries are required to install pollution control devices, such as scrubbers and catalytic converters, to reduce emissions. They must also monitor their emissions regularly and report the data to regulatory agencies. Non-compliance with the standards can result in penalties and legal consequences.

This case study demonstrates the effectiveness of regulation in controlling and reducing industrial air pollution. By imposing emission standards and enforcing compliance, governments can hold industries accountable and encourage the adoption of cleaner technologies. Such regulations play a crucial role in protecting public health, improving air quality, and mitigating the impacts of industrial activities on the environment.

Key Takeaways

- Pollution refers to the introduction of harmful substances or contaminants into the environment, causing adverse effects on living organisms and ecosystems.

- Major types of pollution include air pollution, water pollution, soil pollution, noise pollution, light pollution, thermal pollution, and radioactive pollution.

- Pollution can arise from various sources, including industrial activities, transportation, agricultural practices, waste management, natural processes, and human activities.

- Understanding the types and sources of pollution is crucial for developing effective strategies and policies to prevent or mitigate their negative impacts.

- Regulation and enforcement, technological advancements, sustainable practices, and public awareness are essential in addressing and reducing pollution levels.

Discussion Questions

1. Discuss the primary sources of air pollution in your local area and their potential impacts on human health and the environment.

2. What are the challenges faced in regulating and reducing water pollution caused by agricultural runoff?

3. Investigate a specific case of soil pollution and discuss its causes, impacts, and potential remediation strategies.

4. How does light pollution affect the natural behavior and ecological balance of nocturnal animals in urban areas?

5. Research a notable incident or accident related to radioactive pollution and discuss its long-term environmental and health consequences.

Further Reading

To learn more about pollution and its impacts, you may refer to the following resources:

1. Karr, J. R., & Chu, E. W. (1997). Restoring life in running waters: Better biological monitoring. Island Press.

2. Stumm, W., & Morgan, J. J. (1996). Aquatic chemistry: Chemical equilibria and rates in natural waters. Wiley.

3. United Nations Environment Programme. (2016). Global waste management outlook. UNEP/Earthprint.

4. International Dark-Sky Association. (2020). Light pollution. Retrieved from https://www.darksky.org/light-pollution/

5. Nuclear Energy Agency. (2019). Radioactive waste management glossary. OECD Publishing.

Remember to use these resources critically and cite them appropriately in your own work.

Human Health Impacts

Human health is deeply interconnected with the environment. Our well-being is directly influenced by the quality of the air we breathe, the water we drink, and the food we eat. Environmental pollution and degradation can have significant impacts on human health, leading to a wide range of illnesses and diseases. In this section, we will explore the various ways in which environmental factors affect human health and discuss the importance of addressing these impacts for a greener future.

Link Between Environmental Pollution and Human Health

Environmental pollution is a major contributor to human health problems. Exposure to pollutants in the air, water, and soil can have detrimental effects on various systems within our bodies. For instance, air pollution, often caused by industrial emissions, vehicle exhaust, and burning fossil fuels, has been linked to

1.6.5 THE IMPORTANCE OF BIODIVERSITY

83

respiratory diseases, such as asthma, bronchitis, and lung cancer. Fine particulate matter (PM2.5) and toxic gases, such as nitrogen dioxide (NO2), can penetrate deep into the lungs, causing inflammation and long-term damage.

Water pollution is another significant concern. Contamination of water sources with harmful chemicals, heavy metals, or microbial pathogens can lead to waterborne diseases, including cholera, dysentery, and gastrointestinal infections. Prolonged exposure to polluted water can also result in chronic health issues, such as kidney damage and hormonal disruption.

Contaminated soil poses risks as well. Chemical pollutants, such as pesticides, heavy metals, and industrial waste, can be absorbed by plants and subsequently ingested by humans. These toxic substances can accumulate in our bodies over time, causing various health problems, including neurological disorders, reproductive issues, and cancer.

Impact of Climate Change on Human Health

Climate change is one of the greatest challenges of our time and has significant implications for human health. Rising global temperatures, changing precipitation patterns, extreme weather events, and sea-level rise all contribute to the complex web of impacts on our well-being.

One of the most immediate health risks associated with climate change is heat-related illness. As temperatures continue to rise, particularly in urban areas, heatwaves become more frequent and intense. Heat stress and heatstroke can result in dehydration, cardiovascular strain, and even death, particularly among vulnerable populations, such as the elderly, young children, and those with pre-existing health conditions.

Climate change also affects infectious disease dynamics. Vector-borne diseases, such as malaria, dengue fever, and Lyme disease, are highly sensitive to temperature, precipitation, and ecological changes. As climate conditions shift, the geographic range of these diseases may expand, exposing new populations to infection. Additionally, changes in temperature and rainfall patterns can alter the breeding habitats of disease-carrying mosquitoes, ticks, and other vectors, leading to increased transmission rates.

The impacts of climate change on mental health should not be overlooked either. Natural disasters, displacement, and the loss of livelihoods can cause significant psychological distress and contribute to anxiety, depression, and post-traumatic stress disorder (PTSD). The uncertainty and disruption caused by climate change can have profound effects on the well-being of individuals and communities.

Mitigation and Adaptation Strategies

Addressing the human health impacts of environmental pollution and climate change requires a comprehensive approach that includes both mitigation and adaptation strategies.

Mitigation strategies aim to reduce the sources of pollution and greenhouse gas emissions to prevent or minimize environmental harm. This can be achieved through transitioning to cleaner energy sources, implementing sustainable land and water management practices, promoting waste reduction and recycling, and adopting more sustainable transportation options. By reducing pollution and mitigating climate change, we can protect human health and prevent the onset of related illnesses.

Adaptation strategies, on the other hand, focus on building resilience and capacity to cope with the unavoidable impacts of climate change. This may involve improving public health systems to respond to changing disease patterns, implementing early warning systems for heatwaves and extreme weather events, enhancing urban planning to reduce heat island effects, and providing adequate support and resources to vulnerable communities.

Case Studies

Air Pollution in Delhi, India: Delhi, the capital city of India, has been grappling with severe air pollution for years. The levels of PM2.5 and NO2 regularly exceed the safe limits set by the World Health Organization (WHO), leading to a public health crisis. High air pollution has been linked to increased respiratory illnesses, cardiovascular diseases, and premature deaths. The government and local authorities are implementing various measures to combat air pollution, including promoting the use of clean energy, restricting vehicle emissions, and improving waste management practices.

Impact of Hurricane Maria in Puerto Rico: In 2017, Hurricane Maria devastated Puerto Rico, causing widespread damage and a humanitarian crisis. The aftermath of the hurricane led to disruptions in the healthcare system, lack of clean water, and limited access to essential medications. These conditions significantly impacted the health and well-being of the population, leading to increased rates of infectious diseases, mental health issues, and chronic illnesses. The recovery efforts highlighted the importance of resilient healthcare systems and preparedness for future climate-related events.

1.6.5 THE IMPORTANCE OF BIODIVERSITY

Conclusion

Human health is intricately connected to the environment, and the impacts of pollution and climate change are posing significant challenges. Addressing the human health impacts requires concerted efforts from individuals, communities, governments, and international organizations. By prioritizing the reduction of pollution, transitioning to renewable energy sources, promoting sustainable practices, and strengthening public health systems, we can pave the way for a healthier and greener future. Remember, the health of our planet is closely tied to our own well-being. Let's work together to create a sustainable and healthy environment for ourselves and future generations.

Pollution Prevention and Control Measures

Pollution prevention and control measures play a crucial role in preserving the environment and safeguarding human health. Pollution refers to the introduction of harmful substances or contaminants into the natural environment, which can cause adverse effects on living organisms and ecological systems. By implementing preventive measures and effective control strategies, we can minimize the release and impact of pollutants, thereby mitigating environmental pollution.

Understanding Pollution

To effectively prevent and control pollution, we must first understand the different types and sources of pollution. Pollution can originate from various activities, including industrial processes, transportation, agriculture, and domestic waste disposal. It can manifest as air pollution, water pollution, soil contamination, noise pollution, and even light pollution.

Air pollution is caused by the release of harmful gases, particulate matter, and volatile organic compounds (VOCs) into the atmosphere. Sources of air pollution include vehicle emissions, industrial emissions, power plants, and burning of fossil fuels. These pollutants can contribute to respiratory diseases, smog formation, and climate change.

Water pollution occurs when harmful substances, such as toxic chemicals, heavy metals, sewage, and agricultural runoff, contaminate water bodies like rivers, lakes, and oceans. This can have severe impacts on aquatic ecosystems, leading to the loss of biodiversity and posing risks to human health if contaminated water is consumed.

Soil contamination occurs when soil is contaminated with hazardous substances, such as industrial waste, pesticides, or heavy metals. This can have detrimental effects on soil fertility, crop productivity, and the overall health of ecosystems.

Noise pollution refers to excessive noise that disrupts the natural environment and causes annoyance or harm to humans and wildlife. Common sources of noise pollution include transportation, industrial activities, construction, and recreational activities.

Light pollution occurs when artificial lights interfere with the natural darkness of the night sky. It can disrupt ecosystems, affect wildlife behavior, and impair human sleep patterns.

Prevention Strategies

Preventing pollution is always the most effective approach to protect the environment and human well-being. Here are some key pollution prevention strategies:

1. **Source Reduction:** This involves reducing the production and use of harmful substances at their source. For example, industries can adopt cleaner production techniques to minimize the generation of pollutants. Reducing the use of pesticides in agriculture and promoting organic farming practices are other examples of source reduction.

2. **Waste Minimization:** By minimizing waste generation, we can reduce the potential for pollution. This can be achieved through recycling, reusing materials, and implementing waste reduction strategies at individual, industrial, and societal levels.

3. **Energy Conservation:** Conserving energy reduces the demand for energy production, which often results in pollution. Implementing energy-efficient measures, using renewable energy sources, and promoting sustainable transportation systems are effective ways to conserve energy and reduce pollution.

4. **Green Infrastructure:** Incorporating green infrastructure, such as green roofs, permeable pavements, and urban green spaces, can help mitigate pollution by promoting natural filtration and absorption of pollutants. Green infrastructure also provides multiple benefits like improved air quality, reduced stormwater runoff, and enhanced urban biodiversity.

5. **Environmental Education and Awareness:** Educating individuals and communities about the importance of pollution prevention and sustainable

1.6.5 THE IMPORTANCE OF BIODIVERSITY

practices can drive behavior change. Promoting environmental awareness and empowering people to make informed decisions can significantly contribute to pollution reduction.

Control Measures

While prevention is essential, control measures are necessary to mitigate pollution from existing sources. Here are some common control measures:

1. **Air Pollution Control:** The control of air pollution involves the use of various technologies and strategies to reduce emissions of harmful pollutants. Examples of control measures include the installation of pollution control devices in factories and power plants, implementation of stricter emission standards for vehicles, and promoting cleaner fuel alternatives.

2. **Water Pollution Control:** Water pollution can be controlled through various means, such as wastewater treatment systems, sedimentation ponds, and constructed wetlands. These systems help remove contaminants from wastewater before it is discharged into natural water bodies. Additionally, regulations and policies are crucial in managing and controlling pollution from industrial and agricultural activities by imposing standards and monitoring systems.

3. **Soil Remediation and Land Restoration:** Contaminated soil can be remediated using various techniques like bioremediation, phytoremediation, and soil washing. These methods help to remove, reduce or transform pollutants in soil, making it less harmful to the environment. Restoring degraded land through reforestation, soil erosion control measures, and promoting sustainable land management practices can further prevent pollution.

4. **Noise Pollution Control:** To control noise pollution, regulations and building codes can be implemented to limit noise levels in residential areas, construction sites, and industrial zones. Noise barriers, soundproofing, and noise reduction technologies can also be employed in specific settings to minimize the impact of noise pollution.

5. **Light Pollution Control:** Controlling light pollution involves the use of shielding, directing, or dimming artificial lights. Regulations can be set to limit the intensity and direction of outdoor lighting, especially in areas near

ecologically sensitive regions or astronomical observatories. Using efficient lighting fixtures and promoting responsible lighting practices are also important measures.

Case Study: The Clean Air Act in the United States

One of the most significant success stories in pollution prevention and control is the Clean Air Act (CAA) in the United States. Enacted in 1970, the CAA has played a vital role in addressing air pollution issues and improving air quality across the country. The legislation set emission standards for major pollutants, such as sulfur dioxide, nitrogen oxides, particulate matter, lead, and carbon monoxide.

Through the CAA, the Environmental Protection Agency (EPA) established regulations for various industries to reduce emissions and implemented stricter standards for vehicle exhausts. This has led to significant reductions in air pollutants, resulting in improved air quality and associated health benefits.

The CAA also enabled the implementation of innovative technologies, such as catalytic converters in automobiles, electrostatic precipitators in power plants, and scrubbers in industrial facilities, to control pollution effectively. These measures have contributed to the reduction of smog, acid rain, and other harmful air pollutants.

Conclusion

Pollution prevention and control measures are crucial for maintaining a clean and sustainable environment. By understanding the different types and sources of pollution and implementing preventive strategies, we can effectively reduce pollution at its source. Control measures further help in mitigating pollution from existing sources. The case study of the Clean Air Act in the United States demonstrates the positive impact of comprehensive legislation and effective implementation in improving air quality and protecting human health. It is vital for individuals, industries, and governments to work together to develop and implement pollution prevention and control measures to ensure a greener and healthier future.

Sustainable Development

Sustainable development is a concept that encompasses the integration of economic, social, and environmental aspects to meet the needs of the present generation without compromising the ability of future generations to meet their own needs. It recognizes that economic growth must be balanced with social equity and environmental stewardship to ensure a greener and more sustainable future.

1.6.5 THE IMPORTANCE OF BIODIVERSITY

1. Principles of Sustainable Development:

Sustainable development is guided by several key principles that provide a framework for decision-making and action. These principles include:

1.1 Interdependence: Recognizing the interconnectedness of social, economic, and environmental systems, sustainable development seeks to address the complex interdependencies between these systems.

1.2 Equity: Sustainable development promotes fairness and equality, ensuring that all people have equal access to resources, opportunities, and benefits.

1.3 Long-term thinking: Sustainable development takes a proactive and future-oriented approach, considering the long-term consequences of actions and decisions.

1.4 Conservation and efficiency: Emphasizing the importance of resource conservation and efficiency, sustainable development aims to minimize waste and reduce environmental impacts.

1.5 Collaboration and participation: Sustainable development encourages collaboration and the active involvement of all stakeholders, including governments, businesses, communities, and individuals.

2. Sustainable Development Goals (SDGs):

To guide global efforts towards sustainable development, the United Nations has established a set of Sustainable Development Goals (SDGs). These goals, comprising 17 targets, address a range of environmental, social, and economic challenges, including poverty eradication, climate action, clean energy, sustainable cities, and responsible consumption and production.

2.1 Example: SDG 7 - Affordable and Clean Energy

SDG 7 aims to ensure universal access to affordable, reliable, sustainable, and modern energy for all. It recognizes the importance of transitioning from fossil fuels to renewable energy sources to mitigate climate change and improve energy security.

2.2 Solution: Promoting Renewable Energy

To achieve SDG 7, countries can implement various strategies and policies to promote renewable energy adoption. This includes incentivizing the development of renewable energy projects, such as solar and wind farms, and supporting research and development in energy storage technologies.

For example, countries can introduce feed-in tariffs or tax incentives to encourage investment in renewable energy infrastructure. They can also establish regulations and standards that promote energy efficiency in buildings and appliances.

3. Challenges in Sustainable Development:

While sustainable development offers promising solutions to global challenges, it also faces several significant challenges. These challenges include:

3.1 Trade-offs and conflicts: Balancing economic growth, social equity, and environmental protection often involves making trade-offs and navigating conflicts of interest. For instance, economic development may come at the expense of environmental conservation.

3.2 Limited resources: Meeting the needs of a growing global population while preserving and managing finite resources poses a significant challenge. Sustainable development requires innovative approaches to resource management and efficient use of resources.

3.3 Institutional and systemic barriers: Overcoming entrenched practices, policies, and institutional barriers is crucial for achieving sustainable development. This includes addressing issues such as corruption, inadequate governance, and lack of political will.

3.4 Unequal distribution of resources: Addressing social inequalities and ensuring equitable access to resources and opportunities is an ongoing challenge in sustainable development. Poverty eradication and social inclusivity are critical aspects of achieving sustainable development.

4. Integrating Sustainable Development into Decision-Making:

To effectively integrate sustainable development principles into decision-making processes, various approaches and tools can be employed. These include:

4.1 Life Cycle Assessment (LCA): LCA is a methodology used to assess the environmental impacts of a product or system throughout its life cycle. It helps identify opportunities for improvements and encourages the adoption of more sustainable practices.

4.2 Cost-Benefit Analysis (CBA): CBA is a technique used to evaluate the economic viability of a project or policy by comparing its costs and benefits. Incorporating environmental and social costs and benefits into such analyses can provide a more comprehensive understanding of the true impacts.

4.3 Multi-stakeholder engagement: Involving diverse stakeholders, including communities, businesses, NGOs, and governments, in decision-making processes promotes transparency, inclusivity, and accountability.

4.4 Sustainable development indicators: Developing and utilizing indicators that measure progress towards sustainable development goals can help track performance and inform policy decisions.

5. Unconventional Solution: Sustainable Tourism

Sustainable tourism is a concept that seeks to minimize the negative environmental, social, and cultural impacts of tourism while maximizing the benefits for local communities and the environment. It promotes responsible travel practices and aims to preserve natural and cultural heritage.

1.6.5 THE IMPORTANCE OF BIODIVERSITY

By implementing sustainable tourism practices, destinations can mitigate the degradation of natural resources, protect ecosystems, and enhance the well-being of local communities. This can be achieved through measures such as eco-friendly accommodations, community-based tourism initiatives, and education programs for tourists to raise awareness about sustainable behavior.

Overall, sustainable development is a multi-faceted concept that requires a holistic approach to address the challenges of our time. By integrating economic, social, and environmental considerations, we can pave the way for a greener and more sustainable future.

Principles of Sustainable Development

Sustainable development is a fundamental concept that guides environmental science and strives to meet the needs of the present generation without compromising the ability of future generations to meet their own needs. It is a holistic approach that seeks to balance economic, environmental, and social goals in a way that promotes long-term well-being and preserves the integrity of ecosystems.

1. The Triple Bottom Line Approach: One of the key principles of sustainable development is the triple bottom line approach, which recognizes that economic development, environmental protection, and social equity are interconnected and equally important. This approach aims to achieve a harmonious balance between these three pillars by considering their interdependencies and impacts on each other. It emphasizes the need to go beyond economic growth and take into account environmental and social considerations when making decisions.

2. Intergenerational Equity: Sustainable development also emphasizes intergenerational equity, which means that the needs and rights of future generations should be given equal consideration to the needs and rights of the present generation. This principle acknowledges that the resources and ecological systems we rely on are finite and should be managed responsibly to ensure their availability and quality for future generations.

3. Conservation and Efficient Use of Resources: Another key principle of sustainable development is the conservation and efficient use of resources. This involves minimizing waste, reducing resource extraction, and promoting the use of renewable resources. It also encourages the adoption of cleaner production methods and technologies that minimize environmental impacts while maximizing resource efficiency.

4. Environmental Integration and Consideration: Sustainable development recognizes the importance of integrating environmental considerations into

decision-making processes. It emphasizes the need to assess and minimize potential environmental risks and impacts associated with economic activities. This principle highlights the importance of environmental impact assessments, environmental management systems, and the precautionary principle, which advocates for taking proactive measures to prevent harm to the environment.

5. Social Equity and Justice: Sustainable development promotes social equity and justice by ensuring fair distribution of resources, opportunities, and benefits among all members of society. It aims to address social inequalities, reduce poverty, and improve the livelihoods of marginalized communities. This principle also emphasizes the importance of engaging local communities and stakeholders in decision-making processes to ensure their meaningful participation.

6. Collaboration and Partnerships: Sustainable development recognizes that achieving its goals requires collaboration and partnerships among governments, businesses, civil society organizations, and individuals. It encourages the formation of multi-stakeholder partnerships that bring together diverse perspectives, expertise, and resources to address complex environmental and social challenges. This principle emphasizes the importance of cooperation, dialogue, and knowledge sharing to find innovative and inclusive solutions.

7. Education and Awareness: Sustainable development recognizes the critical role of education and awareness in promoting a sustainable mindset. It emphasizes the need to educate individuals, communities, and decision-makers about the principles and practices of sustainable development. This principle includes raising awareness about the environmental, economic, and social consequences of unsustainable behaviors and promoting the adoption of sustainable lifestyles and consumption patterns.

To illustrate these principles, let's consider the example of sustainable agriculture. Sustainable agriculture aims to meet current and future food needs while minimizing negative environmental impacts, conserving resources, and promoting social equity. It applies the principles of sustainable development by integrating ecological principles, resource conservation, and social considerations into agricultural practices.

For instance, sustainable agriculture emphasizes the conservation and efficient use of resources by promoting practices such as organic farming and permaculture. These approaches minimize the use of synthetic fertilizers and pesticides, reduce water consumption, and improve soil health. By adopting these practices, farmers can enhance the long-term productivity and sustainability of their farms while minimizing negative impacts on the environment.

Additionally, sustainable agriculture recognizes the importance of social equity and justice by promoting fair trade practices, supporting local farmers, and

1.6.5 THE IMPORTANCE OF BIODIVERSITY

ensuring the rights and well-being of agricultural workers. It emphasizes the need to address issues such as food security, access to markets, and fair prices for farmers. By considering social factors in agricultural practices, sustainable agriculture contributes to the overall well-being and resilience of communities.

In conclusion, the principles of sustainable development provide a framework for achieving harmony between economic development, environmental protection, and social equity. By integrating these principles into decision-making processes and practices, we can work towards a greener future that ensures the well-being of both current and future generations.

Sustainable Resource Management

In order to achieve a greener future, it is crucial to adopt sustainable resource management practices. Sustainable resource management refers to the responsible use, conservation, and preservation of natural resources to meet the needs of current and future generations. This involves balancing social, economic, and environmental considerations to ensure long-term ecological integrity and human well-being.

The Importance of Sustainable Resource Management

Natural resources, such as water, land, forests, minerals, and energy sources, are essential for human survival and development. However, their misuse and overexploitation can lead to depletion, environmental degradation, and social conflicts. Sustainable resource management aims to strike a balance between resource use and conservation by minimizing waste, maximizing efficiency, and promoting renewable alternatives.

By adopting sustainable resource management practices, we can address several pressing challenges:

- **Environmental Preservation:** Sustainable resource management helps protect ecosystems, wildlife habitats, and biodiversity. It reduces deforestation, soil erosion, pollution, and habitat destruction, thereby ensuring the sustained functioning of natural systems and the services they provide.

- **Social Equity:** It promotes equitable access to resources and benefits, ensuring that marginalized and vulnerable communities are not left behind. Sustainable resource management initiatives seek to address social disparities, empower local communities, and promote inclusive decision-making processes.

+ **Economic Development:** By managing resources sustainably, we can create economic opportunities while minimizing negative impacts. Sustainable resource management practices stimulate innovation, promote green industries, and foster economic resilience in the face of environmental challenges.

+ **Climate Change Mitigation:** Sustainable resource management plays a crucial role in mitigating climate change. By transitioning to renewable energy sources, optimizing resource use, adopting low-carbon technologies, and conserving forests, we can reduce greenhouse gas emissions and promote a low-carbon economy.

Principles of Sustainable Resource Management

Sustainable resource management is guided by several key principles that help ensure its effectiveness:

1. **Intergenerational Equity:** Resources should be managed in a way that meets the needs of the present without compromising the ability of future generations to meet their own needs. This principle emphasizes long-term thinking and the preservation of natural resources for the benefit of all.

2. **Precautionary Approach:** When there is scientific uncertainty, a precautionary approach should be taken to avoid irreversible or harmful impacts. This principle highlights the importance of proactive decision-making and the consideration of potential risks before proceeding with resource management strategies.

3. **Polluter Pays Principle:** Those who pollute or degrade the environment should bear the costs of mitigation and restoration. This principle promotes accountability and incentivizes responsible behavior among individuals, businesses, and industries.

4. **Integration and Collaboration:** Sustainable resource management requires an interdisciplinary and collaborative approach. It involves the participation of various stakeholders, including government agencies, communities, industries, and non-governmental organizations, working together towards shared goals.

1.6.5 THE IMPORTANCE OF BIODIVERSITY

5. **Adaptive Management:** Resource management strategies and practices should be flexible and adaptive. This principle recognizes the complexity and dynamic nature of ecosystems and the need to adjust approaches based on monitoring, evaluation, and feedback.

6. **Resource Efficiency:** Sustainable resource management aims to optimize resource utilization by minimizing waste, improving efficiency, and promoting circular economy principles. This involves reducing resource extraction, promoting recycling and reuse, and adopting resource-efficient technologies and practices.

Challenges and Solutions in Sustainable Resource Management

Despite the advantages of sustainable resource management, there are several challenges that need to be addressed in order to achieve its full potential:

1. **Limited Awareness and Education:** Many individuals and societies lack awareness and understanding of sustainable resource management practices. Education and awareness campaigns are needed to promote sustainable consumption and production patterns, as well as to encourage behavior change at the individual and community levels.

2. **Policy and Regulatory Barriers:** Inadequate or conflicting policies and regulations can hinder sustainable resource management efforts. Governments and international organizations need to develop and enforce effective policies that incentivize sustainable practices, discourage resource-intensive activities, and create a supportive regulatory framework.

3. **Lack of Financial Resources:** Financing sustainable resource management projects can be a major challenge, especially in developing countries. Innovative financing mechanisms, such as green bonds, public-private partnerships, and international assistance, can help bridge the funding gap and support the implementation of sustainable resource management initiatives.

4. **Globalization and Consumerism:** The globalized nature of the modern economy and consumer-driven societies contribute to unsustainable patterns of resource consumption and waste generation. Addressing these challenges requires a shift towards sustainable consumption and production, as well as promoting alternative economic models that prioritize well-being over endless growth.

5. **Conflict and Governance Issues:** Resource management often involves competing interests and conflicts, particularly in areas where resources are scarce or valuable. Good governance, inclusive decision-making, and conflict resolution mechanisms are essential for sustainable resource management, ensuring that all stakeholders are involved in the decision-making process and benefiting from the resources in a fair and equitable manner.

Examples of Sustainable Resource Management

There are numerous successful examples of sustainable resource management strategies and initiatives implemented around the world. These examples demonstrate the potential for creating positive change and achieving sustainable development:

+ **Water Resource Management in Singapore:** Singapore, a water-scarce country, has implemented a comprehensive and innovative approach to water resource management. Through a combination of water pricing, water conservation measures, wastewater recycling, and the development of alternative water sources, such as desalination and NEWater (reclaimed water), Singapore has achieved water self-sufficiency and reduced reliance on imported water.

+ **Forest Conservation in Costa Rica:** Costa Rica has successfully implemented a payment for ecosystem services (PES) program to protect its valuable forests. Landowners are paid to conserve and restore forests, which provide carbon sequestration, watershed protection, and habitat for biodiversity. This initiative has helped increase forest cover and preserve critical ecosystems.

+ **Community-based Fisheries Management in Belize:** In Belize, a community-based approach to fisheries management has been effective in maintaining healthy fish populations and supporting sustainable livelihoods. Local communities have been empowered to manage their own fishing grounds, implement regulations, and monitor fish stocks, leading to improved resource sustainability and enhanced community resilience.

+ **Renewable Energy Transition in Denmark:** Denmark has made significant strides in transitioning to renewable energy sources, particularly wind power. Through a combination of supportive policies, incentives, and public investment in research and development, Denmark has become a global

1.6.5 THE IMPORTANCE OF BIODIVERSITY

97

leader in wind energy production, reducing its reliance on fossil fuels and contributing to climate change mitigation.

- **Sustainable Agriculture in Cuba:** Cuba's experience with sustainable agriculture provides valuable lessons in resource conservation and food security. Following the collapse of the Soviet Union, Cuba faced severe economic challenges and limited access to external resources, including agrochemicals and fossil fuels. As a result, the country adopted organic and agroecological farming practices, promoting local food production, urban agriculture, and biodiversity conservation.

Conclusion

Sustainable resource management is a vital component of environmental science and the fight for a greener future. By adopting principles of intergenerational equity, precautionary approaches, and resource efficiency, we can address pressing environmental, social, and economic challenges. However, achieving sustainable resource management requires concerted efforts, including raising awareness, developing supportive policies, securing adequate financial resources, and promoting collaboration among various stakeholders. The successful examples of sustainable resource management showcased here demonstrate the potential for positive change and highlight the path toward a more sustainable and resilient future.

Green Technologies and Innovations

Green technologies and innovations play a crucial role in addressing environmental challenges and moving towards a sustainable future. These advancements aim to minimize the negative impacts of human activities on the environment while promoting economic growth and social well-being. In this section, we will explore the principles, applications, and success stories of various green technologies and innovations.

Principles of Green Technologies

Green technologies are designed to reduce the use of non-renewable resources, minimize pollution and waste, and promote efficient and sustainable practices. They are based on the following principles:

1. Renewable Energy: Green technologies focus on harnessing energy from renewable sources such as sunlight, wind, water, and biomass. By shifting from

fossil fuels to clean and sustainable energy, we can significantly reduce greenhouse gas emissions and combat climate change.

2. Energy Efficiency: Improving energy efficiency is an essential aspect of green technologies. This involves enhancing the performance of devices and systems to reduce energy consumption without compromising functionality. Energy-efficient appliances, smart grids, and energy management systems are some examples.

3. Waste Management: Green technologies aim to minimize waste generation and promote proper waste management practices. Recycling, composting, and waste-to-energy conversion technologies help reduce the amount of waste sent to landfills, conserve resources, and mitigate environmental pollution.

4. Sustainable Transportation: Transportation is a significant contributor to carbon emissions. Green technologies in transportation include electric vehicles, hybrid vehicles, and alternative fuels such as biofuels and hydrogen. These technologies reduce air pollution, dependence on fossil fuels, and promote sustainable mobility solutions.

5. Green Building: Green building technologies and practices focus on constructing energy-efficient and environmentally friendly buildings. This includes using renewable materials, incorporating efficient insulation, utilizing solar panels for energy generation, and implementing intelligent building management systems.

Applications of Green Technologies

Green technologies and innovations have found applications in various sectors, including energy, agriculture, transportation, and waste management. Let's explore some notable examples:

1. Renewable Energy Systems: Solar panels and wind turbines are widely used to generate electricity from renewable sources. These technologies have transformed the energy landscape by providing clean and sustainable power, reducing reliance on fossil fuels.

2. Smart Grids: Smart grids integrate advanced technology, including real-time monitoring, automated controls, and demand response systems. They improve the efficiency and reliability of the electrical grid, enable better integration of renewable energy sources, and empower consumers to make informed energy choices.

3. Precision Agriculture: Green technologies are revolutionizing agriculture by optimizing resource use and reducing environmental impacts. Sensors, drones, and satellite imagery help monitor crop growth, soil moisture, and nutrient levels, allowing farmers to apply precise amounts of water, fertilizers, and pesticides.

4. Electric Vehicles (EVs): EVs are a key component of sustainable transportation. With advancements in battery technology and charging

1.6.5 THE IMPORTANCE OF BIODIVERSITY 99

infrastructure, EVs offer a clean and efficient alternative to conventional vehicles, reducing air pollution and dependence on fossil fuels.

5. Waste-to-Energy Systems: Green technologies convert waste into valuable resources and energy. Anaerobic digestion and incineration with energy recovery are examples of waste-to-energy systems that help reduce landfill waste, generate electricity, and produce heat.

Success Stories in Green Technologies

Several success stories demonstrate the potential impact of green technologies and innovations. Here are a few noteworthy examples:

1. The Great Smog of London: In the 1950s, London experienced a severe air pollution event known as the "Great Smog." This event prompted the UK government to introduce the Clean Air Act of 1956, leading to the development of cleaner fuels, emission standards, and catalytic converters, significantly improving air quality in the city.

2. Solar Power in Germany: Germany has made remarkable progress in adopting solar power. Through strong government support, favorable policies, and financial incentives, Germany became a global leader in solar energy production. This success has transformed the energy landscape, reducing greenhouse gas emissions and creating thousands of jobs.

3. Vertical Farming in Singapore: With limited land availability, Singapore has embraced vertical farming, where crops are grown in vertically stacked layers. This innovative approach reduces the need for large agricultural spaces, optimizes resource consumption, and ensures food security in urban areas.

4. Offshore Wind Farms in Denmark: Denmark has been a pioneer in offshore wind energy. Through extensive investment in wind turbine installations off its coastlines, the country now generates a significant portion of its electricity from clean and renewable sources, reducing reliance on fossil fuels.

5. Green Building Practices: The global green building movement has gained momentum, with successful projects worldwide. One notable example is the EDGE (Excellence in Design for Greater Efficiencies) building certification program, which promotes resource-efficient buildings across the globe.

Challenges and Future Directions in Green Technologies

While green technologies have made significant progress, several challenges persist. Overcoming these challenges is crucial for widespread adoption and continued advancements. Here are some key challenges and future directions:

1. Cost and Affordability: Green technologies often come with a higher upfront cost, making them less accessible to certain populations. Continued research and development, along with supportive policies, are necessary to drive down the cost and increase affordability.

2. Infrastructure Development: The transition to renewable energy requires significant infrastructure development, including transmission lines, energy storage systems, and charging stations. Coordinated planning and investment in this infrastructure are essential for scaling up green technologies.

3. Policy and Regulatory Frameworks: Governments play a crucial role in supporting the adoption of green technologies through favorable policies, regulations, and incentives. Strengthening and expanding these frameworks will encourage innovation and investment in green technologies.

4. Resource Management and Circular Economy: Green technologies should focus on efficient resource utilization and circular economy principles. Minimizing resource extraction, optimizing recycling and waste management, and promoting sustainable consumption patterns are key considerations.

5. Research and Development: Continued research and development are necessary to drive innovation in green technologies. This includes exploring new materials, improving efficiency, enhancing energy storage capabilities, and developing breakthrough technologies to address environmental challenges.

The future of green technologies and innovations holds great promise. As we continue to advance in science and engineering, the development and implementation of sustainable solutions will contribute significantly to building a greener and more sustainable future for generations to come.

Summary

Green technologies and innovations are vital in addressing environmental challenges and promoting sustainability. These technologies focus on renewable energy, energy efficiency, waste management, sustainable transportation, and green building practices. Successful implementations of green technologies include solar power in Germany, vertical farming in Singapore, and offshore wind farms in Denmark. However, challenges such as cost, infrastructure, and policy frameworks need to be addressed to ensure wider adoption. Continued research and development will drive innovation and enable the transition to a greener future.

Chapter 2: Restoring Ecosystems

Chapter 2: Restoring Ecosystems

Chapter 2: Restoring Ecosystems

Ecosystems are intricate networks of living organisms and their surrounding environment. They provide numerous benefits to society, including clean air and water, food, medicine, and recreational opportunities. Unfortunately, human activities have caused significant degradation and destruction of many ecosystems worldwide. This chapter focuses on the principles and techniques of ecosystem restoration, which aims to reverse the damage and enhance the resilience of ecosystems.

Definition and Objectives of Ecosystem Restoration

Ecosystem restoration is the process of assisting the recovery of an ecosystem that has been degraded, damaged, or destroyed. It involves activities such as removing or mitigating pollutants, reintroducing native species, and rehabilitating habitats. The primary objective of ecosystem restoration is to regain the functioning and structure of the ecosystem, enabling it to provide its services and support biodiversity.

Restoration projects can vary in scale, from small-scale efforts like reestablishing a wetland pond to large-scale undertakings like restoring a degraded forest ecosystem. The objectives may differ depending on the ecosystem type and the specific goals of the restoration project. Some common objectives of ecosystem restoration include:

- Enhancing biodiversity by reintroducing native species and protecting endangered species

CHAPTER 2: RESTORING ECOSYSTEMS

+ Improving water quality and quantity by restoring wetlands and riparian zones

+ Mitigating the impacts of climate change by restoring carbon-sequestering habitats like forests and peatlands

+ Controlling and minimizing invasive species that threaten the native biodiversity

+ Restoring degraded landscapes for sustainable land use and agriculture

Restoration projects often take a holistic approach, addressing not only the ecological aspects but also social, cultural, and economic aspects. Collaboration and engagement with local communities, stakeholders, and indigenous peoples are essential for the success of restoration initiatives.

Ecological Succession and Ecosystem Development

Understanding ecological succession is crucial for planning and implementing successful ecosystem restoration projects. Ecological succession refers to the predictable sequence of changes in species composition and ecosystem structure over time following a disturbance. It involves the colonization of species adapted to the environmental conditions created by the disturbance, followed by the replacement of these pioneer species by more complex and diverse communities.

Primary succession occurs in environments where no ecosystem or soil exists, such as volcanic slopes or bare glacial moraines. Pioneer species, such as lichens and mosses, are the first to colonize these barren habitats, gradually transforming them into suitable environments for other plants and animals. Over time, shrubs, trees, and a variety of organisms establish themselves, leading to the development of a stable climax community.

Secondary succession, on the other hand, occurs in ecosystems that have been disturbed but still retain soil and some remaining vegetation. Examples include abandoned agricultural fields or areas affected by wildfires. In these cases, the process of succession begins with the reestablishment of early successional species, followed by the gradual recovery of the original vegetation and ecosystem structure.

Understanding the trajectory of ecological succession, including the characteristics of pioneer species, facilitation, and inhibition processes, and the role of environmental factors, is critical for selecting appropriate restoration techniques and determining the expected timeframe for ecosystem recovery.

CHAPTER 2: RESTORING ECOSYSTEMS

Approaches to Ecosystem Restoration

Ecosystem restoration can be approached through various techniques, depending on the specific goals and conditions of the project. Some commonly employed restoration techniques include:

- Habitat restoration: This involves recreating or enhancing the physical features of an ecosystem to support the recovery of its native flora and fauna. It may include activities such as regrading, revegetation, and creating artificial structures like nesting sites or oyster reefs.

- Species reintroduction: Reintroducing locally extinct or endangered species is a crucial strategy to restore the ecological functions and dynamics of an ecosystem. Careful planning, including habitat suitability assessments, predator control, and monitoring, is essential for successful reintroduction efforts.

- Biological control: Invasive species can significantly impact native biodiversity and ecosystem functioning. Biological control involves the intentional introduction of natural enemies or competitors to control the spread and impact of invasive species, thereby promoting the recovery of the native species and ecosystem processes.

- Hydrological restoration: Many ecosystems, such as wetlands, streams, and floodplains, are highly dependent on appropriate water regimes. Restoring natural hydrological patterns, such as water flow, groundwater levels, and flood events, is essential for ensuring the successful recovery of these ecosystems.

- Genetic rescue: In situations where populations of native species have become small and genetically depleted, genetic rescue can be employed. This involves introducing individuals from other populations or closely related species to increase genetic variability and enhance the resilience of the population.

The selection of appropriate restoration techniques depends on various factors, including the type and extent of degradation, the availability of resources, and the involvement of stakeholders and local communities. Restoration projects often require interdisciplinary collaboration among ecologists, hydrologists, landscape architects, community members, and other experts.

104   *CHAPTER 2: RESTORING ECOSYSTEMS*

Case Studies in Ecosystem Restoration

Several remarkable success stories demonstrate the effectiveness of ecosystem restoration in different parts of the world. Let's explore a few of these case studies:

Coral Reef Restoration: Coral reefs, highly diverse and productive marine ecosystems, are under severe threat due to climate change, pollution, and destructive fishing practices. In response, restoration methods such as coral gardening, artificial reef deployment, and habitat restoration have been employed to rehabilitate damaged reefs. Successful restoration efforts have been observed in locations like the Florida Keys in the United States, the Great Barrier Reef in Australia, and the Andaman Sea in Southeast Asia.

Reforestation and Afforestation Projects: Deforestation has had drastic impacts on terrestrial ecosystems, including loss of habitat, soil erosion, and climate change. Reforestation involves replanting forests in areas that were once forested, while afforestation involves creating forests in areas that have not previously supported forest growth. Globally, initiatives like the Bonn Challenge aim to restore millions of hectares of degraded land through reforestation and afforestation efforts, contributing to carbon sequestration, biodiversity conservation, and sustainable land use.

Wetland Restoration: Wetlands provide essential ecosystem services, including water purification, flood mitigation, and habitat provision. However, extensive draining and conversion for agriculture and urban development have led to the degradation and loss of wetland ecosystems. Restoration projects, such as the restoration of the Everglades in Florida, USA, and the Hulun Lake wetland in China, have successfully restored hydrological processes, enhanced biodiversity, and improved water quality.

Urban Ecological Restoration: Urban areas face unique challenges due to the high density of human population and infrastructure. Urban ecological restoration focuses on incorporating nature into urban landscapes, promoting biodiversity, and enhancing the quality of life for residents. For example, the High Line Park in New York City has transformed an abandoned elevated railway into a flourishing green space, providing habitat for pollinators and recreational opportunities for residents.

River and Stream Restoration: Rivers and streams support diverse ecosystems, but their health is often compromised by pollution, channelization, and alteration

CHAPTER 2: RESTORING ECOSYSTEMS

of natural flow regimes. Restoration efforts aim to rejuvenate these water bodies by restoring meandering channels, reintroducing native fish species, and improving water quality. Successful river restoration projects can be found in locations like the Elwha River in Washington State, USA, where dam removal led to the recovery of salmon populations and restoration of natural processes.

These case studies highlight the potential for restoring degraded ecosystems and showcase the remarkable resilience of nature when provided with the opportunity to recover.

Challenges and Future Directions in Ecosystem Restoration

While ecosystem restoration has shown promise in reversing ecosystem degradation, several challenges and future directions must be considered to ensure its long-term success:

- Financial and political barriers: Ecosystem restoration projects often require significant funding and political support. Securing financial resources and garnering political will can be challenging, especially for large-scale restoration initiatives. Innovative financing mechanisms, such as payment for ecosystem services and green bonds, can help overcome these barriers.

- Monitoring and evaluation of restoration projects: It is crucial to monitor the progress and evaluate the effectiveness of restoration projects to ensure that the intended objectives are being met. Developing standardized monitoring protocols and utilizing remote sensing technologies can facilitate accurate and cost-effective monitoring.

- Integrating indigenous knowledge in restoration practices: Indigenous peoples have invaluable traditional ecological knowledge and practices that can significantly contribute to ecosystem restoration. Collaborating with indigenous communities and integrating their knowledge systems into restoration projects can enhance project outcomes and promote cultural sustainability.

- Climate change impacts on restoration efforts: Climate change poses additional challenges to ecosystem restoration. Increasing temperatures, altered rainfall patterns, and extreme weather events can affect the success of restoration projects. Considering climate change resilience in restoration planning, such as selecting climate-adapted species and restoring natural habitats that can sequester carbon, is crucial.

- Scaling up restoration for global impact: The degradation of ecosystems is a global problem that requires large-scale and concerted efforts for restoration. Collaborative initiatives, such as the United Nations Decade on Ecosystem Restoration, aim to mobilize governments, organizations, and communities to restore millions of hectares of degraded lands worldwide, fostering sustainable development and resilience.

In conclusion, ecosystem restoration is a powerful tool to reverse the damage caused by human activities and restore the functionality and diversity of ecosystems. By understanding the principles of ecological succession, applying appropriate techniques, and addressing the challenges, we can pave the way towards a more sustainable and resilient future. Through collective action and collaboration, we can restore the incredible beauty and complexity of the world's ecosystems and ensure their continued benefits for generations to come.

Ecosystem Restoration: Principles and Concepts

Definition and Objectives of Ecosystem Restoration

Ecosystem restoration is a critical field within environmental science that aims to repair and restore natural ecosystems that have been damaged or degraded by human activities. It involves the implementation of various strategies and techniques to reverse the negative impacts caused by factors such as pollution, habitat destruction, and climate change. The main objectives of ecosystem restoration are to enhance the health and functionality of ecosystems, promote biodiversity conservation, and ensure the provision of ecosystem services for present and future generations.

Defining Ecosystem Restoration

Ecosystem restoration can be defined as the deliberate and planned process of assisting the recovery of an ecosystem that has been disturbed or destroyed. It is a proactive approach that goes beyond simply conserving existing ecosystems. Restoration activities focus on repairing the physical structure, ecological processes, and biodiversity of the affected ecosystem. The overall goal is to return the ecosystem to a state that is ecologically stable, resilient, and capable of providing the multiple functions and benefits it originally offered.

ECOSYSTEM RESTORATION: PRINCIPLES AND CONCEPTS

Objectives of Ecosystem Restoration

1. **Restoring Ecological Functionality**: One of the primary objectives of ecosystem restoration is to restore and enhance the ecological functionality of an ecosystem. This involves recreating the natural processes, interactions, and relationships within the ecosystem. For example, in a wetland restoration project, the objective may be to rebuild the natural hydrological and nutrient cycling processes and reintroduce native plant and animal species.

2. **Conserving Biodiversity**: Ecosystem restoration plays a crucial role in conserving biodiversity by rehabilitating habitats and providing suitable conditions for the recovery of native plant and animal species. The objective is not only to recover the population size of endangered or threatened species but also to enhance the overall biodiversity of the ecosystem. This can be achieved through the removal of invasive species, the restoration of native vegetation, and the creation of suitable habitats for wildlife.

3. **Providing Ecosystem Services**: Ecosystem restoration aims to restore and maintain the provision of critical ecosystem services such as clean air and water, soil fertility, pollination, and climate regulation. By restoring the natural functioning of ecosystems, we can ensure the sustainable delivery of these services, which are essential for human well-being and the functioning of societies.

4. **Enhancing Resilience and Adaptation**: Ecosystem restoration also seeks to improve the resilience of ecosystems to cope with ongoing and future environmental changes. By restoring degraded ecosystems, we can enhance their ability to withstand disturbances, such as extreme weather events, and adapt to changing conditions. This is particularly important in the face of climate change, which poses significant challenges to the stability and functioning of ecosystems.

5. **Promoting Environmental Education and Awareness**: Another objective of ecosystem restoration is to raise awareness and educate the public about the importance of ecosystems and their role in maintaining a healthy environment. Restoration projects often involve community engagement and participation, which can lead to increased environmental literacy and a greater sense of responsibility towards the protection and restoration of ecosystems.

Key Principles of Ecosystem Restoration

The practice of ecosystem restoration is guided by several key principles that provide a framework for planning and implementing restoration projects:

1. **Ecological Integrity**: Restoration efforts should prioritize the restoration of ecological integrity, which refers to the health, resilience, and functioning of

ecosystems. This involves ensuring the presence of native species, maintaining natural ecological processes, and preserving the overall structure and composition of the ecosystem.

2.   **Collaboration and Stakeholder Engagement:** Successful restoration projects often require collaboration and active engagement with various stakeholders, including local communities, government agencies, NGOs, and scientists. Involving stakeholders throughout the restoration process helps build support, ensures the incorporation of diverse perspectives, and increases the chances of long-term success.

3. **Adaptive Management:** Ecosystem restoration is an iterative and adaptive process that involves continuous monitoring, evaluation, and adjustment of restoration activities. Adaptive management allows for learning from the outcomes of restoration efforts and enables project managers to refine their approaches over time based on new information and changing circumstances.

4. **Sustainability and Long-Term Planning:** Restoration projects should be designed and implemented with a long-term vision to ensure their sustainability. This involves considering the potential impacts of climate change, establishing realistic goals and timelines, and securing the necessary resources for ongoing monitoring and maintenance.

Example: Mangrove Forest Restoration

To illustrate the principles and objectives of ecosystem restoration, let's consider the restoration of mangrove forests. Mangrove forests are critical coastal ecosystems that provide various ecosystem services such as coastal protection, carbon sequestration, and habitat for unique biodiversity. However, they are highly vulnerable to human activities such as deforestation, aquaculture expansion, and coastal development.

The objectives of mangrove forest restoration may include the following:

1. Restoring Ecological Functionality: The restoration project aims to recreate the natural hydrological patterns within the mangrove ecosystem by restoring tidal flow and water circulation. This will allow for the natural deposition of sediments and nutrients, facilitating the growth of mangrove trees and providing suitable conditions for other associated plant and animal species.

2. Conserving Biodiversity: The restoration project focuses on reintroducing native mangrove species that are well-adapted to the local conditions. By doing so, it provides habitat for various wildlife species, including migratory birds, fish, and crustaceans. This helps to conserve the biodiversity of the mangrove ecosystem and supports the recovery of endangered or threatened species.

ECOSYSTEM RESTORATION: PRINCIPLES AND CONCEPTS

3. Providing Ecosystem Services: The restored mangrove forest will serve as a buffer zone, protecting coastal areas from storm surges, erosion, and sea-level rise. It will also act as a carbon sink, sequestering carbon dioxide from the atmosphere and mitigating climate change. Additionally, the restored mangroves will support local fisheries by providing a nursery for commercially important fish species.

4. Enhancing Resilience and Adaptation: The restoration project considers the anticipated impacts of climate change, such as sea-level rise and increased frequency of extreme weather events. It incorporates adaptive management practices, such as planting a mix of mangrove species with varying salt tolerance, to enhance the resilience of the restored ecosystem in the face of changing environmental conditions.

By implementing the principles of ecosystem restoration and addressing these objectives, the mangrove forest restoration project not only enhances the health and functionality of the ecosystem but also provides valuable benefits to local communities and the broader environment.

Key Resources and Tools

Ecosystem restoration requires a combination of scientific knowledge, technical expertise, and practical tools. Here are some key resources and tools commonly used in ecosystem restoration projects:

1. **Ecological Assessments:** Detailed ecological assessments are conducted to understand the current condition of the ecosystem, identify the main threats and pressures, and determine the appropriate restoration techniques and strategies.

2. **Monitoring and Evaluation:** Robust monitoring and evaluation protocols are essential for assessing the success of restoration projects. These protocols involve collecting data on various ecological indicators, such as species diversity, vegetation cover, and water quality, to track the progress of restoration efforts over time.

3. **Native Plant Nurseries:** The establishment of native plant nurseries is crucial for sourcing and propagating native plant species used in restoration projects. These nurseries provide a controlled environment for seed germination and seedling growth before they are transplanted into the restoration site.

4. **Hydrological and Soil Management Techniques:** In some cases, hydrological and soil management techniques are employed to recreate or enhance the natural water flow and soil conditions in the restored ecosystem. This may involve constructing levees or channels to regulate water levels or improving soil fertility through the addition of organic matter.

5. **GIS and Remote Sensing:** Geographic Information Systems (GIS) and remote sensing technologies are used for mapping and monitoring the spatial

extent and changes in ecosystems. These tools provide valuable information for decision-making during the planning and implementation phases of restoration projects.

6. **Engagement and Education:** Building community awareness and engagement is vital for the success of restoration projects. Educational programs, workshops, and outreach initiatives help to foster a sense of ownership and stewardship among local communities, ensuring the long-term sustainability of restoration efforts.

Conclusion

Ecosystem restoration is an essential approach for addressing the degradation and loss of natural ecosystems caused by human activities. By defining and understanding the objectives of ecosystem restoration, and by adhering to key principles, we can effectively restore and enhance the health, resilience, and functionality of ecosystems. Through successful restoration projects, we can contribute to the conservation of biodiversity, provision of ecosystem services, and the sustainable management of our natural resources.

Ecological Succession and Ecosystem Development

Ecological succession is a natural process through which ecosystems undergo changes in their structure and composition over time. It refers to the predictable and orderly sequence of community development following a disturbance or the initial colonization of a barren area. This process is driven by interactions between the living organisms and the abiotic factors in the environment. Understanding ecological succession is vital for ecosystem management, biodiversity conservation, and restoration efforts.

Primary and Secondary Succession

Ecological succession can be classified into two types: primary succession and secondary succession. Primary succession occurs in areas that are devoid of life, such as bare rock surfaces, lava flows, or newly formed sand dunes. It starts with pioneer species like lichens and mosses that are capable of colonizing the area and breaking down the substrate, gradually forming soil. As the soil develops, herbaceous plants and shrubs establish themselves, paving the way for the growth of trees and the establishment of a forest ecosystem.

Secondary succession, on the other hand, occurs in areas that have been previously occupied by an ecosystem but have undergone disturbance events such

ECOSYSTEM RESTORATION: PRINCIPLES AND CONCEPTS

111

as fire, logging, or agriculture. In secondary succession, the soil is already present, which provides a seed bank and nutrients for plant growth. Common pioneer species in secondary succession include fast-growing herbaceous plants and shrubs. Over time, these pioneer species are replaced by shade-tolerant tree species, leading to the reestablishment of a mature and diverse forest ecosystem.

Mechanisms of Succession

Several mechanisms drive the process of ecological succession. One critical mechanism is facilitation, where early successional species modify the environment, making it more suitable for future species to establish. For example, pioneer plants can improve soil fertility, enhance water retention, and provide shade, creating microhabitats for other species to colonize.

Another mechanism is inhibition, where early colonizers hinder the establishment of other species. These early species may compete for resources, such as sunlight, water, and nutrients, limiting the growth and survival of other plant species. However, as the ecosystem matures, the inhibitory effects decrease, allowing more species to establish and diversify.

Tolerance is another mechanism that influences succession. As the ecosystem develops, species that are more tolerant of the prevailing conditions become dominant. These species are better adapted to the environment and can outcompete and replace the initial colonizers. Tolerant species are often characterized by slower growth rates and longer lifespans, contributing to the stability and longevity of the ecosystem.

The Role of Disturbances

Disturbances play a crucial role in shaping ecological succession. While disturbances can initially disrupt an ecosystem, they also create opportunities for new species to colonize and for existing species to adapt and regenerate. Disturbances can be natural, such as wildfires, hurricanes, or volcanic eruptions, or human-induced, like logging or agriculture.

In some ecosystems, disturbances are an integral part of their natural cycle, promoting species diversity and maintaining ecosystem health. For example, fire-adapted ecosystems rely on periodic fires to regenerate and release nutrients locked in plant biomass. These disturbances reset succession, providing open spaces and reducing competition, enabling assemblages of species that are adapted to post-disturbance conditions.

Succession and Ecosystem Development

Ecological succession is a fundamental process in ecosystem development. As succession progresses, there is an increase in species diversity and biomass accumulation. Initially, the ecosystem may be dominated by a few species with large population sizes. However, over time, the community becomes more complex, consisting of diverse species with varying roles and interactions.

Successional changes not only occur in the plant community but also influence the composition and abundance of animal species. As the ecosystem develops, habitat structure and resource availability change, attracting different animal species and influencing their population dynamics. For instance, early successional vegetation may provide ideal nesting sites for certain bird species, while mature forests provide habitat for large mammals.

Applying Succession in Restoration Ecology

Understanding ecological succession is crucial in restoration ecology, where damaged or degraded ecosystems are rehabilitated. By studying the natural successional patterns of target ecosystems, restoration practitioners can develop strategies to fast-track succession and accelerate the recovery process.

For example, during the restoration of degraded lands, practitioners can introduce pioneer species that facilitate soil formation and nutrient cycling. These pioneer species help establish the foundational elements necessary for the successful establishment of more complex plant communities. By carefully selecting and manipulating the species composition and environmental conditions, restoration efforts can speed up succession and restore the desired ecosystem structure.

Conclusion

Ecological succession is a dynamic and predictable process that shapes the structure and function of ecosystems. It involves a series of changes in the composition and organization of species following a disturbance or colonization event. By understanding the mechanisms and patterns of succession, scientists and practitioners can effectively manage and restore ecosystems, promoting biodiversity conservation and the sustainable use of natural resources. In the next chapter, we will explore case studies of successful ecosystem restoration projects that have utilized the principles of ecological succession.

ECOSYSTEM RESTORATION: PRINCIPLES AND CONCEPTS

Approaches to Ecosystem Restoration

Ecosystem restoration is a complex and multidimensional process that aims to reverse the negative impacts of human activities and restore the ecological structure, function, and resilience of degraded ecosystems. It involves the implementation of various approaches and techniques tailored to specific ecosystem types and restoration goals. In this section, we will explore some of the key approaches commonly employed in ecosystem restoration.

Passive Restoration

Passive restoration, also known as natural regeneration, relies on the natural process of ecological succession to restore degraded ecosystems. In this approach, human intervention is minimal, and the ecosystem is given the space and time to recover on its own. This method is suitable for ecosystems with the capacity for self-recovery, such as forests, wetlands, and grasslands. By removing or reducing the sources of degradation and disturbances, such as invasive species, pollution, and land-use changes, passive restoration allows natural processes to resume and facilitates the reestablishment of native species and ecological interactions.

However, passive restoration can be a slow and uncertain process, especially in heavily degraded ecosystems or those facing multiple stressors. It requires careful monitoring and adaptive management to ensure that the ecosystem is progressing towards recovery. Nevertheless, passive restoration has several advantages, including lower costs, minimal disruption to the ecosystem, and the potential for the development of resilient and self-sustaining ecosystems.

Active Restoration

Active restoration involves direct human intervention to accelerate and guide the recovery of degraded ecosystems. This approach is typically employed in cases where natural recovery is unlikely or insufficient to achieve restoration goals within a reasonable timeframe. Active restoration techniques can include:

- **Tree planting and reforestation:** In deforested or degraded areas, planting native tree species can help restore forest ecosystems. This technique promotes the recovery of forest structure, enhances biodiversity, and provides various ecosystem services such as carbon sequestration, soil erosion prevention, and habitat creation.

- **Wetland creation and enhancement:** Wetlands provide critical habitat for many species, mitigate flooding, improve water quality, and sequester carbon. Active restoration of wetlands involves recreating or enhancing natural wetland features, such as hydrology, vegetation, and soil conditions, to promote the return of wetland functions and values.

- **Riparian zone restoration:** Riparian zones, the areas adjacent to rivers and streams, play a crucial role in enhancing water quality, regulating streamflow, and supporting aquatic and terrestrial biodiversity. Restoration efforts in riparian zones often involve replanting native vegetation, stabilizing stream banks, and reintroducing woody debris to create habitat complexity.

- **Artificial reef installation:** Artificial reefs are human-made structures deployed in marine environments to provide substrate for marine organisms, enhance fish habitats, and promote biodiversity. These structures can be made from various materials, such as concrete, steel, or recycled materials, and are strategically placed to improve the ecological function of degraded or damaged coral reefs or other marine ecosystems.

Active restoration requires careful planning, site preparation, and monitoring to ensure the success of restoration efforts. The choice of techniques depends on the specific ecosystem characteristics, restoration objectives, available resources, and stakeholder involvement. It is essential to consider the potential long-term impacts of active restoration activities on the ecosystem, such as unintended consequences or conflicts with other ecosystem functions or services.

Hybrid Restoration

Hybrid restoration combines elements of both passive and active restoration approaches. It recognizes the importance of natural processes while incorporating targeted interventions to jumpstart ecosystem recovery. This approach aims to harness the benefits of natural regeneration while addressing specific challenges or constraints that hinder ecosystem recovery.

For example, in a hybrid restoration project in a degraded grassland ecosystem, passive restoration techniques may be initially employed, such as controlling grazing pressures to allow for natural seed dispersal and vegetation regrowth. Once the grassland has started to recover, active restoration techniques, such as introducing native plant species or conducting prescribed burns, may be used to enhance the desired ecosystem structure and function.

Hybrid restoration can be particularly effective in situations where limited resources or external factors constrain a complete reliance on either passive or

ECOSYSTEM RESTORATION: PRINCIPLES AND CONCEPTS

active restoration. By combining the strengths of both approaches, hybrid restoration offers a flexible and adaptive strategy for achieving restoration objectives.

Collaborative and Community-based Restoration

Collaborative and community-based restoration approaches recognize the importance of engaging relevant stakeholders, including local communities, landowners, indigenous peoples, and government agencies, in the restoration process. These approaches ensure that diverse perspectives, knowledge, and resources are integrated into restoration planning, implementation, and long-term management.

By involving local communities and promoting their active participation, collaborative restoration can facilitate the transfer of traditional ecological knowledge, promote social and economic benefits, and enhance the long-term sustainability of restoration efforts. It also fosters a sense of ownership and stewardship among local stakeholders, leading to increased success and resilience of restoration projects.

Example: In the Mississippi River Delta, the Louisiana Coastal Master Plan adopts a collaborative restoration approach by leveraging partnerships with local communities and diverse stakeholders. Through joint decision-making processes, the restoration projects aim to rebuild coastal wetlands, enhance flood protection, and sustain livelihoods of communities while conserving biodiversity.

Innovative Restoration Techniques

As the field of ecosystem restoration evolves, new and innovative techniques are being developed to address emerging challenges and enhance restoration outcomes. These techniques incorporate cutting-edge technologies, such as biotechnology, remote sensing, and ecological engineering, to facilitate restoration activities.

For example, biotechnology tools, including genetic engineering and tissue culture, can be used to propagate and reintroduce rare or endangered plant species into degraded ecosystems. Remote sensing techniques, such as aerial imagery and LiDAR, enable accurate mapping and monitoring of restoration sites, allowing for better assessment of the restoration progress and identification of potential issues. Ecological engineering utilizes engineering principles and ecological knowledge to design and construct restoration projects that mimic natural processes, such as constructing artificial reefs or oyster beds to enhance coastal resilience and habitat creation.

The integration of innovative techniques into ecosystem restoration practices presents opportunities for more efficient, cost-effective, and ecologically sustainable restoration. However, it is important to assess the potential risks and unintended consequences associated with these techniques, ensuring that their use aligns with the principles of ecological integrity and long-term ecosystem health and resilience.

In summary, ecosystem restoration involves a range of approaches tailored to specific ecosystems and restoration goals. Passive restoration allows natural processes to drive recovery, while active restoration involves direct human intervention. Hybrid restoration combines elements of both approaches, and collaborative approaches engage stakeholders in the restoration process. Innovative techniques and technologies further enhance restoration outcomes. By applying these diverse approaches, we can increase the success and effectiveness of ecosystem restoration initiatives worldwide.

Case Studies in Ecosystem Restoration

Coral Reef Restoration

The world's coral reefs are facing an unprecedented crisis. These invaluable ecosystems, known for their vibrant colors and diverse marine life, are under threat from various factors such as climate change, pollution, and overfishing. As a result, coral reefs are experiencing widespread bleaching events, where the coral loses its vibrant color and ultimately dies. However, there is hope in the form of coral reef restoration efforts, which aim to rebuild and rehabilitate these fragile ecosystems.

Understanding Coral Reefs

Before delving into coral reef restoration, it is important to understand the basic components and functions of a coral reef ecosystem. Coral reefs are built by tiny marine organisms called coral polyps. These polyps secrete a hard calcium carbonate skeleton, which forms the structure of the reef. Coral reefs are home to a vast array of species, including fish, crustaceans, and other marine invertebrates, making them one of the most biodiverse habitats on the planet.

Coral reefs provide crucial ecosystem services. They act as nurseries for many fish species, provide coastal protection from storms, and support local economies through activities such as tourism and fisheries. Additionally, coral reefs play a significant role in the global carbon cycle by sequestering carbon dioxide from the atmosphere.

CASE STUDIES IN ECOSYSTEM RESTORATION

Challenges Facing Coral Reefs

Coral reefs face numerous threats that have resulted in their decline and degradation. One of the most pressing challenges is climate change, particularly the rising ocean temperatures and increased acidity levels. Elevated sea surface temperatures cause coral bleaching, a phenomenon where the coral expels the colorful algae living within its tissues, leading to its death.

Pollution, including nutrient runoff from agriculture and coastal development, also poses a major threat to coral reefs. Excess nutrients can lead to algal overgrowth, which can smother and kill corals. Additionally, overfishing and destructive fishing practices, such as blast fishing and cyanide fishing, can directly damage coral reefs and disrupt the delicate balance of the ecosystem.

Coral Reef Restoration Techniques

Coral reef restoration aims to reverse the damage done to coral reefs and promote their recovery. Several techniques are employed in coral reef restoration, each tailored to the specific needs and challenges of the target reef.

One common restoration technique is coral transplantation. This involves collecting fragments of healthy coral from donor reefs and attaching them to the degraded reef using underwater adhesives or cement. Over time, these transplanted corals grow and spread, enhancing the overall reef structure and biodiversity.

Another technique is coral larval propagation. This method involves collecting coral larvae during their annual mass spawning events and rearing them in specialized facilities until they can be released onto degraded reefs. By introducing new coral recruits, this technique helps boost the reef's recovery and genetic diversity.

Additionally, artificial reef structures can be deployed to create habitats for coral colonization. These structures can be made from materials such as concrete, steel, or even recycled materials like old ships or sunken structures. They provide a substrate for coral attachment and offer protection for juvenile corals.

Success Stories in Coral Reef Restoration

Coral reef restoration efforts have yielded successful outcomes in various parts of the world. One notable success story is the Mesoamerican Reef in the Caribbean. In this region, local communities, research institutions, and non-profit organizations have collaborated to restore damaged reef areas by implementing coral transplantation and other restoration techniques. These efforts have resulted in the recovery of coral cover and improved ecosystem health.

118   CHAPTER 2: RESTORING ECOSYSTEMS

Another success story is the Great Barrier Reef in Australia. The Reef Restoration and Adaptation Program (RRAP) is an ambitious initiative aiming to scale up coral restoration efforts on a large scale. The program focuses on innovative techniques, such as using underwater robots to disperse coral larvae onto damaged reefs. Early results show promise in enhancing the reef's capacity to recover from disturbances.

Challenges and Future Directions

Despite the successes achieved in coral reef restoration, significant challenges remain. One of the main challenges is the scale of the problem. Coral reefs are vast and complex ecosystems, making restoration efforts logistically challenging and resource-intensive. To overcome this, it is crucial to prioritize areas for restoration based on ecological importance, connectivity, and the potential for success.

Another challenge is securing long-term funding and support for restoration projects. Coral reef restoration requires sustained financial investment, as well as active involvement from local communities, governments, and stakeholders. Developing innovative financing mechanisms and engaging in public-private partnerships can help ensure the long-term success of restoration initiatives.

In the future, advancements in the field of coral microbiology and genetics may play a significant role in coral reef restoration. Understanding the genetic diversity and resilience of coral species can inform the selection of suitable donor colonies and enhance the success rates of restoration efforts. Additionally, research into the interactions between corals and their microbiomes may provide insights into strategies for enhancing coral health and resilience.

In conclusion, coral reef restoration is a vital undertaking to conserve and rebuild these precious ecosystems. By implementing innovative techniques and engaging in interdisciplinary collaborations, restoration efforts hold the potential to reverse the damage done to coral reefs and ensure their long-term survival. However, it requires a concerted global effort to address the underlying causes of coral reef decline, such as climate change and pollution, to truly safeguard these remarkable and invaluable ecosystems for future generations.

Reforestation and Afforestation Projects

Reforestation and afforestation are two important practices in environmental science that aim to restore and increase forest cover. Both processes involve the planting of trees; however, they differ in their specific objectives and the context in which they are implemented. In this section, we will explore the principles and techniques of

CASE STUDIES IN ECOSYSTEM RESTORATION

reforestation and afforestation projects, their significance in combating deforestation and climate change, and some successful case studies.

Deforestation and its Impacts

Before delving into reforestation and afforestation, it is essential to understand the problem they address: deforestation. Deforestation refers to the permanent removal of forest cover, primarily caused by human activities such as logging, agriculture expansion, and infrastructure development. This widespread destruction of forests has severe consequences for both the environment and human society.

One of the significant environmental impacts of deforestation is the loss of biodiversity. Forests are home to millions of plant and animal species, many of which are unique and play critical roles in maintaining the ecosystem's balance. Deforestation disrupts these delicate ecosystems, leading to the extinction of species and the degradation of habitats.

Furthermore, deforestation contributes to climate change. Trees act as carbon sinks, absorbing carbon dioxide from the atmosphere through photosynthesis and storing it in their biomass. When forests are cut down, this stored carbon is released back into the atmosphere, contributing to the greenhouse effect and global warming. Deforestation also disrupts the water cycle, leading to reduced precipitation, altered weather patterns, and increased droughts.

Principles and Objectives of Reforestation

Reforestation is the process of replanting trees in areas where forests have been destroyed or degraded. Its primary objective is to restore forest cover and the associated ecological and environmental functions. Reforestation can take place in both natural and man-made forests, such as plantations or degraded woodland areas. The following principles guide reforestation efforts:

1. **Selection of Suitable Species:** The choice of tree species for reforestation depends on various factors, including soil conditions, climate, and intended objectives. It is crucial to select species that are well-adapted to the site and promote biodiversity, including both native and exotic species if appropriate.

2. **Ecological Succession:** Reforestation should consider the principles of ecological succession, which refers to the natural process of vegetation development over time. Understanding the succession patterns helps ensure the establishment of a diverse and resilient forest ecosystem.

120   CHAPTER 2: RESTORING ECOSYSTEMS

3. **Site Preparation:** Proper site preparation is necessary to create favorable conditions for tree growth. This may include clearing invasive species, tilling the soil, or implementing erosion control measures.

4. **Monitoring and Management:** Regular monitoring is essential to evaluate the success of reforestation projects and make necessary adjustments in management practices. This includes measuring tree growth, assessing biodiversity recovery, and addressing any challenges or threats.

5. **Community Engagement:** Involving local communities in reforestation efforts fosters a sense of ownership and promotes long-term sustainability. Communities can contribute to tree planting, habitat restoration, and monitoring activities, ensuring the success of the project.

Afforestation and its Objectives

Afforestation, on the other hand, refers to the establishment of forests in areas where there were no forests previously. It involves converting non-forest lands, such as agricultural fields or barren lands, into forested areas. Afforestation plays a crucial role in expanding forest cover and has several objectives:

1. **Carbon Sequestration:** Afforestation helps mitigate climate change by sequestering carbon dioxide from the atmosphere. As the newly planted trees grow, they absorb carbon dioxide through photosynthesis, helping to offset greenhouse gas emissions.

2. **Soil Conservation:** Afforestation helps prevent soil erosion and degradation by providing a protective cover of vegetation. Tree roots stabilize the soil, reducing the risk of landslides and improving water infiltration.

3. **Biodiversity Conservation:** Afforestation contributes to the preservation and restoration of biodiversity. By creating new habitats, afforestation provides shelter and resources for various plant and animal species, fostering biodiversity conservation.

4. **Water Management:** Afforestation plays a crucial role in water resource management. Forests act as natural water filters, purifying water and improving its quality. They also regulate water flow, prevent floods, and recharge groundwater.

CASE STUDIES IN ECOSYSTEM RESTORATION

5. **Economic Benefits:** Afforestation projects can provide economic benefits to local communities by creating sustainable livelihood opportunities, such as ecotourism, timber production, and non-timber forest products.

Reforestation and Afforestation Techniques

The success of reforestation and afforestation projects depends on the effective implementation of various techniques. Some common techniques used in these projects include:

1. **Direct Seeding:** This technique involves sowing seeds directly into the ground, either by hand or using machinery. It is cost-effective and suitable for areas with favorable soil conditions and low levels of disturbance.

2. **Seedling Planting:** Planting seedlings is a widely used technique in reforestation and afforestation projects. The seedlings are typically raised in nurseries and carefully transplanted into the target site. This technique allows for better control over species selection and ensures a higher survival rate.

3. **Assisted Natural Regeneration:** In some cases, natural regeneration can be encouraged by improving site conditions and protecting young seedlings. This technique involves removing competing vegetation and providing necessary resources for regeneration to occur.

4. **Agroforestry:** Agroforestry is a sustainable land-use system that combines tree planting with agricultural practices. It integrates trees with crops or livestock, providing multiple benefits such as improved soil fertility, increased biodiversity, and additional income streams.

5. **Silvicultural Techniques:** Silviculture involves the application of scientific and sustainable principles to manage forests. Techniques such as selective cutting, thinning, and coppicing can be used to promote forest growth and diversity.

Case Studies

Let's explore two successful case studies in reforestation and afforestation:

1. **Great Green Wall Initiative:** The Great Green Wall is an ambitious afforestation initiative in Africa that aims to combat desertification and

promote sustainable land management. Spanning across multiple countries, the project involves planting a large belt of trees and vegetation along the southern edge of the Sahara desert. The Great Green Wall not only helps to halt desertification but also provides economic opportunities for local communities and enhances biodiversity.

2. **Eden Reforestation Projects:** The Eden Reforestation Projects is a non-profit organization that has successfully implemented reforestation programs in multiple countries, including Madagascar, Haiti, and Nepal. Their approach involves employing local villagers to plant trees and restore degraded land. This not only helps increase forest cover but also creates employment opportunities and improves the socio-economic conditions of the communities involved.

These case studies highlight the importance of community involvement, long-term monitoring, and a holistic approach to reforestation and afforestation projects.

Conclusion

Reforestation and afforestation projects are crucial tools in combating deforestation, restoring ecosystems, and mitigating climate change. By carefully selecting suitable species, engaging local communities, and implementing effective techniques, these projects can have long-lasting positive impacts. However, it is essential to recognize the challenges and limitations of such projects, including the need for proper monitoring, policy support, and addressing underlying drivers of deforestation. Through collective efforts, reforestation and afforestation can contribute significantly to a greener and more sustainable future.

Wetland Restoration

Wetlands are vital ecosystems that provide numerous ecological benefits, including water filtration, flood control, and habitat for a variety of plant and animal species. However, due to human activities such as drainage for agriculture and urban development, wetlands have been extensively degraded and lost worldwide. Wetland restoration aims to reverse these impacts and restore the ecological functions and services of degraded or destroyed wetlands.

CASE STUDIES IN ECOSYSTEM RESTORATION

Definition and Objectives of Wetland Restoration

Wetland restoration can be defined as the process of returning a wetland ecosystem to its original or near-original condition after disturbances or degradation. The objectives of wetland restoration include:

+ Re-establishing the hydrological characteristics of the wetland by restoring water flow patterns, water levels, and the natural water regime.

+ Restoring the biogeochemical processes in the wetland, such as nutrient cycling and sedimentation, to improve water quality.

+ Re-establishing diverse and functioning wetland vegetation communities.

+ Enhancing wildlife habitat to support a variety of wetland-dependent species, including migratory birds, amphibians, and fish.

Approaches to Wetland Restoration

Wetland restoration projects can employ various approaches depending on the specific goals, site conditions, and available resources. Some common approaches include:

1. **Hydrological restoration:** This approach focuses on re-establishing natural water flow patterns and hydrological conditions within the wetland. It may involve modifying drainage systems, constructing levees or dikes, or restoring natural water channels to restore the water balance of the wetland.

2. **Vegetation restoration:** This approach aims to restore the original vegetation communities or establish new ones that are characteristic of the specific wetland type. It often involves removing invasive plant species, planting native wetland plants, and promoting natural regeneration of vegetation.

3. **Sediment and nutrient management:** Wetland restoration projects may involve managing sedimentation and nutrient inputs to improve water quality and restore the biogeochemical processes within the wetland. Techniques such as sediment traps, nutrient interception basins, and constructed wetlands can be employed to capture and retain sediments and nutrients.

124   CHAPTER 2: RESTORING ECOSYSTEMS

4. **Habitat enhancement:** Wetland restoration efforts can focus on enhancing or creating specific habitats within the wetland to support target species. This may include creating shallow-water areas for nesting waterfowl, constructing fish passages, or installing nesting platforms for birds.

5. **Community involvement and education:** Successful wetland restoration often involves engaging local communities and stakeholders in the planning and implementation process. Community involvement can enhance the long-term success of restoration projects through increased awareness, support, and participation.

Case Studies in Wetland Restoration

Several wetland restoration projects have demonstrated successful outcomes in restoring degraded wetlands and their associated ecological functions. Let's explore a few notable case studies:

The Florida Everglades, USA   The Florida Everglades is a globally significant wetland ecosystem that has been severely impacted by drainage for agriculture and urban development. The restoration efforts in the Everglades aim to restore the historic water flows and re-establish the natural hydrological patterns of the wetland. This involves removing levees and canals, constructing reservoirs, and implementing water management strategies to mimic the historic water flow. The restoration efforts also focus on removing invasive species, enhancing natural water filtration processes, and re-establishing wetland vegetation communities. The restoration of the Everglades not only benefits native flora and fauna but also provides water supply and flood control for the surrounding communities.

The Loess Plateau, China   The Loess Plateau in China is an area characterized by severe soil erosion and degradation due to unsustainable agricultural practices. Wetland restoration efforts in the region have involved the construction of check dams, terraces, and contour trenches to slow down water flow and reduce soil erosion. These measures have helped restore the natural hydrological patterns and improve water retention in the wetland areas. The restoration efforts have also included planting native wetland vegetation to stabilize the soil and enhance biodiversity. This combination of engineering and ecological restoration approaches has transformed the degraded landscapes into productive and ecologically valuable wetlands.

CASE STUDIES IN ECOSYSTEM RESTORATION

Challenges and Future Directions in Wetland Restoration

Despite the successes in wetland restoration, several challenges persist, and future efforts need to address these challenges to ensure sustained restoration outcomes. Some of the key challenges include:

- **Lack of funding:** Wetland restoration projects often require substantial financial resources for planning, implementation, and long-term monitoring. Securing adequate funding remains a significant challenge for many restoration initiatives.

- **Multiple stakeholders and conflicts:** Wetland restoration projects often involve multiple stakeholders, each with their own interests, which can lead to conflicts and delays in decision-making. Effective collaboration and conflict resolution are crucial to ensure the success of restoration efforts.

- **Climate change impacts:** Climate change poses additional challenges for wetland restoration. Rising sea levels, increased frequency of extreme weather events, and altered precipitation patterns can affect the success and long-term viability of restoration projects. Adapting restoration strategies to future climate scenarios is essential.

- **Monitoring and assessment:** Monitoring the success of wetland restoration projects is vital to evaluate the effectiveness of restoration efforts and inform adaptive management. Developing standardized monitoring protocols and long-term monitoring programs can be challenging due to the complexity and variability of wetland ecosystems.

- **Public awareness and engagement:** Increasing public awareness about the value of wetland ecosystems and the importance of restoration is crucial for garnering support and participation. Education and outreach programs are essential to engage local communities and stakeholders in restoration efforts.

In conclusion, wetland restoration plays a critical role in recovering and conserving these valuable ecosystems. By restoring hydrological conditions, vegetation communities, and ecological functions, wetland restoration projects can contribute to the conservation of biodiversity, improvement of water quality, and mitigation of climate change impacts. However, addressing the challenges and adopting innovative approaches will be crucial for the success and long-term sustainability of wetland restoration efforts.

Urban Ecological Restoration

Urban areas are characterized by high population density, extensive infrastructure, and limited green spaces. These environments often suffer from a multitude of environmental problems, such as air and water pollution, habitat loss, and heat island effects. However, through urban ecological restoration, it is possible to transform these spaces into vibrant and sustainable ecosystems that support a diverse range of species and provide numerous benefits to human residents. This section will explore the principles, techniques, and success stories of urban ecological restoration.

Definition and Objectives of Urban Ecological Restoration

Urban ecological restoration is the process of improving the health and functionality of urban ecosystems through deliberate human intervention. It involves enhancing biodiversity, restoring ecosystem services, and creating sustainable urban environments. The objectives of urban ecological restoration include:

- Enhancing biodiversity: Urban areas often experience a loss of native species and habitats. Restoring urban ecosystems aims to increase the diversity and abundance of plants, animals, and microorganisms in urban environments.

- Improving ecosystem services: Urban ecosystems provide important services, such as air purification, water filtration, climate regulation, and recreational opportunities. Restoration efforts aim to enhance these services to benefit both humans and wildlife.

- Creating sustainable urban environments: Urban ecological restoration incorporates sustainable practices into city planning and design. It aims to improve the quality of life for urban residents by creating green spaces, promoting walkability, and reducing environmental impacts.

Approaches to Urban Ecological Restoration

Urban ecological restoration employs a variety of approaches to transform urban landscapes into healthy and resilient ecosystems. Some common approaches include:

1. Habitat restoration: This involves creating or restoring habitats to support native plant and animal species. It may include reintroducing native

CASE STUDIES IN ECOSYSTEM RESTORATION 127

vegetation, constructing wildlife corridors, and restoring wetlands or streams.

2. Green infrastructure: Green infrastructure refers to the use of natural systems, such as street trees, green roofs, and rain gardens, to manage stormwater, reduce urban heat island effects, and improve air quality. These features mimic natural processes and provide multiple benefits to urban ecosystems.

3. Brownfield remediation: Brownfields are abandoned or underutilized sites that are contaminated with pollutants. Urban ecological restoration can involve the remediation of brownfields, transforming them into safe and productive spaces for both humans and wildlife.

4. Community engagement: Urban ecological restoration often involves active participation from the community. Engaging residents in the restoration process encourages a sense of ownership and responsibility, leading to more successful and sustainable outcomes.

Case Studies in Urban Ecological Restoration

High Line Park, New York City, USA: The High Line is a remarkable example of urban ecological restoration. It is a 1.45-mile-long elevated park built on a former elevated railway track. The park incorporates sustainable design elements, native plantings, and diverse habitats. It has become a haven for wildlife and a popular recreational space for residents and visitors, contributing to the overall revitalization of the surrounding neighborhood.

Cheonggyecheon Stream Restoration, Seoul, South Korea: Cheonggyecheon Stream was covered by an elevated highway in the 1960s, leading to severe environmental degradation. In 2005, the stream was restored through the removal of the highway and the creation of a natural waterway. The restoration project improved water quality, increased biodiversity, and provided an attractive public space for recreation and cultural activities.

Renaturation of the Donaukanal, Vienna, Austria: In Vienna, the Donaukanal, a canalized section of the Danube River, underwent a restoration project to enhance its ecological value and recreational potential. The project involved the creation of diverse habitats, such as wetlands and meadows, along the canal's banks. The restored area now serves as an important corridor for wildlife and a popular destination for residents and tourists.

Challenges and Future Directions in Urban Ecological Restoration

While urban ecological restoration presents immense opportunities, it also faces several challenges. Some of the key challenges include:

- Limited space: Urban areas often have limited available space for restoration projects. Maximizing the impact of restoration efforts requires innovative design and utilization of underutilized spaces such as rooftops or vacant lots.

- Fragmented habitats: Urban ecosystems are often fragmented due to infrastructure and human activities. Creating connected habitat patches and green corridors is crucial for supporting wildlife populations and promoting ecological resilience.

- Pollution and contamination: Urban environments are frequently polluted and contaminated, posing significant challenges to restoration efforts. Addressing pollution sources and implementing remediation measures are essential for successful restoration outcomes.

- Funding and resources: Urban ecological restoration projects require adequate funding and resources for planning, implementation, and long-term maintenance. Securing financial support from various stakeholders is a critical challenge.

To overcome these challenges and advance urban ecological restoration, several future directions should be considered:

- Incorporating nature-based solutions: Nature-based solutions, such as green infrastructure, can be integrated into urban planning to enhance the resilience of cities to climate change and improve overall environmental quality.

- Engaging the community: Active engagement of local communities in restoration projects fosters a sense of ownership and stewardship. Education and awareness programs can help individuals understand the importance of urban ecology and actively participate in restoration efforts.

- Collaboration and partnerships: Successful urban ecological restoration often requires collaboration among multiple stakeholders, including government agencies, non-profit organizations, and the private sector. Building strong partnerships can leverage resources and expertise for more effective outcomes.

CASE STUDIES IN ECOSYSTEM RESTORATION

- Policy support: Policy frameworks that promote urban ecological restoration and incorporate sustainability principles into urban planning are essential. Governments at various levels should develop and enforce policies that encourage restoration initiatives and prioritize green spaces in urban development.

In conclusion, urban ecological restoration offers a valuable opportunity to transform urban areas into thriving ecosystems that benefit both humans and nature. By employing strategic approaches, learning from successful case studies, and addressing the challenges ahead, cities can become sustainable, resilient, and harmonious environments that promote the well-being of all inhabitants.

River and Stream Restoration

Rivers and streams are vital components of Earth's ecosystems, providing essential habitat for countless species and serving as sources of freshwater for human populations. However, these water bodies face numerous threats including pollution, habitat degradation, and altered flow regimes due to human activities. River and stream restoration aims to address these challenges and restore the health and functioning of these aquatic ecosystems.

Principles and Concepts

River and stream restoration is guided by several key principles and concepts that contribute to successful outcomes. These principles include:

- **Hydrologic connectivity:** Restoring the natural flow regime and maintaining hydrologic connectivity within river networks is crucial for supporting healthy ecosystems. This involves preserving or recreating the natural pathways for water movement, including reconnecting fragmented sections of rivers.

- **Riparian zone restoration:** The riparian zone, the area alongside rivers and streams, plays a critical role in maintaining water quality and providing habitat. Restoration efforts often focus on revegetating and stabilizing riparian areas, enhancing their ability to filter pollutants, control erosion, and provide habitat for wildlife.

- **Channel morphology:** Understanding the natural dynamics of river channels and their response to changes in flow and sediment transport is essential for effective restoration. Restoration projects may involve modifying or reshaping channels to restore natural channel forms and processes, such as meandering patterns and pools.

+ **Enhancing habitat diversity:** Increasing habitat complexity and diversity is a common objective in river and stream restoration. This may involve creating instream structures like boulders or log jams to provide shelter and hiding places for aquatic organisms, as well as restoring side channels and floodplain connectivity to expand available habitat.

+ **Stakeholder engagement:** Successful restoration projects often involve collaboration and engagement with stakeholders, including local communities, government agencies, and environmental organizations. Engaging stakeholders throughout the restoration process helps build support, promote long-term sustainability, and ensure the inclusion of diverse perspectives and expertise.

Approaches to River and Stream Restoration

River and stream restoration can be approached using various techniques depending on the specific goals and challenges of each project. Some common approaches include:

+ **Natural channel design:** This approach emphasizes working with natural processes to restore river channels and reestablish their dynamic equilibrium. It involves using natural materials and methods to create stable and self-sustaining channels that mimic natural conditions as closely as possible.

+ **Streambank stabilization:** Unstable streambanks can contribute to excessive erosion and sedimentation, affecting water quality. Restoration efforts may involve techniques such as bioengineering, which uses vegetation and natural materials to stabilize streambanks and reduce erosion while maintaining ecological functions.

+ **Fish passage restoration:** Many rivers and streams contain barriers, such as dams or culverts, that obstruct fish migration and disrupt their life cycles. Restoring fish passage involves modifying or removing barriers to allow fish to move freely between different habitats, promoting population connectivity and restoring natural ecological processes.

CASE STUDIES IN ECOSYSTEM RESTORATION
131

- **Floodplain restoration:** Floodplains are vital components of river ecosystems, providing habitats for both aquatic and terrestrial species and serving as important natural reservoirs during floods. Restoration efforts often aim to reconnect rivers with their floodplains, allowing for controlled flooding and restoring the natural floodplain dynamics.

- **Water quality improvement:** Addressing pollution sources is crucial for restoring the health of rivers and streams. Restoration strategies may focus on reducing point-source pollution, such as treating wastewater, as well as implementing measures to control non-point source pollution from agricultural runoff or urban stormwater.

Case Studies in River and Stream Restoration

Several successful river and stream restoration projects from around the world serve as inspiring examples of what can be achieved through dedicated efforts. Let's explore a few notable case studies:

Los Angeles River, California, USA The Los Angeles River is an iconic urban waterway that had been severely channelized and degraded for flood control purposes. In recent years, significant restoration efforts have been made to enhance the ecological functioning and recreational value of the river. This includes removing sections of concrete channel to restore natural stream channels, creating wetlands, and providing habitat for fish and migrating birds within the urban landscape.

Thames River, London, UK The Thames River restoration project aimed to reverse the impact of historical engineering works on the river system and promote ecological recovery. Efforts focused on creating natural meanders, reconnecting side channels, and restoring wetlands and floodplains. These measures have not only improved the river's biodiversity and water quality but also enhanced the resilience of adjacent communities to flooding.

Elwha River, Washington, USA The removal of two dams on the Elwha River represents one of the largest dam removal and river restoration projects in history. After the dams were removed, salmon populations rebounded, and the river's ecosystem began to recover. The restoration of the Elwha River demonstrates the power of removing barriers and allowing natural processes to restore rivers to their original state.

Challenges and Future Directions

While river and stream restoration projects have achieved notable successes, they also face several challenges and uncertainties. These challenges include:

+ **Limited funding and resources:** Restoration projects require significant financial resources and long-term commitment. Securing adequate funding and resources remains a challenge, particularly for large-scale restoration initiatives.

+ **Legacy impacts and contaminants:** Many rivers and streams are affected by historical pollution and contamination that can persist for decades. Restoration efforts must address these legacy impacts and manage contaminants to ensure the long-term health of aquatic ecosystems.

+ **Climate change impacts:** Climate change poses additional challenges to river and stream restoration. Rising temperatures, altered precipitation patterns, and increased frequency of extreme events can affect the success of restoration projects and require adaptation strategies to account for changing conditions.

+ **Complexity of river systems:** Rivers and streams are complex ecosystems influenced by multiple interacting factors. Understanding and managing these complexities, such as the interactions between hydrology, ecology, and geomorphology, is crucial for effective restoration.

Looking ahead, future directions in river and stream restoration will likely involve:

+ **Integration of ecosystem services:** Recognizing and quantifying the multiple benefits provided by restored rivers and streams, such as water purification, flood mitigation, and recreational opportunities, can further support and justify restoration efforts.

+ **Adaptive management and monitoring:** Implementing adaptive management strategies that allow for ongoing learning and adjustment based on monitoring data will enhance the success and effectiveness of restoration projects. Monitoring can help evaluate project outcomes, detect potential issues early on, and guide future restoration efforts.

+ **Engagement of local communities:** Involving local communities and stakeholders in decision-making processes and restoration activities is

SUCCESS STORIES IN ECOSYSTEM RESTORATION

crucial. Empowering local communities to actively participate in and benefit from restoration projects fosters stewardship and sustainability.

- **Collaboration and knowledge exchange:** Sharing knowledge, experiences, and best practices among researchers, practitioners, and communities globally can advance the field of river and stream restoration. Collaboration can help overcome common challenges, promote innovation, and maximize the impact of restoration efforts.

In conclusion, river and stream restoration is a critical component of environmental science, aiming to restore and preserve the ecological integrity and functionality of these vital freshwater ecosystems. By applying principles such as hydrologic connectivity, riparian zone restoration, and enhancing habitat diversity, along with various restoration approaches, we can address the challenges and work towards creating healthier and more sustainable rivers and streams for future generations.

Success Stories in Ecosystem Restoration

Loess Plateau Restoration in China

The Loess Plateau is a vast area located in northwestern China, covering approximately 640,000 square kilometers. It is characterized by its unique landscape, consisting of deep gullies, steep slopes, and bare exposed soil. This region has been facing severe ecological degradation and soil erosion for centuries, resulting in the loss of fertile land, declining agricultural productivity, and increased water pollution downstream.

The Problem: Ecological Degradation and Soil Erosion

The ecological degradation of the Loess Plateau can be attributed to a combination of natural factors and human activities. The region experiences a dry and arid climate, which makes it susceptible to erosion. Additionally, the traditional farming practices, such as overgrazing, deforestation, and inappropriate land use, have accelerated soil erosion processes. Moreover, the steep slopes of the plateau exacerbate the erosion problems, as heavy rainfall during the monsoon season leads to the formation of deep gullies.

As a result of these factors, millions of cubic meters of soil are washed away every year, leading to land degradation, loss of biodiversity, and sedimentation in downstream rivers and reservoirs. This not only affects food production and local ecosystems but also impacts the livelihoods of the people living in the region.

Restoration Strategies: The "Grain for Green" Program

To address the ecological degradation and soil erosion in the Loess Plateau, the Chinese government implemented the "Grain for Green" program in 1999. This large-scale ecological restoration initiative aims to convert steep slopes and marginal farmland into forest and grassland areas.

Under the program, farmers are encouraged to voluntarily participate in land conversion projects. They are provided with financial incentives, such as subsidies and compensation for income loss. The program also employs local communities in afforestation and reforestation activities, creating new jobs and income sources.

Implementation and Results

The implementation of the "Grain for Green" program involved several key steps. First, a comprehensive land survey was conducted to identify priority areas for restoration. Then, appropriate tree and grass species were selected based on soil conditions and local climate. Local farmers and communities were trained in techniques for land restoration, including soil conservation and water management.

Over the past two decades, the "Grain for Green" program has achieved remarkable results. More than 35 million hectares of land have been restored, primarily through afforestation and the establishment of grasslands. This has helped stabilize soil, control erosion, and improve water resources in the region.

The restoration activities have also had positive ecological and socio-economic impacts. The reforestation has increased forest cover, promoting biodiversity conservation and providing habitat for wildlife. The regenerated grasslands have improved forage resources for livestock, leading to increased agricultural productivity. Furthermore, the program has created employment opportunities and improved the living standards of local communities.

Challenges and Lessons Learned

The restoration of the Loess Plateau has not been without challenges. One major challenge has been the survival rate of newly planted trees, particularly in arid conditions. The lack of water availability and the vulnerability of young trees to pests and diseases have affected the success of afforestation efforts.

Additionally, the participation and cooperation of local communities are crucial for the long-term success of ecological restoration initiatives. The "Grain for Green"

SUCCESS STORIES IN ECOSYSTEM RESTORATION   135

program has been successful in engaging farmers and incentivizing their involvement through financial support. However, continuous monitoring and support are needed to ensure the sustained commitment of communities.

The restoration of the Loess Plateau serves as a valuable lesson for other regions facing similar ecological challenges. It demonstrates the importance of implementing comprehensive and integrated approaches to address soil erosion and ecological degradation. The involvement of local communities and the provision of long-term support are vital for the success of large-scale restoration projects.

Conclusion

The restoration of the Loess Plateau in China through the "Grain for Green" program showcases the power of ecological restoration to address environmental degradation and improve the livelihoods of local communities. The program has not only stabilized soil and controlled erosion but also promoted biodiversity conservation and created sustainable income sources.

The success of the Loess Plateau restoration provides inspiration and valuable lessons for other regions facing similar challenges. It demonstrates the importance of combining scientific knowledge, community participation, and government support in achieving long-term ecological restoration goals. By implementing sustainable land management practices, we can restore degraded ecosystems and create a greener and more sustainable future.

Oostvaardersplassen Nature Reserve in the Netherlands

The Oostvaardersplassen Nature Reserve in the Netherlands is a unique and remarkable example of successful ecosystem restoration. The reserve, located in the province of Flevoland, was created in the 1960s on reclaimed land that was once part of the Zuiderzee, a large saltwater inlet. Today, it serves as a haven for a wide range of plant and animal species, demonstrating the power of ecological restoration in bringing back biodiversity and creating a thriving ecosystem.

Background

The history of the Oostvaardersplassen Nature Reserve is closely tied to the ambitious land reclamation project known as the Zuiderzee Works. In the early 20th century, the Dutch government decided to reclaim large areas of the Zuiderzee to address flooding issues and create new farmland. The construction of

136 CHAPTER 2: RESTORING ECOSYSTEMS

dikes and the draining of the water transformed the landscape and created an opportunity for ecological restoration.

Principles of Ecosystem Restoration

Ecosystem restoration aims to repair, recreate, or create functional ecosystems that have been degraded or destroyed. The restoration process typically involves ecological research, planning, and active management to enhance natural processes and bring back a diverse array of species. The principles of ecosystem restoration can be summarized as follows:

1. Identify restoration goals: Determine the desired ecological outcomes and define specific goals for the restoration project. In the case of Oostvaardersplassen, the goal was to create a dynamic and self-regulating wetland ecosystem.

2. Understand the ecological context: Conduct a thorough assessment of the site's historical and current ecological conditions, including the identification of key species and habitats. This information provides a baseline for measuring the success of the restoration project.

3. Utilize natural processes: Allow natural ecological processes to guide the restoration process as much as possible. This involves reintroducing key species, promoting natural succession, and ensuring the availability of essential resources such as water and nutrients.

4. Consider connectivity: Enhance connectivity within and beyond the restoration site to promote genetic exchange and species movement. This can involve creating corridors or stepping stones to connect fragmented habitats.

5. Monitor and adapt: Regularly monitor the progress of the restoration project and make adjustments as needed. This adaptive management approach ensures that the project remains on track and allows for continuous learning and improvement.

Restoration Efforts

The restoration efforts at Oostvaardersplassen began in the 1980s and involved the transformation of the former agricultural land into a diverse and dynamic wetland ecosystem. Here are some key steps taken during the restoration process:

1. Habitat Creation: Multiple types of habitats were created within the reserve, including open water, reed beds, grasslands, and deciduous forests. These habitats were carefully designed to support a wide range of plant and animal species.

2. Wildlife Management: To kickstart the ecosystem, a variety of animal species were reintroduced to Oostvaardersplassen, including Heck cattle, Konik horses, and

SUCCESS STORIES IN ECOSYSTEM RESTORATION

red deer. These herbivorous animals play a crucial role in shaping the landscape through grazing and trampling.

3. Natural Succession: The restoration process allowed for natural succession to take place, where pioneer species colonized the bare areas and eventually gave way to more diverse plant communities. This process helped create a mosaic of habitats that support different species.

4. Monitoring and Research: Extensive monitoring and research efforts have been carried out at Oostvaardersplassen to assess the ecological changes and evaluate the success of the restoration project. This data has been instrumental in refining management strategies and guiding future restoration initiatives.

Successes and Benefits

The Oostvaardersplassen Nature Reserve has achieved remarkable successes in ecosystem restoration and has become home to a rich and diverse array of plant and animal species. Some notable successes and benefits include:

1. Biodiversity: The reserve now supports over 600 plant species, more than 450 bird species, and numerous mammals, amphibians, and insects. It has become an important refuge for both resident and migratory species.

2. Ecological Dynamics: Oostvaardersplassen exhibits a remarkable balance between grazing and vegetation growth. The large herbivores play a vital role in shaping the landscape, while the changing vegetation provides food and shelter for a wide range of organisms.

3. Scientific Value: The reserve serves as an outdoor laboratory for studying ecological processes and the interactions between species. Researchers from around the world visit Oostvaardersplassen to conduct studies on topics such as trophic interactions, ecosystem dynamics, and landscape ecology.

4. Education and Recreation: Oostvaardersplassen has also become an important educational and recreational resource for local communities and visitors. Numerous trails, bird hides, and visitor centers provide opportunities for people to learn about and appreciate the natural wonders of the reserve.

Challenges and Future Directions

Despite its successes, the Oostvaardersplassen Nature Reserve also faces challenges and ongoing management considerations. Some key challenges include:

1. Population Management: Maintaining a balance between herbivore populations and available resources is an ongoing challenge. In some years, the

population density becomes too high, leading to concerns about animal welfare and the availability of food.

2. Public Perception: The management practices at Oostvaardersplassen, such as allowing natural processes to take their course, have generated controversy and debate. Balancing the interests and expectations of various stakeholders is essential for the long-term sustainability of the reserve.

3. Climate Change: The impacts of climate change, including changes in precipitation patterns and increasing temperatures, pose future challenges for the restoration efforts at Oostvaardersplassen. Adaptation strategies may be needed to ensure the resilience of the ecosystem.

In conclusion, the Oostvaardersplassen Nature Reserve in the Netherlands stands as a testament to the power of ecosystem restoration. Through careful planning, active management, and the integration of scientific knowledge, this once barren landscape has been transformed into a thriving wetland ecosystem, teeming with biodiversity. The successes and lessons learned from Oostvaardersplassen provide valuable insights into the principles and practices of ecosystem restoration, offering hope for the conservation and restoration of ecosystems around the world.

Rwandan Ecotourism and Conservation Initiatives

Rwandan ecotourism and conservation initiatives have gained significant attention in recent years, showcasing the country's commitment to preserving its natural resources and promoting sustainable development. This section explores the unique conservation efforts and ecotourism practices in Rwanda, highlighting the success stories and the challenges faced in this pursuit.

Background

Rwanda, located in East Africa, is renowned for its diverse landscapes, including lush rainforests, majestic mountains, and expansive savannahs. The country is home to a wealth of biodiversity, including iconic species like the mountain gorillas, golden monkeys, and numerous bird species. However, decades of civil unrest and habitat destruction have posed significant threats to these invaluable ecosystems.

In response to these challenges, Rwanda has made substantial efforts to protect its environment and promote sustainable tourism. The government, in collaboration with local communities, international organizations, and conservationists, has implemented various initiatives to conserve wildlife, restore habitats, and develop responsible tourism practices. These efforts have not only

SUCCESS STORIES IN ECOSYSTEM RESTORATION 139

contributed to biodiversity conservation but have also provided income-generating opportunities for local communities.

Volcanoes National Park and Gorilla Conservation

One of Rwanda's most notable success stories in conservation is the Volcanoes National Park, located in the Virunga Massif. This park is home to the endangered mountain gorillas, which are found in only three countries worldwide. Rwanda's conservation efforts have led to a significant increase in the population of mountain gorillas in recent years.

The government has implemented strict regulations to protect the gorillas and their habitat within the national park. These regulations include limiting the number of visitors, enforcing guidelines for gorilla treks, and establishing revenue-sharing programs with local communities. These measures have helped minimize the negative impacts of tourism on the gorillas while providing economic benefits to the communities living near the park.

Tourism revenue generated from gorilla trekking permits is used to fund conservation activities, including anti-poaching patrols, habitat restoration, and community development projects. This integrated approach has not only ensured the long-term survival of gorillas but has also contributed to poverty reduction and community empowerment.

Akagera National Park and Ecosystem Restoration

Akagera National Park, located in eastern Rwanda, is a vital conservation area supporting a diverse range of ecosystems, including savannahs, wetlands, and lakes. However, like many other protected areas, Akagera faced significant challenges such as poaching, habitat loss, and encroachment by grazing animals.

To address these issues, the government of Rwanda, in partnership with African Parks, implemented a comprehensive management and restoration program for Akagera National Park. This initiative focused on law enforcement, habitat rehabilitation, reintroduction of endangered species, and community engagement.

Anti-poaching measures, including increased patrols and support for local law enforcement, have significantly reduced poaching incidents within the park. Habitat restoration efforts, involving tree planting, invasive species control, and wetland rehabilitation, have helped to improve the overall health and resilience of the ecosystem.

140 CHAPTER 2: RESTORING ECOSYSTEMS

A notable success story is the reintroduction of lions and black rhinos to Akagera National Park. After a 15-year absence, lions were reintroduced in 2015, marking a significant milestone in the park's restoration. Subsequently, in 2017, black rhinos were also reintroduced, further enhancing the park's biodiversity and ecological processes.

Conservation-Based Community Development

Rwanda's conservation efforts go beyond protecting wildlife and ecosystems; they also aim to foster sustainable development and empower local communities. Several initiatives have been implemented to promote conservation-based community development projects, enabling local residents to benefit from ecotourism activities.

For instance, the revenue-sharing program associated with gorilla tourism in Volcanoes National Park has provided communities with incentives to protect natural resources and participate in sustainable livelihood opportunities. Through this program, local communities receive a portion of the revenue generated from tourism activities, which is used for community development projects such as education, healthcare, and clean energy initiatives.

Additionally, community-based ecotourism initiatives have been established, allowing visitors to engage with local communities and experience Rwanda's cultural heritage while contributing to conservation efforts. These initiatives include homestay programs, craft cooperatives, and cultural tourism experiences, providing economic benefits to the communities and promoting cultural preservation.

Challenges and Future Directions

Despite the successes in Rwandan ecotourism and conservation initiatives, several challenges remain. Human-wildlife conflicts, particularly between farmers and wildlife, pose a significant threat to both livestock and wildlife populations. Encroachment on protected areas for agriculture and settlement also continues to undermine conservation efforts.

To address these challenges, it is crucial to develop sustainable solutions that integrate the needs of local communities with conservation goals. This can be achieved through community-led initiatives, increased collaboration between stakeholders, and the implementation of innovative conservation strategies.

Furthermore, it is vital to continue investing in capacity building and education programs to enhance environmental awareness and conservation practices among local communities and tourists. By fostering a greater understanding of the value of Rwanda's natural heritage, individuals can become stewards of the environment and actively participate in conservation efforts.

Rwanda's commitment to ecotourism and conservation serves as an inspiring example of how a developing nation can achieve environmental sustainability and biodiversity conservation. Through a combination of strong governance, community involvement, and innovative approaches, Rwanda is leading the way in promoting a greener future.

Conclusion

Rwandan ecotourism and conservation initiatives exemplify the power of integrated approaches to environmental sustainability. By combining conservation efforts with responsible tourism practices, Rwanda has not only protected its natural resources but also provided economic opportunities for local communities and promoted cultural preservation.

The success stories in Volcanoes National Park and Akagera National Park highlight the importance of strong governance, community engagement, and innovative restoration techniques in achieving sustainable outcomes. These initiatives have not only brought back endangered species but have also improved overall ecosystem health and resilience.

However, challenges such as human-wildlife conflicts and habitat encroachment persist, requiring ongoing efforts and collaboration to find sustainable solutions. By investing in education, capacity building, and community-led initiatives, Rwanda can continue to be a global leader in ecotourism and conservation.

Rwandan ecotourism and conservation initiatives provide valuable lessons for other countries and communities seeking to balance economic development with environmental stewardship. With continued commitment and innovation, Rwanda's journey towards a greener future can serve as a significant inspiration for the world.

Challenges and Future Directions in Ecosystem Restoration

Financial and Political Barriers

In the pursuit of ecosystem restoration, various challenges can arise from both financial and political barriers. These obstacles often hinder or delay restoration

142   CHAPTER 2: RESTORING ECOSYSTEMS

efforts, making it essential for environmental scientists to find ways to overcome them. In this section, we will explore some of the common financial and political barriers that can impede ecosystem restoration projects, as well as strategies to address them effectively.

Financial Barriers

1. Limited Funding: One of the primary financial barriers in ecosystem restoration is the lack of adequate funding. Restoration projects often require significant investments in resources, including equipment, materials, and skilled labor. However, acquiring sufficient funding can be challenging, particularly for large-scale restoration initiatives.

Example: The restoration of the Everglades in Florida faced financial barriers due to the extensive scale of the project. Securing funds for land acquisition, infrastructure development, and ecosystem rehabilitation required collaboration between multiple stakeholders, including government agencies, non-profit organizations, and private investors.

2. Cost-effectiveness Concerns: Another financial barrier is the perception that ecosystem restoration projects are expensive and lack immediate economic benefits. Some may argue that allocating funds to other societal needs, such as healthcare or education, would provide a more tangible return on investment.

Example: The restoration of the Aral Sea faced cost-effectiveness concerns due to the massive scale of the project. Critics questioned the justification of allocating significant financial resources to revive an ecosystem that had already undergone severe degradation.

3. Uncertain Return on Investment: Restoration projects often involve long-term commitments, and the benefits may not be immediate or easily quantifiable. This uncertainty can deter potential investors and make it challenging to secure funding for restoration initiatives.

Example: Wetland restoration projects face challenges in demonstrating a clear return on investment as the benefits, such as improved water quality and increased biodiversity, may take years to materialize.

To address these financial barriers, several strategies can be employed:

- Seeking Multiple Funding Sources: Ecosystem restoration projects can explore various funding options, including government grants, philanthropic contributions, corporate sponsorships, and public-private partnerships. Diversifying funding sources can enhance the financial stability of restoration initiatives.

CHALLENGES AND FUTURE DIRECTIONS IN ECOSYSTEM RESTORATION

- Cost-sharing and Collaboration: Collaborating with different stakeholders, such as NGOs, academic institutions, and local communities, can help distribute the financial burden. Sharing costs through partnerships can make restoration projects more feasible and sustainable.

- Demonstrating Economic Benefits: Highlighting the potential economic returns associated with ecosystem restoration is crucial. Emphasizing the value of restored ecosystems in supporting tourism, recreation, and local economies can garner support and funding from stakeholders concerned with financial gain.

Political Barriers

1. Lack of Political Will: Political will is essential in driving ecosystem restoration initiatives. However, it can be challenging to generate and sustain political support, as competing priorities and short-term political cycles may overshadow the long-term benefits of restoration projects.

Example: Reforestation efforts in some regions face political barriers as authorities prioritize land-use policies that promote agriculture or urban development over restoration activities.

2. Complex Regulatory Frameworks: Regulations and policies related to environmental conservation and restoration can be complex and vary across different jurisdictions. These complexities can create bureaucratic hurdles and slow down the approval processes for restoration projects.

Example: Stream restoration projects can encounter political barriers when there is confusion about the jurisdictional authority responsible for granting permits, leading to project delays and increased costs.

3. Conflicting Stakeholder Interests: Ecosystem restoration often involves multiple stakeholders with diverse interests and priorities. Conflicts can arise when these interests clash, leading to opposition or resistance towards restoration projects.

Example: Coastal restoration projects may face political barriers when the interests of commercial fisheries, tourism industries, and coastal residents are not aligned, and competing stakeholders block or impede restoration efforts.

Strategies to address political barriers:

- Building Awareness and Public Support: Generating public awareness about the importance of ecosystem restoration and its benefits can help create the necessary political will. Engaging communities through education and outreach programs can rally support and mobilize public pressure on policymakers.

- Stakeholder Engagement and Collaboration: Facilitating open dialogue and collaboration among stakeholders can help identify shared interests and common goals. Finding win-win solutions that address the concerns of diverse stakeholders can reduce opposition and create paths for political cooperation.

- Streamlining Regulatory Processes: Working towards simplifying and streamlining the regulatory frameworks can help reduce bureaucratic delays and make it easier for restoration projects to navigate the approval processes. Enhancing coordination between different agencies and clarifying jurisdictional responsibilities can mitigate political barriers.

By addressing financial and political barriers effectively, ecosystem restoration projects can overcome impediments and pave the way for successful restoration efforts. The collaboration between environmental scientists, policymakers, communities, and stakeholders is crucial in surmounting these challenges and achieving long-lasting ecological benefits.

Key Takeaways

- Financial barriers in ecosystem restoration include limited funding, cost-effectiveness concerns, and uncertain return on investment.

- Overcoming financial barriers can be achieved through seeking multiple funding sources, cost-sharing, and demonstrating economic benefits.

- Political barriers in ecosystem restoration stem from the lack of political will, complex regulatory frameworks, and conflicting stakeholder interests.

- Strategies to address political barriers involve building awareness and support, stakeholder engagement and collaboration, and streamlining regulatory processes.

- Collaboration and cooperation among environmental scientists, policymakers, communities, and stakeholders are crucial in overcoming financial and political barriers for successful ecosystem restoration.

CHALLENGES AND FUTURE DIRECTIONS IN ECOSYSTEM RESTORATION

Monitoring and Evaluation of Restoration Projects

Monitoring and evaluation are essential components of any restoration project. They provide feedback on the progress and effectiveness of restoration efforts, helping to identify areas that require adjustment or improvement. In this section, we will explore the importance of monitoring and evaluation in restoration projects and discuss the key methodologies and techniques used in these processes.

Importance of Monitoring and Evaluation

Monitoring and evaluation serve several important purposes in restoration projects. Firstly, they allow project managers and stakeholders to assess whether restoration goals and objectives are being met. By collecting and analyzing relevant data, project teams can determine whether the implemented restoration strategies are achieving the desired outcomes.

Secondly, monitoring and evaluation help to identify any unexpected ecological or environmental changes that may arise during the restoration process. These changes could be positive or negative and may have significant implications for the success of the project. By closely monitoring the project site, restoration practitioners can respond promptly to any emerging issues and adapt their strategies accordingly.

Furthermore, monitoring and evaluation enable project teams to learn from their experiences and improve future restoration efforts. By documenting and analyzing the data collected during the monitoring process, valuable insights can be gained, leading to more informed decision-making and more effective restoration practices.

Methodologies and Techniques

There are several methodologies and techniques available for monitoring and evaluating restoration projects. The selection of specific methods depends on the goals of the project, the resources available, and the ecological characteristics of the site. Here are some commonly used approaches:

- **Baseline Data Collection:** Before initiating the restoration project, it is crucial to gather baseline data about the site's ecological conditions. This provides a reference point for later comparisons and helps in setting realistic restoration goals. Baseline data can include information about species composition, vegetation cover, soil attributes, water quality, and other relevant parameters.

146   *CHAPTER 2: RESTORING ECOSYSTEMS*

- **Field Surveys:** Field surveys involve direct observations and measurements in the project area. These surveys can include techniques such as vegetation sampling, biodiversity assessment, soil sampling, and water quality testing. Field surveys provide valuable information about the progress of the restoration and any changes happening in the ecosystem.

- **Remote Sensing and GIS:** Remote sensing techniques, such as satellite imagery and aerial photography, can be used to monitor large-scale restoration projects. These methods allow for the assessment of vegetation cover changes, habitat fragmentation, and land use changes over time. Geographic Information Systems (GIS) provide a powerful tool for data integration, visualization, and spatial analysis, enabling a comprehensive evaluation of restoration efforts.

- **Data Analysis:** The collected data needs to be analyzed to derive meaningful insights and assess the progress of the restoration project. Statistical analysis techniques can be used to identify patterns, trends, and relationships in the data. This analysis can help in identifying factors that contribute to the success or failure of restoration efforts.

- **Stakeholder Engagement:** Involving local communities and stakeholders in monitoring and evaluation processes can provide valuable perspectives and insights. Their knowledge and observations can complement scientific data and contribute to a more comprehensive understanding of the project's impact.

Case Studies

Let's explore two case studies that demonstrate effective monitoring and evaluation practices in restoration projects.

Case Study 1: Coral Reef Restoration in the Great Barrier Reef

The Great Barrier Reef, located off the northeast coast of Australia, is one of the most biodiverse and ecologically significant coral reef ecosystems in the world. Over the years, it has faced multiple threats, including climate change, pollution, and coral bleaching. To address these challenges, restoration projects have been initiated to enhance coral reef resilience and promote ecosystem recovery.

Monitoring and evaluation efforts in the Great Barrier Reef restoration projects involve a combination of techniques. Satellite imagery is used to assess changes in coral cover and identify areas of coral bleaching. Divers conduct underwater surveys to monitor coral growth and track the establishment of new

CHALLENGES AND FUTURE DIRECTIONS IN ECOSYSTEM RESTORATION

coral colonies. Water quality measurements are taken to evaluate the health of the ecosystem and the effectiveness of conservation actions. Additionally, local communities and indigenous groups actively participate in monitoring activities, providing valuable insights into the social and cultural dimensions of restoration.

Case Study 2: Forest Restoration in the Amazon Rainforest

The Amazon rainforest has been subjected to significant deforestation and habitat degradation due to agricultural expansion and logging activities. Forest restoration projects have been implemented to restore the ecological integrity and biodiversity of the region.

Monitoring and evaluation of the Amazon forest restoration projects involve a combination of field surveys, remote sensing, and participatory approaches. Field surveys are conducted to assess tree growth, biomass accumulation, and the recovery of plant and animal species. Remote sensing techniques, such as LiDAR and aerial photography, are used to estimate forest canopy cover and track changes in vegetation structure. Indigenous communities play a crucial role in monitoring the restoration efforts, providing traditional knowledge and insights into ecosystem dynamics.

Challenges and Future Directions

Monitoring and evaluation of restoration projects face several challenges that need to be addressed for more effective outcomes:

* **Long-term Monitoring:** Restoration projects often require long-term monitoring to evaluate the success and sustainability of the implemented strategies. However, long-term monitoring can be costly and resource-intensive, posing challenges for project continuity.

* **Data Standardization and Sharing:** Standardizing data collection protocols and sharing monitoring data across projects and organizations can enhance the comparability and reliability of restoration assessments. Efforts are underway to develop common frameworks and data repositories for better data management and accessibility.

* **Incorporating Socioeconomic Factors:** Restoration projects should not only focus on ecological aspects but also consider socioeconomic factors. Assessing the social and economic impacts of restoration initiatives can provide a more comprehensive understanding of their overall effectiveness and sustainability.

* **Adaptive Management:** Restoration projects need to be adaptive, allowing for flexibility and responsiveness to changing environmental conditions and

148   *CHAPTER 2: RESTORING ECOSYSTEMS*

emerging challenges. Incorporating adaptive management strategies can enhance the resilience and success of restoration initiatives.

In conclusion, monitoring and evaluation are critical for the success of restoration projects. By collecting and analyzing data, project teams can assess progress, identify challenges, and make informed decisions to improve restoration strategies. As restoration efforts continue to grow, it is essential to develop standardized methodologies, engage local communities, and promote adaptive management practices to ensure the long-term success of ecosystem restoration.

Exercises

1. Select a restoration project in your local area and research the monitoring and evaluation techniques used. Discuss their effectiveness and suggest any improvements or alternative approaches.

2. Conduct a field survey in a nearby natural area and collect data on vegetation cover, species diversity, or water quality. Analyze the collected data using appropriate statistical techniques and interpret the results in the context of restoration efforts.

3. Explore the role of indigenous knowledge in monitoring and evaluation of restoration projects. Discuss its importance and the potential challenges and benefits of incorporating traditional ecological knowledge into scientific monitoring approaches.

Integrating Indigenous Knowledge in Restoration Practices

Restoring ecosystems is a complex and challenging task that requires a deep understanding of the ecological processes at play. In recent years, there has been growing recognition of the importance of integrating indigenous knowledge into restoration practices. Indigenous peoples have lived in close harmony with nature for centuries and possess a wealth of traditional knowledge about the land, biodiversity, and ecosystem dynamics. Incorporating this knowledge into restoration efforts can enhance the effectiveness and sustainability of restoration projects.

Understanding Indigenous Knowledge

Indigenous knowledge refers to the knowledge, practices, and beliefs developed by indigenous communities through their interactions with the environment over generations. It is often passed down orally, from one generation to another, and encompasses a deep understanding of local ecosystems, natural resources, and traditional management practices.

Indigenous knowledge is holistic in nature, recognizing the interconnectedness of all living beings and the environment. It acknowledges the importance of maintaining balance and harmony between humans and nature. This knowledge is rooted in cultural practices, spiritual beliefs, and a profound respect for the land and its resources.

Benefits of Integrating Indigenous Knowledge

Integrating indigenous knowledge into restoration practices offers several key benefits. First, it enhances the accuracy and comprehensiveness of ecological assessments. Indigenous peoples have accumulated detailed knowledge about local flora and fauna, including their medicinal properties, ecological niches, and interactions. This knowledge can supplement scientific data, providing a more comprehensive understanding of ecosystem dynamics.

Second, indigenous knowledge contributes to the selection of appropriate restoration techniques. Indigenous communities have developed sustainable land management practices over generations, such as rotational farming, agroforestry, or controlled burning. These traditional practices can inform the selection of restoration methods that are culturally appropriate, environmentally sustainable, and respectful of traditional values.

Third, integrating indigenous knowledge fosters community engagement and empowerment. By involving indigenous communities in restoration projects, their cultural heritage and traditional practices are valued and respected. This creates a sense of ownership and stewardship among community members, leading to greater long-term commitment and sustainability of restoration efforts.

Challenges and Considerations

While integrating indigenous knowledge in restoration practices offers numerous benefits, it also poses certain challenges and considerations. One crucial aspect is the need for effective collaboration and respectful engagement with indigenous communities. It is essential to establish partnerships based on trust, mutual respect, and shared decision-making. Indigenous communities should be involved throughout the restoration process, from planning to monitoring and evaluation, ensuring their voices are heard and their knowledge is valued.

150 CHAPTER 2: RESTORING ECOSYSTEMS

Another challenge is the need for cultural sensitivity and appropriate protocols. Indigenous knowledge is often deeply intertwined with spiritual beliefs and cultural practices. Researchers and restoration practitioners must approach this knowledge with humility, respecting the intellectual property rights, protocols, and traditions of indigenous communities. Permission should be sought, and benefit-sharing agreements should be established to ensure the equitable distribution of benefits derived from the use of indigenous knowledge.

Furthermore, integrating indigenous knowledge requires a recognition of its dynamism and adaptation to changing environmental conditions. Indigenous communities have historically adapted their practices to environmental changes, and this flexibility should be acknowledged and incorporated into restoration strategies. This includes recognizing the impacts of climate change and other external factors and adapting restoration approaches accordingly.

Example: Traditional Fire Management in Australia

An example of how integrating indigenous knowledge can enhance restoration practices is the traditional fire management practices of Indigenous Australians. Indigenous communities in Australia have used controlled burning techniques for thousands of years to manage the landscape and promote biodiversity.

These traditional practices involve deliberate, low-intensity burning during cooler months to reduce fuel loads, control the spread of wildfires, and promote the regeneration of vegetation. The intricate knowledge held by Indigenous Australians about fire behavior, timing, and intensity allows for precise management of ecosystems, promoting the growth of fire-dependent plant species and maintaining a balance between grasslands, woodlands, and forests.

Recently, there has been a growing recognition of the importance of incorporating Indigenous fire practices into contemporary land management strategies in Australia. Research has shown that incorporating traditional fire management techniques can lead to reduced wildfire risks, increased biodiversity, and enhanced ecosystem resilience.

Conclusion

Integrating indigenous knowledge in restoration practices is a crucial step towards more holistic and sustainable approaches to ecosystem restoration. The traditional knowledge held by indigenous communities can significantly contribute to the success and effectiveness of restoration projects by providing valuable insights into local ecosystems, traditional land management practices, and cultural values.

CHALLENGES AND FUTURE DIRECTIONS IN ECOSYSTEM RESTORATION

However, it is imperative to approach the integration of indigenous knowledge with respect, humility, and cultural sensitivity. Collaboration and mutual learning between scientists, restoration practitioners, and indigenous communities are essential to ensure the success of such initiatives. By embracing and integrating indigenous knowledge, we can strive for a more inclusive, culturally diverse, and sustainable approach to ecosystem restoration.

Climate Change Impacts on Restoration Efforts

Climate change is one of the most pressing challenges facing our planet and has significant implications for ecosystem restoration efforts. As global temperatures rise and weather patterns become more erratic, ecosystems around the world are being subjected to new and unprecedented pressures. These changes have far-reaching consequences for the success and effectiveness of restoration projects.

Understanding the Impacts

One of the key impacts of climate change on ecosystem restoration is the alteration of habitats and ecosystems. Rising temperatures can lead to shifts in the geographic distribution of species, as they seek more suitable conditions. In turn, this can affect the composition and structure of restored ecosystems, potentially making them less resilient and less able to support biodiversity.

Changes in precipitation patterns also have important implications for restoration efforts, especially in water-stressed regions. Increased droughts and intense rainfall events can alter soil moisture levels, affecting the survival and growth of restored plants. In some cases, restoration projects may need to reconsider the choice of plant species and prioritize those that are more tolerant to these changing conditions.

Climate change can also exacerbate existing pressures on ecosystems, making restoration more challenging. For example, invasive species that thrive in warmer climates may become more difficult to manage, potentially hindering the recovery of native species. Additionally, extreme weather events, such as hurricanes and wildfires, can cause severe damage to restored ecosystems, undoing years of restoration efforts.

Adapting Restoration Strategies

In order to overcome the challenges posed by climate change, it is necessary to incorporate adaptive strategies into ecosystem restoration projects. This means taking into account the potential impacts of climate change and actively adjusting restoration approaches accordingly.

One important consideration is the selection of species for restoration. In the face of changing environmental conditions, it is essential to choose species that are more resilient to future climate scenarios. This may involve identifying and utilizing local plant populations that have already demonstrated resilience to changing conditions, or selecting species that have a broader climatic tolerance.

Furthermore, restoration projects need to be designed with flexibility in mind. Rather than focusing on restoring ecosystems to a specific historical state, it may be more effective to aim for functional ecosystems that are capable of adapting to changing conditions. This can involve creating diverse habitats that provide a variety of resources for different species, allowing them to persist and thrive under different climate scenarios.

Partnerships and Collaboration

Addressing the impacts of climate change on ecosystem restoration requires collaboration across disciplines and sectors. Scientists, restoration practitioners, policymakers, and local communities need to work together to develop and implement effective strategies.

Partnerships with indigenous communities can be particularly valuable, as they often possess unique knowledge and understanding of local ecosystems and have a vested interest in their preservation. Their traditional ecological knowledge can inform restoration strategies and help identify species and techniques that are better suited to changing climatic conditions.

It is also important to engage with policymakers and advocate for policies that support climate change mitigation and adaptation. By actively participating in policy discussions, restoration practitioners can help shape decisions that prioritize the long-term resilience of ecosystems.

Real-World Example

The Great Barrier Reef in Australia provides a compelling example of the impact of climate change on restoration efforts. Rising ocean temperatures and ocean acidification, both consequences of climate change, have led to widespread coral bleaching and the decline of coral reefs. These changes have made the restoration of damaged reefs more challenging, as bleached corals are less likely to recover.

In response, restoration initiatives have shifted their focus towards building the resilience of remaining coral populations. This involves strategies such as selective breeding of heat-tolerant corals and the establishment of nurseries to support the growth of reef-building organisms.

However, it is important to note that restoration alone cannot solve the challenges posed by climate change. Climate change mitigation efforts, such as reducing greenhouse gas emissions, are essential to address the root causes of climate change and minimize its impacts on ecosystems.

Conclusion

Climate change poses significant challenges to ecosystem restoration efforts, but it also presents opportunities for innovation and collaboration. By understanding and adapting to the impacts of climate change, restoration projects can become more resilient and effective in the face of future challenges. Through partnerships and a holistic approach, we can work towards restoring and conserving ecosystems for generations to come.

Scaling up Restoration for Global Impact

Scaling up ecosystem restoration efforts is crucial to address the widespread degradation of natural habitats and the associated loss of biodiversity. It requires an integrated and coordinated approach involving governments, non-governmental organizations (NGOs), local communities, and international bodies. This section explores the challenges and potential solutions for achieving global impact in ecosystem restoration.

Current State of Ecosystem Restoration

Despite significant efforts in ecosystem restoration, the scale of degradation still outweighs the restoration activities taking place. Many ecosystems, such as forests, wetlands, and coral reefs, continue to decline in quality and extent due to factors like deforestation, pollution, and climate change. As a result, the need to scale up restoration efforts has become increasingly urgent.

Challenges in Scaling up Restoration

1. **Funding and Resources:** One of the major challenges in scaling up restoration is the limited availability of funding and resources. Restoration projects often require substantial financial investments, infrastructure, and skilled human resources. Without adequate funding, it is difficult to implement large-scale restoration initiatives that can have a significant impact.

154   CHAPTER 2: RESTORING ECOSYSTEMS

2. **Institutional Coordination:** Effective coordination among various institutions and stakeholders is essential for successful ecosystem restoration at a global scale. However, different organizations may have divergent objectives, approaches, and priorities, making collaboration and coordination a complex task.

3. **Scientific Knowledge and Innovation:** Scaling up restoration efforts requires access to up-to-date scientific knowledge and innovative approaches. It is essential to continually refine and adapt restoration techniques based on scientific research and best practices. However, the dissemination of this knowledge and the adoption of innovative methods can face various barriers.

4. **Socio-political Context:** The socio-political context of different regions can significantly impact the success of ecosystem restoration. Factors like governance, policy frameworks, land rights, and conflicts can either support or hinder restoration efforts. Addressing these socio-political challenges is crucial for achieving meaningful restoration on a global scale.

Strategies for Scaling up Restoration

1. **Mobilizing Financial Resources:** Increasing financial investments in restoration projects is crucial for scaling up efforts. Governments, development agencies, and private investors need to prioritize funding for restoration initiatives. Exploring innovative financing mechanisms, such as impact investment and payment for ecosystem services, can also expand the pool of available resources.

2. **Enhancing International Cooperation:** Collaboration and cooperation among countries, international organizations, and NGOs are crucial for scaling up restoration globally. International agreements, such as the United Nations Convention on Biological Diversity and the Ramsar Convention on Wetlands, provide a framework for cooperation and support in ecosystem restoration.

3. **Building Capacity and Knowledge-sharing:** Enhancing capacity-building programs and knowledge-sharing platforms can facilitate the adoption of best practices and innovative techniques. Training restoration practitioners, scientists, and local communities on restoration principles and methodologies can empower them to implement effective restoration projects.

4. **Integrating Restoration into Policy and Planning:** Incorporating restoration goals and targets into national policies, land-use planning, and development projects can mainstream restoration efforts. Integrating restoration into broader frameworks, such as sustainable development goals and climate change mitigation plans, can increase the visibility and impact of restoration initiatives.

CHALLENGES AND FUTURE DIRECTIONS IN ECOSYSTEM RESTORATION

Case Studies

1. **The Great Green Wall Initiative:** The Great Green Wall initiative, led by the African Union, aims to restore degraded landscapes across the Sahara and Sahel regions of Africa. The project involves the planting of millions of trees, sustainable land management practices, and the creation of jobs for local communities. The initiative exemplifies a large-scale, transnational ecosystem restoration effort.

2. **Bonn Challenge:** The Bonn Challenge is a global effort to restore 350 million hectares of degraded land by 2030. It brings together governments, NGOs, and private sector actors to commit to restoring degraded and deforested areas. The initiative emphasizes the importance of political will, finance, and capacity building for successful restoration at a global scale.

Future Directions

1. **Holistic Approaches:** Future restoration efforts should adopt holistic approaches that consider the interconnectedness of ecosystems. Rather than focusing on single species or habitats, restoration should aim to restore ecological processes and functions across landscapes.

2. **Resilience and Climate Change Adaptation:** The restoration of ecosystems should also address the challenges posed by climate change. Restoration projects should aim to enhance the resilience of ecosystems to changing climatic conditions and consider climate change adaptation strategies.

3. **Indigenous and Local Knowledge:** Integrating indigenous and local knowledge systems into restoration practices can enhance the effectiveness and sustainability of restoration efforts. Traditional ecological knowledge often holds valuable insights into ecosystem dynamics and restoration techniques.

4. **Public Engagement and Awareness:** Raising public awareness about the importance of ecosystem restoration and involving local communities in restoration projects are crucial for long-term success. Engaging with local communities and stakeholders can build support, enhance project ownership, and ensure the sustainability of restoration efforts.

In conclusion, scaling up ecosystem restoration for global impact requires addressing financial, institutional, scientific, and socio-political challenges. By mobilizing resources, enhancing cooperation, building capacity, and integrating restoration into policies, it is possible to achieve large-scale restoration. Case studies like the Great Green Wall initiative and the Bonn Challenge provide valuable insights into successful restoration efforts. Moving forward, holistic approaches, climate change adaptation, incorporation of indigenous knowledge, and public engagement will contribute to the effectiveness and sustainability of large-scale ecosystem restoration.

Chapter 3: Conservation Biology and Wildlife Management

Chapter 3: Conservation Biology and Wildlife Management

Chapter 3: Conservation Biology and Wildlife Management

In this chapter, we will explore the fascinating field of conservation biology and wildlife management. We will delve into the fundamental principles of biodiversity conservation, the various threats facing species, and the strategies and techniques used to mitigate these threats. Conservation biology is an interdisciplinary science that combines elements of ecology, genetics, and natural resource management to protect and restore the Earth's biological diversity.

Foundations of Conservation Biology

Conservation biology is based on the concept of biodiversity, which refers to the variety of life on Earth at all levels, from genes to ecosystems. Biodiversity is essential for the functioning of ecosystems and provides important ecosystem services such as pollination, soil fertility, and climate regulation.

The endangerment and extinction of species are major concerns in conservation biology. A species is considered endangered if it is at risk of extinction throughout all or a significant portion of its range. Factors contributing to species endangerment include habitat loss, overexploitation, pollution, introduced species, and climate change.

CHAPTER 3: CONSERVATION BIOLOGY AND WILDLIFE MANAGEMENT

Conservation genetics is a field that focuses on the genetic management of small populations to prevent inbreeding and the loss of genetic diversity. Genetic techniques such as DNA analysis and captive breeding programs are used to assess the genetic health of populations and develop effective strategies for species conservation.

Wildlife Management Techniques

Wildlife management is a crucial component of conservation biology as it aims to maintain populations of wild species in their natural habitats while also ensuring the sustainable use of natural resources. Several techniques are employed in wildlife management:

1. **Wildlife population assessment:** This involves estimating population size, monitoring population trends, and understanding factors affecting population dynamics. Techniques such as mark-recapture studies, camera trapping, and satellite tracking are used for population assessment.

2. **Habitat management and restoration:** This focuses on maintaining and enhancing suitable habitats for wildlife. Habitat management involves activities such as controlled burning, reforestation, and creation of artificial nesting sites. Habitat restoration aims to restore degraded habitats to their original condition.

3. **Species reintroduction and translocation:** Reintroduction programs involve releasing captive-bred or relocated individuals back into their natural habitats. Translocation refers to moving individuals from areas of high population density to areas with low population density or establishing new populations in suitable habitats.

Case Studies in Conservation Biology

Let's explore some notable case studies that highlight the application of conservation biology principles:

1. **Giant Panda Conservation in China:** The giant panda is an iconic symbol of conservation. China has implemented a range of measures, including protected areas, habitat restoration, and captive breeding programs, to conserve this endangered species.

CHAPTER 3: CONSERVATION BIOLOGY AND WILDLIFE MANAGEMENT

2. **African Elephant Conservation and Anti-Poaching Efforts:** African elephants face threats from poaching and habitat loss. Conservation organizations work towards protecting elephant populations through anti-poaching initiatives, community-based conservation programs, and international collaborations.

3. **California Condor Recovery Program:** The California condor, one of the world's most endangered birds, faced extinction in the 1980s due to factors like lead poisoning and habitat loss. Intensive captive breeding and release programs have successfully increased the population of this species.

4. **Sea Turtle Conservation and Protection:** Various species of sea turtles are under threat from habitat destruction, pollution, and fishing practices. Conservation efforts involve beach protection, regulating fishing practices, and public awareness campaigns to reduce threats to these ancient mariners.

5. **Orangutan Conservation in Borneo and Sumatra:** Orangutans, highly intelligent primates, are critically endangered due to habitat loss from deforestation for palm oil production. Conservation efforts include establishing protected areas, promoting sustainable palm oil production, and rehabilitating orphaned orangutans.

Success Stories in Conservation Biology

Conservation biology has achieved remarkable success in protecting and restoring biodiversity around the world. Here are some notable success stories:

1. **Galapagos Islands Conservation and Ecotourism:** The Galapagos Islands' unique biodiversity and delicate ecosystem have been safeguarded through strict conservation measures, including limits on tourism, invasive species control, and scientific research.

2. **Wolf Reintroduction in Yellowstone National Park:** The reintroduction of gray wolves in Yellowstone National Park in the United States has had significant ecological benefits, including a cascading effect on the ecosystem by controlling herbivore populations and restoring natural balance.

3. **Indigenous-led Conservation Initiatives:** Indigenous communities around the world play a vital role in protecting biodiversity through their traditional knowledge and sustainable resource management practices. Their conservation efforts demonstrate the importance of bridging traditional knowledge with modern conservation practices.

4. **Marine Protected Areas and Biodiversity Conservation:** The establishment of marine protected areas (MPAs) has proven effective in conserving marine biodiversity and restoring degraded habitats. MPAs restrict human activities, such as fishing and oil drilling, and allow marine ecosystems to recover.

5. **Conservation of Endangered Birds in New Zealand:** New Zealand has implemented intensive predator control programs to protect native bird species, such as the kiwi and kakapo, from invasive predators like rats and stoats. These efforts have led to population increases and the recovery of endangered bird populations.

Challenges and Future Directions in Conservation Biology

While conservation biology has made significant progress, numerous challenges persist:

1. **Human-Wildlife Conflicts:** As human populations expand and habitats shrink, conflicts between humans and wildlife arise. Finding ways to mitigate conflict while ensuring species conservation is a ongoing challenge.

2. **Illegal Wildlife Trade:** The illegal trade in wildlife products, such as ivory and rhino horn, poses a significant threat to many endangered species. Collaborative efforts by governments, law enforcement agencies, and conservation organizations are necessary to combat this global issue.

3. **Habitat Fragmentation and Loss:** Habitat destruction and fragmentation due to agriculture, urbanization, and infrastructure development continue to degrade ecosystems and threaten biodiversity. Balancing economic development with conservation is a complex challenge.

4. **Climate Change and Species Conservation:** Climate change poses a significant threat to biodiversity, causing shifts in species' ranges and disrupting ecological interactions. Conservation strategies need to incorporate climate change adaptation and mitigation measures.

5. **Engaging Local Communities in Conservation Efforts:** Involving local communities in conservation initiatives is crucial for ensuring sustainable and long-term success. Building partnerships, promoting awareness, and addressing local concerns are essential for effective conservation.

FOUNDATIONS OF CONSERVATION BIOLOGY

In conclusion, conservation biology and wildlife management play vital roles in protecting and restoring biodiversity. Through the application of scientific knowledge and collaborative efforts, we can address the challenges facing our planet's species and ecosystems and work towards a more sustainable future.

Foundations of Conservation Biology

The Concept of Biodiversity

Biodiversity is a fundamental concept in conservation biology and a key focus of environmental science. It refers to the variety of living organisms and ecosystems found on Earth. The term "biodiversity" is derived from the words "biological" and "diversity," emphasizing the multitude of species, genes, and ecosystems that exist in nature. This section will explore the importance of biodiversity, the different levels of biodiversity, and the threats that biodiversity faces.

The Importance of Biodiversity

Biodiversity is crucial for maintaining the stability and resilience of ecosystems. It provides a wide range of ecosystem services, which are the benefits that humans receive from the natural environment. These services include food production, air and water purification, carbon sequestration, and climate regulation.

One of the primary reasons biodiversity is important is its role in supporting food security. The variety of plant species found in ecosystems ensures a diverse and nutritious diet for humans. In addition, insects and other animals play a crucial role in pollinating crops, contributing to their productivity and sustainability. Loss of biodiversity can lead to decreased food production and increased vulnerability to pests and diseases.

Biodiversity also has cultural significance as it encompasses the richness of traditional knowledge, practices, and beliefs tied to the natural world. Indigenous communities, for example, have a deep understanding of their local environments and rely on biodiversity for their livelihoods and cultural identity. Preserving biodiversity is therefore essential for maintaining cultural diversity and promoting social well-being.

Furthermore, biodiversity is a source of inspiration for scientific and medical advancements. Many important drugs and therapies have been derived from natural compounds found in plants, animals, and microorganisms. The loss of biodiversity could mean a loss of potential future discoveries that could revolutionize medicine and technology.

Levels of Biodiversity

Biodiversity can be examined at different levels: genetic diversity, species diversity, and ecosystem diversity.

Genetic diversity refers to the variation in genes within and among populations of the same species. It is responsible for the adaptability of species to changing environments. Genetic diversity increases the likelihood that some individuals within a population will possess traits that enable them to survive and reproduce in the face of environmental challenges, such as disease outbreaks or climate change. Higher genetic diversity also enhances the potential for future evolution and adaptation.

Species diversity refers to the variety of species present in a particular area or on the entire planet. It includes the number of species, their relative abundance, and their distribution patterns. Species diversity is a measure of the ecological complexity and stability of an ecosystem. High species diversity indicates a healthy and resilient ecosystem, while low species diversity can be an indication of ecological disturbance or degradation.

Ecosystem diversity encompasses the range of different habitats, communities, and ecological processes present in a given region. It reflects the variety of ecosystems, such as forests, wetlands, grasslands, and coral reefs. Ecosystem diversity is critical for maintaining key ecological functions, such as nutrient cycling, water filtration, and climate regulation. It also provides a wide array of habitats and resources for different species, supporting their survival and promoting interactions between organisms.

Threats to Biodiversity

Despite its importance, biodiversity faces numerous threats, many of which are anthropogenic (human-caused). These threats include habitat loss and fragmentation, overexploitation of species, pollution, climate change, and invasive species.

Habitat loss is the primary driver of biodiversity loss worldwide. The conversion of natural habitats for agriculture, urban development, logging, and infrastructure projects reduces the available space and resources for species. This leads to the displacement and extinction of many organisms.

Overexploitation occurs when species are harvested or collected from the wild at rates that exceed their ability to reproduce. Examples include overfishing, illegal wildlife trade, and the hunting of endangered species for their body parts or

FOUNDATIONS OF CONSERVATION BIOLOGY

medicinal value. Overexploitation can lead to population declines and the loss of important ecological functions.

Pollution from industrial and agricultural activities poses a significant threat to biodiversity. Chemical contaminants can accumulate in ecosystems, affecting the health and reproduction of organisms. Pollution can also degrade and destroy habitats, as well as alter the composition of communities.

Climate change is a global phenomenon that alters temperature patterns, precipitation, and sea levels. These changes can disrupt ecosystems and affect the distribution and abundance of species. Climate change also exacerbates other threats, such as habitat loss and the spread of invasive species.

Invasive species are non-native organisms that are introduced into new habitats, often due to human activities. Invasive species can outcompete and displace native species, disrupt ecosystem dynamics, and reduce biodiversity. They can also introduce new diseases and parasites that native species are not adapted to.

Conservation of Biodiversity

Conserving biodiversity is essential for maintaining the health and functioning of ecosystems, ensuring sustainable development, and preserving the countless benefits it provides to humans and the planet. Effective conservation strategies require a combination of scientific research, policy development, and community engagement.

Protected areas, such as national parks and nature reserves, play a crucial role in the conservation of biodiversity. These areas provide habitats for a wide range of species and act as safe havens for vulnerable and endangered populations. Protecting key habitats and preventing further habitat loss is a priority in biodiversity conservation efforts.

In addition to protected areas, conservation efforts often focus on habitat restoration, endangered species recovery, conservation planning, and sustainable land and resource management. It is important to involve local communities in conservation initiatives, as their knowledge and cooperation are essential for the long-term success of conservation projects.

Education and raising awareness about biodiversity and its importance are also vital. By promoting a deeper understanding of the value and fragility of biodiversity, individuals and communities can be empowered to make environmentally conscious decisions and support conservation efforts.

Case Study: Coral Reefs and Biodiversity

Coral reefs are incredibly diverse ecosystems known for their high biological productivity and critical ecological functions. They support an estimated 25% of all marine species and provide habitats for numerous organisms, including fish, mollusks, and corals.

Coral reefs face numerous threats, including rising sea temperatures, ocean acidification, pollution, overfishing, and destructive fishing practices. These threats have led to widespread coral bleaching, where corals expel their symbiotic algae and turn white. Bleached corals are more susceptible to disease and death.

Efforts to conserve coral reef biodiversity include the establishment of marine protected areas, implementing sustainable fishing practices, reducing pollution and sediment runoff, and promoting climate change mitigation and adaptation strategies. Additionally, research and monitoring initiatives help scientists better understand the complex interactions within coral reef ecosystems and develop strategies for their conservation.

Conclusion

Biodiversity is a key aspect of environmental science and conservation biology. It encompasses the variety of species, genes, and ecosystems found on Earth. Biodiversity provides numerous ecosystem services, supports food security, has cultural significance, and offers potential for scientific and medical advancements.

Biodiversity exists at different levels, including genetic diversity, species diversity, and ecosystem diversity. Each level contributes to the overall stability and resilience of ecosystems.

However, biodiversity faces significant threats from habitat loss, overexploitation, pollution, climate change, and invasive species. Conservation efforts are essential to mitigate these threats and ensure the long-term survival of biodiversity. Protected areas, habitat restoration, endangered species recovery, and community engagement are key strategies in conservation initiatives.

Case studies, such as coral reefs, illustrate the importance of biodiversity and highlight the ongoing efforts to protect and restore these fragile ecosystems.

Preserving biodiversity is not only an ethical responsibility but also essential for the well-being of our planet and future generations. By understanding the concept of biodiversity and the threats it faces, we can work towards a greener and more sustainable future.

FOUNDATIONS OF CONSERVATION BIOLOGY

Extinction and Endangerment of Species

The concept of extinction refers to the complete disappearance of a particular species from the Earth. This occurs when there are no surviving individuals of that species. Throughout the history of Earth, there have been several mass extinctions, where a significant number of species disappeared within a relatively short span of time. The most well-known mass extinction event is the one that led to the extinction of the dinosaurs approximately 65 million years ago.

Endangerment of species, on the other hand, refers to the state in which a species is at risk of extinction. This can be due to a variety of factors, such as habitat loss, overexploitation, introduction of invasive species, pollution, and climate change. When a species becomes endangered, it means that its population has declined to a critical level and it is facing imminent extinction if the factors threatening its survival are not addressed.

Conservation biologists and environmental scientists study the patterns and causes of extinction and endangerment in order to understand the underlying processes and develop strategies for preventing or mitigating them. This field of research is crucial for preserving biodiversity and protecting the delicate balance of ecosystems.

Causes of Extinction and Endangerment

There are several key factors that contribute to the extinction and endangerment of species. Understanding these causes is essential for developing effective conservation strategies. Some of the main causes include:

1. Habitat Loss: One of the leading causes of extinction and endangerment is the destruction and fragmentation of natural habitats. The conversion of forests into agricultural or urban areas, for example, reduces the available habitat for many species and disrupts their natural ecological processes.

2. Overexploitation: Unsustainable hunting, fishing, and trading of wildlife can lead to the depletion of populations and eventual extinction. This often occurs when species are targeted for their valuable parts, such as ivory or fur, or when they are subjected to excessive fishing pressure.

3. Introduction of Invasive Species: Non-native species that are introduced into new habitats can have devastating impacts on native species, leading to their decline or extinction. Invasive species can outcompete native species for resources or introduce new diseases and predation pressures, which the native species may not be able to cope with.

CHAPTER 3: CONSERVATION BIOLOGY AND WILDLIFE MANAGEMENT

4. Pollution: Pollution, such as chemical contamination of water and air, can have detrimental effects on both terrestrial and aquatic species. It can disrupt ecosystems, harm reproductive abilities, and weaken immune systems, making species more vulnerable to extinction.

5. Climate Change: The rapidly changing climate due to human activities is a significant threat to many species. Rising temperatures, altered weather patterns, and sea-level rise can force species to shift their ranges, impacting their survival and ability to adapt.

Examples of Extinction and Endangered Species

The impact of human activities on species extinction and endangerment can be observed globally. Here are a few notable examples:

1. The Dodo: The dodo, a flightless bird endemic to the island of Mauritius, became extinct in the 17th century due to habitat destruction and hunting by humans and introduced species. Its extinction serves as a classic example of human-induced extinction.

2. Sumatran Orangutan: The Sumatran orangutan, found in Indonesia, is critically endangered due to widespread deforestation for palm oil plantations. The loss of their natural habitat has resulted in severe population decline, making them highly vulnerable to extinction.

3. Black Rhinoceros: The black rhinoceros, native to Africa, is critically endangered due to poaching for their horns, which are highly valued in traditional medicine. Despite conservation efforts, their populations continue to decline, posing a serious threat to their survival.

4. Loggerhead Sea Turtle: The loggerhead sea turtle, a migratory species found in oceans around the world, is listed as endangered due to habitat destruction, pollution, and accidental capture in fishing gear. Climate change and rising sea levels also pose a significant threat to their nesting sites.

5. Polar Bear: The polar bear, an iconic species of the Arctic, is considered vulnerable due to the loss of sea ice habitat caused by climate change. As the Arctic warms, the polar bear's ability to hunt and reproduce is compromised, leading to population decline.

FOUNDATIONS OF CONSERVATION BIOLOGY

Conservation Efforts

Efforts to prevent species extinction and protect endangered species are crucial for maintaining biodiversity and ecosystem functionality. Conservation organizations, governments, and local communities employ various strategies to address the causes of extinction and endangerment. Some of these strategies include:

1. Protected Areas: Establishing protected areas, such as national parks and nature reserves, helps to safeguard habitats and provide refuge for endangered species. These areas can also serve as important research and education sites.

2. Species Recovery Programs: Programs focused on the recovery and reintroduction of endangered species play a vital role in preventing extinction. These programs involve breeding in captivity, habitat restoration, and reducing threats to the species' survival.

3. Legislation and Policy: Laws and regulations aimed at protecting endangered species and their habitats provide a legal framework for conservation efforts. International agreements, such as the Convention on International Trade in Endangered Species (CITES), help regulate the trade of endangered species and their parts.

4. Public Awareness and Education: Raising awareness about the importance of biodiversity and the need for species conservation is crucial for generating public support and participation. Education programs help instill a sense of responsibility and empower individuals to take action.

5. Sustainable Resource Management: Promoting sustainable practices in areas such as forestry, fisheries, and agriculture is essential for minimizing habitat destruction, overexploitation, and pollution. By ensuring the sustainable use of resources, we can reduce the negative impacts on species and ecosystems.

The Role of Citizen Science

Citizen science, the involvement of public citizens in scientific research, is playing an increasingly important role in understanding and addressing extinction and endangerment. This approach allows individuals to contribute to data collection, monitoring, and conservation efforts. Citizen science projects, such as bird-watching programs or species monitoring initiatives, provide valuable information to scientists and organizations working to protect species and ecosystems.

By engaging citizens in the scientific process, citizen science promotes a sense of ownership and responsibility towards conservation. It also enables collaboration between scientists and local communities, making conservation efforts more effective and sustainable.

Conclusion

The alarming rate of species extinction and endangerment underscores the pressing need for conservation efforts worldwide. Understanding the causes and consequences of extinction, as well as implementing effective strategies, is essential for preserving biodiversity and maintaining the health of our planet. By addressing habitat loss, overexploitation, invasive species, pollution, and climate change, we can work towards a sustainable future where species thrive and ecosystems flourish.

Conservation Genetics

Conservation genetics is a field of study that focuses on the genetic diversity and population dynamics of endangered species. It combines principles from genetics, ecology, and evolutionary biology to inform conservation strategies and ensure the long-term survival of species. By understanding the genetic factors affecting populations, conservation geneticists can make informed decisions about management practices, such as promoting genetic diversity, reducing inbreeding, and mitigating the risks of population decline.

The Importance of Genetic Diversity

Genetic diversity refers to the number and variety of different genes in a population. It is crucial for the long-term adaptability and survival of a species. A diverse gene pool provides the raw material for natural selection to act upon, allowing populations to respond to environmental changes, resist diseases, and adapt to new conditions. Additionally, genetic diversity promotes reproductive and evolutionary potential, as it reduces the negative effects of inbreeding, such as reduced fertility and increased susceptibility to genetic disorders.

In conservation genetics, understanding the genetic diversity of endangered species is essential for developing effective management strategies. By assessing the genetic structure of populations, scientists can identify individuals or populations with unique genetic traits and prioritize their conservation to maintain overall genetic diversity. This information helps to prevent further loss of genetic

FOUNDATIONS OF CONSERVATION BIOLOGY

variation, which is particularly important in small and isolated populations that are prone to genetic drift and inbreeding.

Genetic Techniques in Conservation Biology

Conservation geneticists employ various techniques to investigate the genetic makeup and population dynamics of endangered species. These techniques enable them to study genetic diversity, gene flow, population structure, and the impact of human activities on natural populations. Here are some commonly used techniques in conservation genetics:

- **Genetic Markers:** DNA markers are used to assess genetic variation within and between populations. These can be specific DNA sequences that vary among individuals and species, such as microsatellites or single nucleotide polymorphisms (SNPs). By analyzing these markers, researchers can determine genetic relatedness, population structure, and patterns of gene flow.

- **Population Genomics:** Population genomics combines traditional genetic markers with large-scale genomic sequencing to study the genetic variation within and between populations. This approach provides a comprehensive view of the genome, allowing researchers to identify specific genes or genomic regions associated with adaptation, disease resistance, or other important traits.

- **DNA Barcoding:** DNA barcoding is a technique used to identify and distinguish species based on a short DNA sequence. By comparing the DNA barcode of an unknown sample to a reference database, researchers can quickly and accurately identify species, including endangered or elusive ones.

- **Non-Invasive Sampling:** Non-invasive sampling techniques, such as collecting feces, hair, or feathers, allow researchers to obtain genetic material without disturbing or capturing the organisms. This approach is particularly useful for rare, elusive, or endangered species, as it reduces the stress and potential harm associated with traditional sampling methods.

- **Parentage Analysis:** Parentage analysis involves reconstructing the pedigree or family tree of individuals within a population. By genotyping individuals and their offspring, researchers can identify parent-offspring relationships,

Case Studies in Conservation Genetics

Conservation genetics has played a crucial role in the management and recovery of several endangered species. Let's explore a few case studies to understand how genetic techniques have contributed to their conservation efforts:

1. **Florida Panther:** The Florida panther, a subspecies of the mountain lion, faced severe population decline due to habitat loss and genetic isolation. Conservation geneticists used genetic markers to assess the level of genetic diversity and inbreeding in the population. To increase genetic variation, researchers translocated individuals from Texas to Florida, resulting in a genetic rescue and population recovery.

2. **Black-footed Ferret:** The black-footed ferret, one of North America's most endangered mammals, faced a population bottleneck due to habitat loss and a decline in its primary prey, the prairie dog. Conservation geneticists implemented a captive breeding program to maintain genetic diversity and prevent inbreeding. By carefully managing the breeding pairs and incorporating genetic information, they successfully reintroduced captive-bred individuals into the wild, boosting wild populations.

3. **Iberian Lynx:** The Iberian lynx, the most endangered wild cat species, experienced a rapid decline in population due to habitat fragmentation and reduced prey availability. Conservation geneticists used genetic techniques to study the genetic diversity and structure of the population. This information guided the development of captive breeding programs and informed the selection of individuals for release into the wild, maximizing genetic diversity and facilitating population recovery.

Challenges and Future Directions in Conservation Genetics

While conservation genetics has significantly contributed to the conservation and management of endangered species, several challenges and future directions exist. Some of these include:

+ **Genomic Data Analysis:** With advancements in genomic sequencing, the generation of vast amounts of genetic data has become more accessible.

WILDLIFE MANAGEMENT TECHNIQUES

However, analyzing and interpreting these data present computational and analytical challenges. Developing robust analytical tools and approaches to effectively handle large genomic datasets is crucial for advancing conservation genetics.

- **Ethical Considerations:** Conservation genetics often involves making decisions that directly impact the management and conservation of species. Ethical considerations, such as determining the appropriate use of genetic information and the handling of captive breeding programs, require ongoing discussions and collaborations among researchers, conservationists, policymakers, and local communities.

- **Integration with other Disciplines:** Conservation genetics can benefit from closer integration with other fields, such as ecology, population biology, and social sciences. Collaborative efforts and interdisciplinary research can provide a more comprehensive understanding of the ecological and social context of conservation issues, enabling more effective management strategies.

In summary, conservation genetics plays a critical role in understanding the genetic diversity and population dynamics of endangered species. By employing genetic techniques, scientists can identify and conserve genetic variation, mitigate the risks of inbreeding, and implement effective management strategies. Despite the challenges, ongoing advancements in genetic technologies and interdisciplinary collaborations hold promise for addressing current and future conservation issues.

Wildlife Management Techniques

Wildlife Population Assessment

In conservation biology and wildlife management, it is crucial to have accurate and reliable data on wildlife populations. Wildlife population assessment involves the collection and analysis of data to estimate the abundance, density, and distribution of different species in their respective habitats. This information is essential for understanding population dynamics, making informed management decisions, and implementing effective conservation strategies.

Methods of Wildlife Population Assessment

There are various methods used for wildlife population assessment, each with its advantages and limitations. Here, we will discuss some commonly used techniques:

1. **Direct Count Methods:** These methods involve visually counting individuals in a defined area. For species that are easily observable and have distinct markings or features, direct counts can provide accurate population estimates. For example, researchers may use transect surveys, where they walk along predetermined paths and record the number of individuals observed within a specified distance on either side of the path. This method is particularly useful for estimating populations of large mammals like elephants or giraffes.

2. **Indirect Count Methods:** Indirect count methods are used when direct observation is difficult or impractical. These methods rely on detecting signs or traces of wildlife presence rather than observing the animals themselves. Examples of indirect count methods include:

- *Track Counts:* Tracks, footprints, or other signs left by animals are counted and used to estimate population size. This method is commonly used for species that leave clear tracks, such as large cats or ungulates.

- *Scat Surveys:* The collection and analysis of animal feces can provide valuable information about population size, diet, health, and reproductive status. DNA analysis of scat samples can also help identify individual animals. This method is useful for studying elusive or nocturnal species like wolves or bears.

- *Nest Site Surveys:* Counting active nests or breeding sites can provide estimates of population size for species that construct nests or display site fidelity. This method is commonly used for birds, turtles, or insects.

3. **Mark-Recapture Methods:** Mark-recapture methods involve capturing and marking a subset of individuals, releasing them back into the population, and then recapturing a second sample at a later time. The number of marked individuals recaptured in the second sample allows estimation of the total population size. This method assumes that the proportion of marked individuals in the second sample represents the proportion of marked individuals in the entire population. Mark-recapture methods can be used for a wide range of species, from small mammals to amphibians and fish.

Challenges and Considerations

Wildlife population assessment poses several challenges and requires careful consideration of various factors. Some key challenges and considerations include:

WILDLIFE MANAGEMENT TECHNIQUES

1. Sampling Bias: Biases can occur when sampling is not representative of the entire population. For example, if surveys are conducted only in easily accessible areas, the resulting population estimates may not accurately reflect the entire species' distribution. Efforts should be made to minimize sampling bias by employing random or systematic sampling techniques.

2. Detection Bias: Certain species may be difficult to detect or observe, leading to an underestimation of their populations. This can be problematic for cryptic or elusive species, especially those active during specific times or in dense habitats. Researchers must account for detection bias when analyzing population data and consider using complementary methods to improve accuracy.

3. Spatial and Temporal Variation: Wildlife populations can exhibit spatial and temporal variations in abundance and distribution. Changes in habitat quality, climate, or other environmental factors can influence population dynamics. It is important to consider these variations when designing and interpreting population assessments and to conduct long-term monitoring to account for these dynamics adequately.

4. Ethics and Animal Welfare: Conducting wildlife population assessments raises ethical considerations. Researchers must ensure that their methods and procedures prioritize the welfare of animals and minimize any potential harm or stress. Institutional ethics committees and relevant wildlife authorities often review and approve research protocols to ensure compliance with ethical guidelines.

Example: Estimating Tiger Population

Let's consider an example of using wildlife population assessment to estimate the population of tigers in a particular reserve. Due to the tiger's elusive nature, direct counts may not be feasible. Instead, researchers may employ a combination of techniques, including camera traps and scat surveys.

Camera traps are remotely activated cameras that capture images or videos when triggered by an animal's movement. By analyzing the images captured over a defined period, researchers can identify individual tigers based on their unique stripe patterns. The number of identified individuals, along with capture rates and recapture rates, can be used to estimate tiger density and population size using mark-recapture methods.

Simultaneously, scat surveys can be conducted to collect feces samples for DNA analysis. DNA profiling can help determine the number of unique individuals in the population and provide insights into genetic diversity, relatedness, and overall population health.

CHAPTER 3: CONSERVATION BIOLOGY AND WILDLIFE MANAGEMENT

To account for spatial and temporal variation, researchers may set up camera traps and collect scat samples at different locations within the reserve and repeat the assessments over multiple seasons or years. This longitudinal approach enables them to monitor changes in tiger abundance, assess habitat usage and movement patterns, and inform conservation interventions.

Conclusion

Wildlife population assessment plays a crucial role in conservation biology and wildlife management. Through the use of direct count, indirect count, and mark-recapture methods, researchers can estimate population sizes, monitor population trends, and make informed decisions to protect and manage wildlife species. However, overcoming challenges such as biases, detection limitations, and temporal variations is crucial to obtain accurate and reliable estimates. Continuous improvements in assessment techniques and the integration of advanced technologies will contribute to better understanding and conservation of wildlife populations.

Habitat Management and Restoration

Habitat management and restoration are key strategies in wildlife management and conservation biology. They involve actively modifying and enhancing natural habitats to create favorable conditions for target species or ecosystems. This section will explore the principles, techniques, and case studies related to habitat management and restoration.

Principles of Habitat Management

Habitat management aims to maintain or improve habitat quality to support target species populations or ecological processes. To achieve this, several principles should be considered:

+ **Habitat selection:** Understanding the habitat requirements and preferences of target species is crucial for effective management. Different species have specific needs for food, water, shelter, and breeding sites. Managers must identify and provide suitable habitats to support their target species.

+ **Habitat structure:** Habitat structure refers to the physical characteristics of the environment, such as vegetation density, canopy cover, and the presence of water bodies or deadwood. It influences species diversity and abundance

WILDLIFE MANAGEMENT TECHNIQUES

by providing different niches and resources. Habitat management should aim to create or maintain a diverse range of habitat structures to support a variety of species.

+ **Habitat connectivity:** Habitat fragmentation is a pervasive threat to biodiversity. Isolated patches of habitat can restrict species movement, gene flow, and access to resources. Habitat management should focus on enhancing connectivity by creating corridors or wildlife-friendly landscapes that facilitate species movement between habitats.

+ **Invasive species control:** Invasive species can outcompete native species, degrade habitat quality, and disrupt ecological processes. Habitat management should include strategies for identifying and controlling invasive species through removal, containment, or biological control methods.

Habitat Restoration Techniques

Habitat restoration aims to reverse habitat degradation and improve ecological functionality. It involves the following techniques:

+ **Ecological site assessment:** Before initiating habitat restoration, an ecological site assessment is conducted to evaluate the current condition of the habitat, identify the causes of degradation, and determine the restoration goals. This assessment helps inform the selection of appropriate restoration techniques.

+ **Restoration of physical components:** Physical components of a habitat, such as soil, water bodies, and topography, may be restored to their natural state or modified to enhance ecological processes. For example, wetland restoration may involve reestablishing hydrological regimes, removing drainage ditches, and reintroducing native plant species.

+ **Revegetation and reforestation:** Restoring vegetation is a common strategy in habitat restoration. It may involve planting native plant species, controlling invasive plants, or assisting natural regeneration processes through seed dispersal or selective thinning of vegetation.

+ **Creation of artificial structures:** In some cases, artificial structures like birdhouses, bat roosts, or artificial reefs may be installed to compensate for the loss of natural structures or provide additional resources to targeted species.

CHAPTER 3: CONSERVATION BIOLOGY AND WILDLIFE MANAGEMENT

- **Species reintroduction:** In situations where a species has been locally extirpated or endangered, habitat restoration efforts may include captive breeding programs and subsequent reintroduction of individuals into restored habitats. This technique aims to reestablish viable populations and restore ecological interactions.

Case Studies in Habitat Management and Restoration

Let's explore two case studies that demonstrate the success of habitat management and restoration efforts:

Case Study 1: Riparian Habitat Restoration Riparian habitats are critical ecosystems along the banks of rivers and streams. They support a variety of plant and animal species and play a crucial role in water quality and erosion control. However, riparian habitats are often degraded or lost due to agriculture, urbanization, and other human activities.

In the Pacific Northwest of the United States, the "Streamside Stewardship Program" was implemented to restore riparian habitat along streams and rivers. The program focused on planting native vegetation, such as willows and alders, to stabilize stream banks, provide shade to maintain cooler water temperatures, and create wildlife habitat. The program also involved removing invasive species and fencing off riparian areas to prevent livestock access.

As a result of these restoration efforts, the water quality improved, and the riparian habitat became more resilient. Fish populations, including endangered salmon species, rebounded, and bird diversity increased. The success of this case study highlights the importance of riparian habitat restoration in preserving aquatic ecosystems.

Case Study 2: Grassland Restoration Grasslands are globally significant ecosystems that support diverse plant and animal communities. However, they are under threat due to habitat fragmentation, conversion to agriculture, and inappropriate land management practices.

The "Prairie Seed Program" in the Midwest region of the United States aimed to restore native grasslands by reintroducing locally adapted plant species and managing invasive species. The program collected seeds from remnant prairie patches, propagated them in nurseries, and then spread the seeds on restored sites. Prescribed fires were also used to simulate natural disturbance regimes and promote the growth of native grasses.

WILDLIFE MANAGEMENT TECHNIQUES

Over time, the restored grasslands showed increased plant diversity, including the return of rare and endemic species. The restored habitats attracted a variety of grassland-dependent birds, butterflies, and other pollinators. This case study demonstrates the importance of restoring native plant communities in grassland ecosystems for conserving biodiversity.

Challenges and Future Directions

Despite the successes in habitat management and restoration, several challenges persist:

+ **Limited resources:** Habitat management and restoration projects require significant financial and technical resources. Securing funding and long-term commitment is often a challenge, especially for large-scale restoration initiatives.

+ **Climate change impacts:** Climate change poses new challenges for habitat management and restoration. Rising temperatures, altered precipitation patterns, and increased frequency of extreme weather events make it necessary to consider climate resilience in restoration strategies.

+ **Socioeconomic factors:** Habitat management and restoration must consider socioeconomic factors such as land use conflicts, stakeholder engagement, and local community involvement. Collaboration among different stakeholders is crucial to ensure the success of restoration efforts.

To address these challenges, future directions in habitat management and restoration should focus on:

+ **Research and innovation:** Continued research and innovation are essential for developing new restoration techniques, optimizing resource allocation, and monitoring the long-term outcomes of restoration projects.

+ **Collaborative approaches:** Collaboration among scientists, land managers, policymakers, and local communities should be fostered to ensure holistic and sustainable habitat management practices. Indigenous knowledge and traditional land management practices should also be integrated into restoration efforts.

+ **Ecosystem-based approaches:** Restoration efforts should embrace ecosystem-based approaches that consider the interconnectedness of ecological processes and address multiple issues simultaneously. This approach can lead to more resilient and self-sustaining ecosystems.

In conclusion, habitat management and restoration are vital tools for conserving biodiversity and maintaining ecosystem services. By implementing effective management strategies and restoration techniques, we can enhance habitat quality, support vulnerable species, and contribute to the overall health of our natural environments.

Species Reintroduction and Translocation

Species reintroduction and translocation are conservation strategies aimed at restoring or establishing populations of endangered or extinct species in their historical or suitable habitats. These strategies have gained significant attention and success in recent years, contributing to the recovery of several species around the world.

Background

Many species face declining populations or even extinction due to various threats such as habitat loss, overhunting, and climate change. In response to these threats, conservation biologists have developed strategies to restore and protect species at risk. Reintroduction and translocation are two such strategies that focus on moving individuals from existing populations to new areas or reintroducing them into areas where they have become locally extinct.

Principles of Reintroduction and Translocation

The process of reintroduction involves carefully selecting and moving individuals or populations from donor sites to suitable habitats. Translocation, on the other hand, involves moving individuals or populations to areas where the species did not previously exist but may now thrive.

Both strategies are guided by several key principles:

1. **Thorough Planning:** Reintroduction and translocation efforts require extensive planning and assessment of the target species and the release site. Factors such as habitat suitability, population dynamics, and potential threats need to be considered.

WILDLIFE MANAGEMENT TECHNIQUES

2. **Source Population Selection:** Individuals for reintroduction or translocation are usually sourced from healthy populations with characteristics suitable for survival in the new environment. Genetic diversity and adaptation to local conditions are important considerations.

3. **Gradual Release:** Reintroduced or translocated individuals are often released in small numbers and monitored closely to gauge their survival and adaptation. This approach allows for adjustments in case unforeseen challenges arise.

4. **Post-release Monitoring and Management:** Monitoring the reintroduced or translocated population is crucial to assess their progress, address any issues, and make necessary adjustments to increase their chances of success.

5. **Community Engagement:** Involving local communities and stakeholders in the reintroduction or translocation process can foster support, increase awareness, and reduce potential conflicts.

Examples

Several successful reintroduction and translocation efforts have been carried out worldwide. Let's look at a couple of notable examples:

California Condor (Gymnogyps californianus): With a population reduced to only 27 individuals in the 1980s, the California Condor became critically endangered. Intensive conservation efforts involving captive breeding and reintroduction programs have helped increase the population to over 400 birds today. The release sites were carefully selected, and constant monitoring and management have ensured the success of the program.

Black-footed Ferret (Mustela nigripes): The black-footed ferret was once considered extinct until a small population was discovered in 1981. A captive breeding program was initiated, and reintroduction efforts began in 1991. Carefully selected release sites and ongoing monitoring have facilitated the recovery of the species, although ongoing challenges such as disease outbreaks and habitat loss remain.

Challenges and Considerations

Reintroduction and translocation efforts are not without challenges and potential drawbacks. Some of the key considerations include:

CHAPTER 3: CONSERVATION BIOLOGY AND WILDLIFE
MANAGEMENT

- **Habitat Suitability:** Ensuring that the release site has suitable habitat and sufficient resources to support the reintroduced or translocated species is crucial for their survival and successful establishment.

- **Genetic Considerations:** Maintaining genetic diversity is important to prevent inbreeding depression and enhance the resilience of the population. Careful selection of source populations and ongoing genetic monitoring are essential.

- **Predation and Competition:** Reintroduced or translocated species may face predation or competition from other species in the new area. Preparing for and mitigating these challenges is vital for their long-term survival.

- **Disease Transmission:** The introduction of individuals from different populations may increase the risk of disease transmission, especially if the new area harbors pathogens to which the reintroduced species has no immunity. Disease monitoring and preventive measures are necessary.

- **Human-Animal Conflict:** Reintroduced or translocated species may come into conflict with human activities or interests, such as agriculture or infrastructure development. Engaging local communities and implementing conflict resolution strategies are crucial for minimizing conflicts.

Resources and Additional Reading

For those interested in further exploring the topic of species reintroduction and translocation, here are some recommended resources:

- Sodhi, Navjot S., and Paul R. Ehrlich. *Conservation Biology for All.* Oxford: Oxford University Press, 2010. (Chapter 16: Reintroductions and Other Conservation Translocations)

- Armstrong, Doug P., and Reed F. Noss, eds. *Conservation Reintroduction and Other Conservation Translocations.* Washington, DC: Island Press, 2010.

- Olney, Matthew. *Reintroduction of Captive-Bred Wildlife: A Global Review.* Surrey, UK: Surrey Wildlife Trust, 2016.

- IUCN/SSC Reintroduction Specialist Group. Guidelines for Reintroductions and Other Conservation Translocations. Gland, Switzerland: IUCN, 2013.

Conclusion

Species reintroduction and translocation play a critical role in conserving endangered or locally extinct species by restoring populations and expanding their ranges. While these strategies come with challenges, proper planning, monitoring, and community engagement can significantly increase their chances of success. As our understanding of ecological dynamics improves, reintroduction and translocation efforts will continue to evolve, offering hope for the recovery of more species in the future.

Case Studies in Conservation Biology

Giant Panda Conservation in China

The conservation of giant pandas (Ailuropoda melanoleuca) in China has been an extraordinary success story that showcases the power of conservation biology in saving endangered species. The giant panda is an iconic symbol of wildlife conservation and is highly valued for its cultural and ecological significance. This section will delve into the various aspects of giant panda conservation, including the efforts made to preserve their habitat, increase their population, and promote public awareness and involvement.

The Importance of Giant Panda Conservation

Giant pandas are endemic to China and are primarily found in mountainous regions, particularly in the Sichuan, Shaanxi, and Gansu provinces. They are known for their distinctive black and white fur, bamboo diet, and docile nature. However, due to habitat loss, poaching, and a low reproduction rate, giant pandas faced a severe decline in their population, making them one of the most endangered species in the world.

Conservation efforts for giant pandas are crucial not only to ensure their survival but also to preserve the biodiversity and ecological integrity of their habitats. Pandas play a vital role in maintaining the balance of bamboo forests, as they help disperse seeds and promote new growth. By protecting pandas and their habitats, we also safeguard other endangered species that share the same ecosystems.

Conservation Strategies

To tackle the challenges faced by the giant panda population, conservationists implemented a range of strategies aimed at habitat preservation, population

CHAPTER 3: CONSERVATION BIOLOGY AND WILDLIFE
MANAGEMENT

management, and community engagement.

Habitat Preservation: One of the primary focuses of giant panda conservation has been the preservation and restoration of panda habitats. This involves establishing protected areas and corridors to connect fragmented habitats and minimize human-wildlife conflicts. The Chinese government has designated Nature Reserves specifically for giant pandas, such as the Wolong Nature Reserve and the Chengdu Research Base of Giant Panda Breeding, which provide safe havens for pandas to thrive.

Population Management: The reproductive biology of giant pandas presents a unique challenge. Female pandas are only receptive to mating for a brief period each year, and their overall reproductive rate is low. To increase the population, scientists have employed techniques such as artificial insemination and captive breeding programs. These programs aim to maximize breeding opportunities and enhance genetic diversity to ensure the long-term viability of the population.

Community Engagement: Engaging local communities in conservation efforts is crucial for the success of any conservation program. In the case of giant pandas, local communities have been involved in initiatives such as ecotourism, sustainable agriculture, and community-based conservation projects. These initiatives not only provide economic incentives for the communities but also raise awareness and foster a sense of stewardship toward panda conservation.

Successes and Challenges

The efforts invested in giant panda conservation have yielded significant successes and brought the species back from the brink of extinction.

Population Recovery: Thanks to conservation measures, the giant panda population has shown a positive trajectory in recent years. The International Union for Conservation of Nature (IUCN) downgraded the giant panda's conservation status from "endangered" to "vulnerable" in 2016, reflecting the population's recovery.

Habitat Protection: The establishment of protected areas and the implementation of strict regulations against deforestation and illegal hunting have contributed to the preservation of panda habitats. The creation of corridors

CASE STUDIES IN CONSERVATION BIOLOGY

between fragmented areas has allowed pandas to move freely and promote gene flow, strengthening the long-term survival prospects of the species.

Despite these successes, challenges remain that require continuous attention and adaptive management.

Habitat Fragmentation: Fragmentation of giant panda habitats due to human activities, such as infrastructure development and agriculture expansion, continues to pose a threat to their long-term survival. Maintaining and improving connectivity between habitats is essential to ensure the genetic viability and resilience of panda populations.

Climate Change: The impact of climate change on giant panda habitats and their primary food source, bamboo, is a growing concern. Changes in temperature and precipitation patterns can disrupt bamboo growth, affecting the availability and quality of food for pandas. Conservation efforts must incorporate climate change adaptation strategies to mitigate the potential negative impacts.

Conclusion

Giant panda conservation in China stands as a shining example of successful species recovery through dedicated efforts in habitat preservation, population management, and community involvement. The positive trajectory of the panda population and the downgrading of their conservation status demonstrate the effectiveness of conservation biology in ensuring the survival of endangered species.

However, the conservation journey is far from over. Ongoing challenges such as habitat fragmentation and climate change necessitate the continued commitment and innovation of scientists, policymakers, local communities, and the public. By learning from the successes and challenges of giant panda conservation, we can apply valuable lessons to other wildlife conservation efforts and create a greener and more sustainable future for all species.

African Elephant Conservation and Anti-Poaching Efforts

African elephants (Loxodonta africana and Loxodonta cyclotis) are majestic creatures found across the African continent. Known for their large size, intelligent behavior, and unique social structures, these iconic animals are a symbol of Africa's natural heritage. Unfortunately, they face numerous threats, including habitat loss, human-wildlife conflicts, and most significantly, poaching for their ivory tusks. In

CHAPTER 3: CONSERVATION BIOLOGY AND WILDLIFE MANAGEMENT

this section, we will explore the conservation efforts and anti-poaching initiatives aimed at protecting African elephants.

Background and Importance of African Elephant Conservation

African elephants play a crucial role in maintaining the ecological balance of their habitats. As both herbivores and seed dispersers, they are vital for the regeneration and sustainability of the African savannah and forest ecosystems. Moreover, they contribute to tourism revenue, providing economic benefits to local communities and national economies. Preserving their populations is not only essential for ecological reasons but also for socio-economic stability in the regions where they reside.

Poaching Crisis and Ivory Trade

The primary threat to African elephants is poaching for their valuable ivory tusks. The global demand for ivory, particularly in Asian markets, drives the illegal trade, resulting in a rapid decline in elephant populations. Ivory is sought after for its use in traditional medicine, decorative items, and as a status symbol. The high profits associated with the illegal ivory trade incentivize poachers, organized crime syndicates, and armed militia groups to engage in wildlife trafficking.

Conservation Strategies

Efforts to conserve African elephants and combat poaching involve various strategies, including law enforcement, community participation, habitat protection, and international cooperation. Some key approaches are highlighted below:

1. **Anti-Poaching Units and Wildlife Law Enforcement** Establishing anti-poaching units and strengthening wildlife law enforcement agencies is crucial for deterring poachers. Training rangers with advanced skills in tracking, surveillance, and firearm handling helps ensure effective protection. Additionally, equipping them with modern technology such as drones, thermal cameras, and GPS tracking devices enhances their monitoring and response capabilities.

2. **Community Engagement and Alternative Livelihoods** Engaging local communities and providing alternative livelihood options is essential for reducing human-wildlife conflicts and gaining their support for elephant conservation. Sustainable agriculture, eco-tourism, and crafts-based enterprises are some

CASE STUDIES IN CONSERVATION BIOLOGY

185

examples of alternative income-generating activities that can alleviate poverty while promoting conservation.

3. Protected Areas and Corridor Conservation Creating protected areas and establishing wildlife corridors are vital for ensuring the long-term survival of African elephants. Protected areas provide secure habitats, while corridors facilitate the movement of elephants between different landscapes, maintaining genetic diversity and promoting ecological balance.

4. International Cooperation and Policy Advocacy Collaboration between countries, international organizations, and NGOs is key to tackling the transnational nature of wildlife smuggling and ivory trade. This includes sharing intelligence, harmonizing legislation, and raising awareness globally. Advocacy efforts aim to increase political will, strengthen regulations, and influence consumer behavior to reduce the demand for illegal ivory.

Case Study: Garamba National Park, Democratic Republic of Congo

Garamba National Park in the Democratic Republic of Congo (DRC) has been a significant battleground in the fight against elephant poaching. The park has faced extreme challenges due to armed conflicts, political instability, and weak law enforcement. However, concerted efforts have been made to protect Garamba's elephant population:

Park Management and Security Garamba National Park authorities, along with partner organizations and the Congolese army, have worked relentlessly to improve park management and security. This includes increasing patrols, implementing aerial surveillance, and using sniffer dogs to detect illegal wildlife products.

Community Engagement Engagement with local communities bordering Garamba National Park has been crucial for gaining their support. Projects focusing on education, healthcare, and sustainable livelihoods have helped reduce poaching incidents by providing alternatives to illegal activities.

International Support The international community has extended support to the conservation efforts in Garamba National Park. Financial aid, technical assistance, and capacity building initiatives have been instrumental in improving park infrastructure, strengthening law enforcement, and enhancing anti-poaching measures.

Despite the progress made in Garamba National Park, ongoing efforts are required to ensure the sustainable conservation of African elephants and the protection of their habitats across the continent. Vigilance needs to be maintained to address emerging challenges and adapt conservation strategies accordingly.

Challenges and Future Directions

While significant progress has been made in African elephant conservation, several challenges persist:

1. Poverty and Governance Limited economic opportunities, political instability, and weak governance contribute to poaching and illegal wildlife trade. Addressing poverty and promoting good governance are crucial for achieving sustainable conservation outcomes.

2. Demand Reduction Reducing the demand for ivory and other wildlife products is critical to curbing poaching. Public awareness campaigns, along with policy interventions and targeted law enforcement, are needed to change consumer behavior and attitudes.

3. Transnational Cooperation International collaboration and cooperation between countries affected by the illegal wildlife trade are essential. Harmonizing legislation, sharing intelligence, and supporting enforcement efforts can help disrupt criminal networks and dismantle the ivory trade.

4. Climate Change Climate change poses additional challenges to elephant conservation. It disrupts ecosystems, alters migration patterns, and affects the availability of key resources. Conservation strategies need to take into account the ecological impacts of climate change and incorporate adaptive measures.

In conclusion, African elephant conservation and anti-poaching efforts are crucial for preserving these iconic species and the ecosystems they inhabit. By implementing comprehensive strategies that involve law enforcement, community engagement, habitat protection, and international cooperation, we can work towards a future where African elephants continue to roam freely and thrive in their natural habitats. However, addressing challenges such as poverty, governance, demand reduction, and climate change is critical for ensuring

CASE STUDIES IN CONSERVATION BIOLOGY   187

long-term success. Only through our collective efforts can we secure a brighter future for African elephants and foster a greener and more sustainable world.

Resources: - African Wildlife Foundation: https://www.awf.org/ - Convention on International Trade in Endangered Species of Wild Fauna and Flora (CITES): https://cites.org/ - Save the Elephants: https://www.savetheelephants.org/ - TRAFFIC: https://www.traffic.org/

Exercises: 1. Analyze the socio-economic impacts of African elephant conservation on local communities. 2. Research and evaluate the effectiveness of international policies, such as the ivory trade ban, in combating elephant poaching. 3. Design a community-based initiative to promote alternative livelihoods and reduce human-elephant conflicts in an elephant habitat region. 4. Investigate the role of technology, such as artificial intelligence and satellite tracking, in strengthening anti-poaching efforts. 5. Discuss the ethical considerations involved in relocating elephants to protected areas outside their natural range.

Tricks and Caveats: - When researching or reporting on African elephant conservation, ensure you access up-to-date information as the situation and conservation strategies may evolve over time. - Be mindful of cultural differences and sensitivities when engaging with local communities involved in elephant conservation initiatives. - Check the legality and ethical considerations when using any products derived from elephants, such as ivory or other elephant by-products.

Unconventional Insight: African elephants have a deep cultural significance and play a vital role in maintaining ecological balance. By highlighting the importance of these majestic animals in storytelling, folklore, and art, we can foster a deeper appreciation and respect for their conservation. Incorporating traditional knowledge and narratives into conservation strategies can create a more holistic and community-centered approach, ensuring the long-term survival of African elephants.

California Condor Recovery Program

The California Condor Recovery Program is a conservation initiative aimed at preventing the extinction of the critically endangered California condor (Gymnogyps californianus). The program, which began in the 1980s, has been one of the most successful examples of captive breeding and reintroduction efforts for a threatened species. In this section, we will explore the background of the program, its key principles and strategies, the challenges it faces, and its remarkable achievements.

CHAPTER 3: CONSERVATION BIOLOGY AND WILDLIFE MANAGEMENT

Background

The California condor, with its distinctive black feathers and large wingspan, is a symbol of the American West. However, by the early 1980s, the species was on the brink of extinction, primarily due to habitat loss, lead poisoning from ingesting bullet fragments, and low reproductive rates. In 1982, only 27 individuals remained in the wild, prompting the launch of the California Condor Recovery Program.

Principles and Strategies

The California Condor Recovery Program is based on several key principles and strategies aimed at ensuring the survival and recovery of the species. These include:

1. **Captive Breeding:** The program established captive breeding facilities to breed California condors in a controlled environment. This approach allowed for the management of breeding pairs, genetic diversity, and the ability to provide specialized care for individuals requiring medical intervention.

2. **Reintroduction:** Once the captive-bred condors reach maturity, they are released into the wild in carefully selected sites. Reintroduction efforts focus on areas with suitable habitat and reduced threats, such as lead poisoning, and involve close monitoring to ensure the survival and integration of the released condors into existing wild populations.

3. **Lead Ammunition Reduction:** A major cause of mortality for California condors is lead poisoning, often resulting from the ingestion of lead ammunition in carcasses. The recovery program has collaborated with hunters, government agencies, and organizations to promote lead-free ammunition and reduce the risk of lead exposure to condors and other wildlife.

4. **Monitoring and Research:** The program employs rigorous monitoring and research techniques to assess the health and behavior of the condors both in captivity and in the wild. This information allows scientists to better understand the species' needs, identify threats, and adjust management strategies accordingly.

5. **Public Education and Outreach:** The recovery program recognizes the importance of raising public awareness about the California condor and its conservation. Through educational programs, visitor centers, and public

CASE STUDIES IN CONSERVATION BIOLOGY

outreach campaigns, efforts are made to engage and inspire people to support the recovery program and take actions to protect the species and its habitat.

Challenges and Achievements

The California Condor Recovery Program has faced numerous challenges throughout its implementation. Some of the main challenges include:

- **Small Population Size:** The small number of individuals left in the wild posed a significant challenge, as genetic diversity was limited, making the condors more susceptible to diseases and environmental changes.

- **Lead Poisoning:** The persistence of lead ammunition in certain areas continued to pose a threat to the condors, despite efforts to promote lead-free alternatives. Ongoing monitoring and outreach are crucial in mitigating this risk.

- **Habitat Restoration:** Ensuring the availability of suitable habitat for California condors is essential for their long-term survival. Habitat loss and degradation remain ongoing challenges, requiring collaboration with landowners and implementing habitat restoration projects.

Despite these challenges, the California Condor Recovery Program has achieved remarkable success in bringing the species back from the brink of extinction. As of *insert current year*, the wild population has increased to over 400 birds, with approximately half of them flying freely in their natural range. The captive breeding efforts have resulted in a self-sustaining population, with breeding pairs successfully raising chicks in the wild.

The program's achievements include:

- **Population Growth:** The recovery program has successfully increased the California condor population, demonstrating the effectiveness of captive breeding and reintroduction strategies.

- **Species Range Expansion:** Reintroduction efforts have facilitated the expansion of the condor's range, allowing the species to occupy historical territories and explore new areas.

- **International Collaboration:** The recovery program has fostered collaboration with international partners, sharing knowledge and best practices for conserving endangered species and their habitats.

CHAPTER 3: CONSERVATION BIOLOGY AND WILDLIFE
MANAGEMENT

+ **Public Support and Engagement:** The program has generated significant public support and engagement, with individuals and organizations actively contributing to condor conservation efforts through donations, volunteer work, and advocacy.

Unconventional Approach: Condor Cams

A particularly innovative and unconventional aspect of the California Condor Recovery Program is the use of "condor cams." These live video feeds set up at nest sites allow researchers, conservationists, and the public to observe and learn about the behavior of these magnificent birds. Condor cams provide a unique opportunity for people to connect with the species, fostering a sense of empathy and a deeper understanding of the challenges they face.

Conclusion

The California Condor Recovery Program serves as a shining example of successful species recovery through a combination of captive breeding, reintroduction efforts, lead ammunition reduction, monitoring, research, and public engagement. Its accomplishments highlight the importance of collaborative conservation and the dedication required to save species on the brink of extinction.

As the program continues, ongoing efforts to address existing and emerging challenges are essential. The California Condor Recovery Program demonstrates that with collective action, innovative strategies, and public support, it is possible to reverse the course of extinction and secure a brighter future for endangered species.

Sea Turtle Conservation and Protection

Sea turtles are magnificent creatures that have been swimming in our oceans for millions of years. These endangered species play a crucial role in maintaining the health of marine ecosystems. However, they face numerous threats that have led to a decline in their populations. In this section, we will explore the challenges faced by sea turtles and the success stories in their conservation and protection.

The Importance of Sea Turtles

Sea turtles are important indicators of the health of marine environments. As keystone species, they help maintain the balance of ecosystems by influencing the abundance of other species. Here are some key roles played by sea turtles:

CASE STUDIES IN CONSERVATION BIOLOGY

- **Maintaining seagrass beds:** Sea turtles feed on seagrass, which keeps it healthy and promotes oxygen production. Healthy seagrass beds, in turn, support a variety of marine species.

- **Controlling jellyfish population:** Sea turtles are voracious jellyfish eaters. By keeping jellyfish populations in check, they prevent jellyfish blooms that can negatively impact fisheries and tourism.

- **Distributing nutrients:** Sea turtles transport nutrients from the ocean to beaches through their nesting activities. This benefits coastal vegetation and supports the overall ecosystem.

- **Enhancing coral reef growth:** Sea turtles feed on sponges, which often compete with corals for space. By reducing sponge populations, they create opportunities for coral reef expansion and diversity.

Threats to Sea Turtles

Sea turtles face numerous threats throughout their life cycle, from their early years as hatchlings to their adult journeys across vast oceans. Some of the major threats include:

- **Coastal development:** Urbanization and tourism lead to habitat loss and degradation, as coastal areas are converted into resorts, homes, and infrastructure.

- **Climate change:** Rising temperatures and sea levels, as well as changes in ocean currents, affect sea turtle nesting sites, hatchling survival, and food availability.

- **Pollution:** Marine pollution, particularly plastic debris, poses significant threats to sea turtles. They can mistake plastic bags for jellyfish, their favorite prey, leading to ingestion and suffocation.

- **Fishing activities:** Sea turtles are at risk of being accidentally caught in fishing nets and lines. This bycatch can result in injuries or death.

- **Poaching:** Sea turtles have been hunted for their meat, eggs, shells, and leather. Despite legal protections, illegal trade still persists in some regions.

CHAPTER 3: CONSERVATION BIOLOGY AND WILDLIFE MANAGEMENT

Conservation Efforts and Success Stories

Conservation organizations, governments, and local communities have been working tirelessly to protect and conserve sea turtles. Their efforts have led to many success stories. Here are a few notable examples:

- **Archie Carr National Wildlife Refuge, Florida, USA:** Established in 1991, this refuge provides critical nesting habitat for loggerhead, green, and leatherback turtles. Strict regulations and monitoring have helped stabilize and increase nesting populations.

- **Gahirmatha Marine Sanctuary, India:** This sanctuary, established in 1997, is a major nesting site for olive ridley turtles. Strict patrolling, community involvement, and awareness programs have contributed to a significant rise in nesting numbers.

- **Great Barrier Reef Marine Park, Australia:** As part of the broader conservation efforts for the Great Barrier Reef, measures have been taken to protect and monitor the nesting and foraging habitats of green turtles and loggerhead turtles.

- **Tortuguero National Park, Costa Rica:** This park is home to one of the largest green turtle nesting sites in the world. Conservation efforts, such as patrolling, research, and community engagement, have helped in the recovery of green turtle populations.

Challenges and Future Directions

While significant progress has been made in sea turtle conservation, challenges remain. Continued efforts are required to address the following issues:

- **Habitat protection:** It is essential to conserve and restore critical nesting beaches, seagrass beds, and foraging habitats to provide safe spaces for sea turtles.

- **Reducing pollution:** Implementing effective waste management systems, reducing plastic use, and raising awareness about the impacts of pollution are crucial steps in protecting sea turtles from the dangers of marine debris.

- **Climate change adaptation:** Developing strategies to mitigate the effects of climate change on sea turtle habitats and ensuring their long-term survival are critical for their conservation.

CASE STUDIES IN CONSERVATION BIOLOGY

- **Sustainable fishing practices:** Implementing turtle-friendly fishing techniques, such as the use of turtle excluder devices (TEDs) in fishing nets, can significantly reduce bycatch and prevent the accidental capture of sea turtles.

- **Community engagement and education:** Involving local communities in conservation efforts, raising awareness about the importance of sea turtles, and promoting responsible tourism can play a vital role in their protection.

Conclusion

Sea turtles are extraordinary creatures that embody the beauty and diversity of our oceans. Their conservation is not only crucial for their own survival but also for the health and balance of marine ecosystems. By recognizing the importance of sea turtles and taking concrete actions to conserve and protect them, we can ensure their survival for future generations to enjoy. Let us join hands in this endeavor and make a positive impact on the lives of these magnificent creatures and the world they inhabit.

Orangutan Conservation in Borneo and Sumatra

Orangutans, the iconic great apes found in the rainforests of Borneo and Sumatra, are among the most endangered species on our planet. Their populations have been severely affected by habitat loss, illegal hunting, and the pet trade. In this section, we will explore the challenges faced by orangutans and the conservation efforts aimed at protecting and preserving these incredible creatures.

The Importance of Orangutans

Orangutans play a crucial role in the ecosystems of Borneo and Sumatra. As the largest arboreal mammals, they act as key seed dispersers, helping to maintain the biodiversity of their habitat. Additionally, they contribute to forest regeneration by creating nests and travel paths in the forest canopy. These activities have a positive impact on the overall health and sustainability of the rainforest ecosystem.

Threats to Orangutan Survival

Orangutans face numerous threats to their survival, primarily due to human activities. Deforestation, driven by the expansion of agriculture and palm oil plantations, is the leading cause of habitat loss for orangutans. As their forest homes are destroyed, orangutans are forced into smaller and fragmented areas, making them more vulnerable to poaching and other dangers.

Illegal hunting and the pet trade also pose significant threats to orangutans. Baby orangutans are often taken from their mothers and sold as pets, both in local markets and internationally. This cruel practice not only harms individual orangutans but also disrupts their social structure and reproductive patterns.

Conservation Strategies for Orangutans

Conserving orangutans requires a multi-faceted approach that addresses both direct threats and the underlying causes of their decline. Here are some key strategies employed in orangutan conservation:

1. **Habitat Protection:** Protecting the remaining orangutan habitat is vital for their survival. Collaboration between governments, NGOs, and local communities is essential in establishing protected areas and managing them effectively. This includes enforcing laws against illegal logging and encroachment and promoting sustainable land use practices.

2. **Community Engagement:** Engaging and empowering local communities is crucial for successful orangutan conservation. By involving communities in conservation efforts, their knowledge and support can be leveraged to protect orangutan habitat, combat illegal hunting, and develop sustainable livelihood options that are compatible with conservation objectives.

3. **Anti-Poaching Measures:** Combating the illegal hunting and pet trade of orangutans requires strong law enforcement and increased penalties for offenders. Efforts should focus on disrupting the supply chains, raising awareness about the negative impacts of the pet trade, and providing alternative sources of income for local communities to reduce their reliance on wildlife exploitation.

4. **Raising Awareness:** Increasing public awareness about the importance of orangutan conservation is vital for generating support and garnering resources for conservation initiatives. Education campaigns, both locally and globally, can help foster a sense of connection and responsibility towards orangutans and their habitats.

5. **Scientific Research:** Conducting scientific research is crucial for understanding orangutan ecology, behavior, and population dynamics. This information is essential for developing effective conservation strategies,

CASE STUDIES IN CONSERVATION BIOLOGY

tracking the success of ongoing efforts, and identifying new challenges that may arise.

Success Stories in Orangutan Conservation

Despite the challenges, there have been notable success stories in orangutan conservation. Some of these include:

1. **Tanjung Puting National Park, Borneo:** Tanjung Puting National Park is one of the largest protected areas for orangutans in Borneo. This park has successfully implemented anti-poaching measures, engaged local communities in conservation efforts, and established rehabilitation centers for orphaned and confiscated orangutans. These efforts have led to an increase in orangutan populations and a decrease in illegal activities.

2. **Gunung Leuser National Park, Sumatra:** Gunung Leuser National Park is a UNESCO World Heritage site and home to one of the most significant populations of Sumatran orangutans. Conservation organizations and local communities have collaborated to protect this important habitat, conduct research, and promote sustainable tourism. As a result, orangutan populations in the park have shown signs of recovery.

3. **International Collaboration:** The conservation of orangutans has also benefited from international collaboration. Organizations such as the Orangutan Foundation International and the Borneo Orangutan Survival Foundation work in partnership with local governments, NGOs, and research institutions to implement comprehensive conservation programs. These partnerships allow for the sharing of knowledge, resources, and expertise, ultimately contributing to the conservation of orangutans.

Challenges and Future Directions

While progress has been made in orangutan conservation, significant challenges remain. Here are some key challenges and future directions for orangutan conservation:

+ **Palm Oil Industry:** The expansion of palm oil plantations continues to be a major threat to orangutan habitat. Addressing this challenge requires a multi-stakeholder approach that involves governments, the palm oil industry, consumers, and conservation organizations. Sustainable palm oil

CHAPTER 3: CONSERVATION BIOLOGY AND WILDLIFE MANAGEMENT

certification and increased consumer demand for palm oil from responsible sources can incentivize industry practices that are less harmful to orangutan habitats.

- **Climate Change:** Climate change poses additional challenges for orangutan conservation. Rising temperatures, changing rainfall patterns, and increased frequency of extreme weather events can impact the availability of food and habitat for orangutans. Integrating climate change adaptation strategies into conservation efforts is crucial for the long-term survival of orangutans.

- **Illegal Wildlife Trade:** The illegal wildlife trade, including the smuggling of orangutans, remains a persistent challenge. Strengthening law enforcement, increasing penalties for offenders, and enhancing international cooperation are essential in combatting this illegal activity.

- **Engaging Indigenous Communities:** Indigenous communities have a deep connection to orangutan habitats and possess traditional knowledge that can contribute to conservation efforts. Engaging these communities as partners and decision-makers in orangutan conservation is vital for creating sustainable and inclusive conservation strategies.

- **Education and Advocacy:** Continued efforts in raising awareness, both locally and globally, are essential for mobilizing support and resources for orangutan conservation. Education initiatives targeting schools, local communities, and consumers can play a significant role in fostering a culture of conservation and making sustainable choices.

In conclusion, orangutan conservation in Borneo and Sumatra is a complex challenge that requires a comprehensive and interdisciplinary approach. By addressing the key threats, engaging local communities, and promoting sustainable practices, we can ensure the survival of orangutans and protect the invaluable ecosystems they inhabit.

Additional Resources:

1. World Wildlife Fund (WWF): *Orangutan Conservation* - `https://www.worldwildlife.org/species/orangutan`

2. Orangutan Foundation International - `https://orangutan.org/`

3. Borneo Orangutan Survival Foundation - `https://orangutan.or.id/`

Exercises:

1. Research and identify one successful case of community-led orangutan conservation. Describe their approaches and outcomes.

2. Discuss the ethical implications of using captive orangutans for rehabilitation and reintroduction programs. What are the potential benefits and challenges?

3. Investigate the impact of palm oil consumption on orangutan habitat destruction. Identify ways in which individuals can reduce their palm oil footprint.

Remember, orangutan conservation is not just about saving a species; it is about safeguarding the delicate balance of our ecosystems and preserving the richness of our natural world for generations to come.

Success Stories in Conservation Biology

Galapagos Islands Conservation and Ecotourism

The Galapagos Islands, located in the Pacific Ocean off the coast of Ecuador, are renowned for their unique ecosystems and incredible biodiversity. The islands are a UNESCO World Heritage site and have been designated a national park. The conservation of the Galapagos Islands and the sustainable management of its natural resources are of utmost importance to protect and preserve this extraordinary ecosystem for future generations.

Conservation Challenges

The Galapagos Islands face numerous conservation challenges due to their fragile and sensitive ecosystems. These challenges include invasive species, overfishing, habitat degradation, and climate change.

Invasive species, such as rats, goats, and ants, have been introduced to the islands by human activities. These invasive species pose a significant threat to the native flora and fauna because they compete for resources and disrupt the natural balance of the ecosystem. For example, the introduction of goats led to the destruction of vegetation, which endangered the survival of several endemic species.

Overfishing is another major challenge in the Galapagos Islands. The islands' waters are home to a diverse range of marine species, but unsustainable fishing practices have depleted fish populations. This not only disrupts the natural food chain but also affects the livelihoods of local communities who rely on fishing for their subsistence.

CHAPTER 3: CONSERVATION BIOLOGY AND WILDLIFE MANAGEMENT

Habitat degradation is a consequence of increasing human activities such as tourism, urbanization, and agriculture. These activities contribute to the destruction of native vegetation and the fragmentation of habitats, leading to the loss of biodiversity and the disruption of ecological processes.

Climate change poses a significant threat to the Galapagos Islands. Rising sea temperatures, ocean acidification, and increased frequency of extreme weather events are impacting the delicate balance of the islands' ecosystems. These changes have negative effects on coral reefs, marine species, and terrestrial organisms, further exacerbating the conservation challenges faced by the islands.

Conservation Strategies

To address these conservation challenges, various strategies and initiatives have been implemented in the Galapagos Islands.

One of the key strategies is the eradication of invasive species. Conservation organizations and local authorities have launched programs to eliminate invasive species from the islands. For instance, the Galapagos National Park has successfully eradicated goats from several islands, allowing native vegetation to recover and supporting the resurgence of endemic species.

Another important conservation strategy is the establishment and management of marine protected areas (MPAs). These MPAs help regulate fishing activities and protect critical habitats for marine species. The Galapagos Marine Reserve, one of the largest MPAs in the world, has played a vital role in conserving the islands' marine biodiversity.

Sustainable tourism and ecotourism have emerged as essential tools for conservation in the Galapagos Islands. Strict regulations and guidelines are in place to ensure that tourism activities are carried out in an environmentally responsible manner. This includes limiting the number of visitors, controlling access to sensitive areas, and promoting education and awareness about the islands' unique ecosystems. Local communities are actively involved in ecotourism initiatives, fostering a sense of ownership and responsibility for the conservation of their natural heritage.

Additionally, scientific research and monitoring programs are crucial for understanding the ecological processes of the Galapagos Islands and providing valuable data for conservation efforts. Researchers from around the world collaborate with local institutions to study and monitor the islands' ecosystems, enabling evidence-based decision-making for effective conservation strategies.

SUCCESS STORIES IN CONSERVATION BIOLOGY

Success Stories

The conservation efforts in the Galapagos Islands have led to several success stories, demonstrating the effectiveness of sustainable management and community involvement.

The restoration of Floreana Island's ecosystem is one such success story. The island was heavily impacted by invasive species, with many native species on the brink of extinction. Through a comprehensive eradication program and the involvement of local communities, invasive rats and other pests were successfully removed from the island. As a result, the native flora and fauna have started to recover, and several endangered species have shown signs of population growth.

The promotion of sustainable fisheries is another notable success in the Galapagos Islands. Strict regulations and monitoring systems have been put in place to prevent overfishing and ensure the long-term sustainability of marine resources. These measures have led to the recovery of fish populations, benefiting both marine ecosystems and local fishing communities.

The establishment of the Charles Darwin Research Station on Santa Cruz Island is a significant achievement for scientific research and conservation in the Galapagos Islands. The research station conducts vital studies on the islands' flora and fauna, develops conservation strategies, and collaborates with local communities and international organizations. The station's work has contributed to the conservation and management of the islands' biodiversity.

Challenges and Future Directions

Despite the conservation successes in the Galapagos Islands, numerous challenges remain, and continuous efforts are needed to ensure the long-term sustainability of the ecosystem.

One of the ongoing challenges is the effective control of invasive species. Constant monitoring, rapid response protocols, and public awareness campaigns are essential to prevent the reintroduction of invasive species and their potential impacts on native biodiversity.

Climate change poses a significant threat to the Galapagos Islands, and adaptation strategies are crucial for the survival of the ecosystems. This includes the restoration of degraded habitats, the development of resilient marine protected areas, and the promotion of sustainable practices in agriculture and tourism.

Engaging local communities in conservation efforts and providing alternative livelihood options are essential for the long-term success of conservation initiatives.

CHAPTER 3: CONSERVATION BIOLOGY AND WILDLIFE MANAGEMENT

Empowering local residents through education and training programs can foster a sense of stewardship and promote sustainable practices.

Collaboration between government agencies, conservation organizations, researchers, and local communities is vital for effective conservation planning and implementation. Integrated approaches that consider social, economic, and environmental factors will be critical for addressing the complex challenges faced by the Galapagos Islands.

In conclusion, the conservation of the Galapagos Islands and its unique ecosystems requires a multi-faceted approach involving the eradication of invasive species, the establishment of marine protected areas, sustainable tourism practices, scientific research, and community involvement. The success stories in the Galapagos Islands demonstrate the positive outcomes that can be achieved through dedicated conservation efforts. However, ongoing challenges such as invasive species control and climate change necessitate continuous monitoring and adaptive management strategies. Only through sustained commitment and collaboration can we ensure the long-term preservation of this remarkable natural heritage.

Wolf Reintroduction in Yellowstone National Park

Yellowstone National Park, located in the United States, is not only known for its breathtaking natural beauty but also for its remarkable success story in wolf reintroduction. In the 1930s, wolves were completely eradicated from the park due to aggressive predator control programs. However, recognizing the important role of wolves in maintaining a balanced ecosystem, efforts were made to bring them back.

Ecosystem Dynamics and the Role of Wolves

To understand the significance of wolf reintroduction in Yellowstone, it is essential to grasp the dynamics of the ecosystem. Yellowstone National Park is a complex ecosystem consisting of various interconnected components, including the food web, vegetation, and physical environment.

Before the reintroduction of wolves, the absence of their presence had profound effects on the ecosystem. Elk populations, which serve as a crucial prey source for wolves, increased dramatically. Overgrazing by elk led to the depletion of willows and aspen trees, causing a decline in their populations. This, in turn, had detrimental effects on the habitats and biodiversity of other species, such as beavers, songbirds, and fish.

SUCCESS STORIES IN CONSERVATION BIOLOGY

The reintroduction of wolves aimed to restore the natural balance by controlling the elk population and subsequently allowing the recovery of vegetation and other animal species. Through predation, wolves regulate the distribution and behavior of ungulate prey, preventing overgrazing and promoting habitat diversity.

Wolf Reintroduction and Challenges

The process of wolf reintroduction in Yellowstone National Park was a complex and challenging endeavor. In 1995, following a lengthy debate and scientific study, 31 gray wolves from healthy populations in Canada were reintroduced into the park. The primary aims were to reestablish a sustainable wolf population and restore ecological balance.

One of the main challenges faced during the reintroduction process was addressing concerns from stakeholders, including ranchers and hunters, who were worried about potential conflicts with wolves. Conservation organizations and park management implemented extensive public outreach programs to educate and address these concerns, demonstrating the importance of wolves for both the ecosystem and the tourism industry.

Another challenge was the selection of appropriate release sites for the reintroduced wolves. Careful consideration was given to factors such as habitat suitability, prey availability, and potential human-wolf conflicts. Through comprehensive research and consultation, suitable locations within the park were identified, ensuring the highest chance of successful reintroduction.

Monitoring and Outcomes

Monitoring the reintroduced wolf population and its impact on the ecosystem has been a vital aspect of the Yellowstone wolf project. Conservation biologists and ecologists have employed various techniques such as radio telemetry, DNA analysis, and aerial surveys to track and study wolf movements, population dynamics, and interactions with other species.

The outcomes of wolf reintroduction in Yellowstone have been remarkable. The reintroduced wolves adapted well to their new environment, establishing territories, forming packs, and successfully reproducing. Over time, the elk population has undergone significant changes, displaying altered behavior and distribution patterns in response to predation pressure from wolves.

The recovery of vegetation, particularly willows and aspen trees, has also been observed. This recovery has reinstated suitable habitats for beavers, songbirds, and fish species, enhancing overall ecosystem health and biodiversity. The presence of wolves has even affected the behavior of herbivores, indirectly benefiting small mammals and birds by reducing grazing pressure.

Lessons Learned and Future Implications

Wolf reintroduction in Yellowstone National Park not only restored a keystone species but also provided valuable lessons for conservation biology and ecosystem management. The success of this endeavor demonstrates the potential for ecological restoration and the importance of maintaining balanced trophic interactions.

The reintroduction of wolves also highlighted the need for collaboration and communication among diverse stakeholders. By involving local communities, ranchers, scientists, and policymakers, the Yellowstone wolf project fostered a shared understanding and support for wolf conservation efforts.

The story of wolf reintroduction in Yellowstone can serve as an inspiration for similar restoration projects worldwide. However, it is important to consider the unique ecological and social contexts of each restoration effort. Adaptive management strategies and ongoing monitoring will be crucial for ensuring the long-term success and sustainability of such projects.

In conclusion, the reintroduction of wolves in Yellowstone National Park has proven to be a triumph in environmental science and conservation biology. By understanding the role of wolves in maintaining ecosystem balance and addressing the concerns of stakeholders, this initiative has restored not only a keystone species but also the ecological dynamics of one of America's most iconic national parks.

Indigenous-led Conservation Initiatives

Indigenous-led conservation initiatives play a crucial role in the protection and preservation of biodiversity worldwide. Indigenous communities have valuable traditional knowledge and practices that have been passed down through generations, enabling them to effectively manage and sustain their land, water, and natural resources. These initiatives incorporate indigenous peoples' deep connection to the land, their holistic worldview, and their sustainable practices. In this section, we will explore the principles, challenges, and success stories of indigenous-led conservation initiatives.

Principles of Indigenous-led Conservation

Indigenous-led conservation initiatives are guided by several key principles that reflect the cultural values and practices of indigenous communities:

SUCCESS STORIES IN CONSERVATION BIOLOGY

1. **Respect for Traditional Knowledge:** Indigenous communities possess a wealth of traditional knowledge about their ecosystems, including detailed knowledge of local flora and fauna, ecological processes, and sustainable resource management. This knowledge is integrated into conservation efforts, ensuring the long-term viability of ecosystems.

2. **Holistic Approach:** Indigenous cultures often embrace a holistic worldview that recognizes the interconnectedness of all living beings. Indigenous-led conservation initiatives consider the ecological, cultural, and spiritual dimensions of biodiversity conservation, recognizing that human well-being is intimately linked to the health of ecosystems.

3. **Community-based Governance:** Indigenous-led conservation is grounded in local self-governance and decision-making processes, with communities actively participating in the management of their ancestral lands. This community-based approach promotes the empowerment and autonomy of indigenous peoples.

4. **Conservation Through Traditional Practices:** Indigenous peoples have developed sustainable practices and resource management systems over centuries. Indigenous-led conservation initiatives integrate traditional practices such as rotational farming, sustainable harvesting techniques, and prescribed burning to maintain ecological balance.

5. **Protection of Sacred Sites:** Many indigenous communities consider certain areas as sacred, embodying their cultural heritage and spiritual beliefs. Indigenous-led conservation initiatives prioritize the protection of these sites, recognizing their significance for biodiversity conservation and the preservation of cultural identity.

Challenges Faced by Indigenous-led Conservation Initiatives

Despite the significant contributions of indigenous-led conservation initiatives, they face numerous challenges that need to be addressed:

1. **Lack of Legal Recognition:** Indigenous communities often lack legal recognition of their land rights, making their territories vulnerable to encroachment and exploitation by external actors. Strengthening legal frameworks that recognize indigenous land rights is crucial for the success of these initiatives.

CHAPTER 3: CONSERVATION BIOLOGY AND WILDLIFE MANAGEMENT

2. **Lack of Resources and Capacity:** Indigenous communities usually have limited financial resources and capacity to address complex conservation challenges. Adequate funding, technical support, and capacity-building programs are needed to strengthen their conservation efforts and enhance their ability to respond to environmental threats.

3. **Conflicts with Development Initiatives:** Indigenous lands are often targeted for industrial development, such as mining, logging, and infrastructure projects. These activities can lead to habitat destruction, pollution, and displacement of indigenous peoples, undermining their conservation efforts. Balancing economic development with environmental and social sustainability is crucial in achieving harmonious coexistence.

4. **Marginalization and Discrimination:** Indigenous communities often face marginalization, discrimination, and the loss of cultural identity. Addressing these social injustices is essential for empowering indigenous peoples and ensuring their active and equal participation in conservation decision-making processes.

5. **Climate Change Impacts:** Indigenous communities are disproportionately affected by climate change due to their close relationship with the environment. Rising temperatures, changing rainfall patterns, and more frequent natural disasters pose significant challenges to indigenous-led conservation initiatives. Climate resilience strategies and support are essential to safeguard their ecosystems and traditional practices.

Success Stories of Indigenous-led Conservation

Indigenous-led conservation initiatives have achieved remarkable success in various regions across the globe. Here are a few notable examples:

1. **Guna Yala Indigenous Reserve, Panama:** The Guna people in Panama have created the Guna Yala Indigenous Reserve, a sustainable management model that balances cultural preservation and ecosystem protection. Their traditional practices, such as rotational fishing and agroforestry, have led to the recovery of fish populations and the restoration of degraded ecosystems.

2. **Namibian Community Conservancies, Namibia:** Indigenous communities in Namibia have established community conservancies, where they are actively engaged in wildlife conservation and eco-tourism. These conservancies have seen the recovery of endangered species, such as black

SUCCESS STORIES IN CONSERVATION BIOLOGY

rhinos and desert elephants, while generating economic benefits for the communities through responsible tourism.

3. **Jukajärvi Village, Finland:** The indigenous Sámi community in Jukajärvi, Finland, has implemented traditional reindeer herding practices that maintain the health of the Arctic tundra ecosystem. Their sustainable land use practices have contributed to the conservation of biodiversity, while preserving their cultural traditions and livelihoods.

Resources and Knowledge Exchange

Indigenous-led conservation initiatives can benefit greatly from knowledge exchange and collaboration with scientists, policymakers, and other stakeholders. Such partnerships can foster the integration of scientific and traditional knowledge systems, enabling a comprehensive understanding of ecosystems and effective conservation strategies. Platforms, such as indigenous-led conferences, community-based workshops, and participatory research, facilitate knowledge exchange while respecting indigenous protocols and intellectual rights.

Exercises

1. Research and identify an indigenous-led conservation initiative in your region or country. Describe their key conservation practices, challenges they face, and any success stories they have achieved.

2. Discuss the importance of culturally sensitive conservation approaches and the challenges of integrating traditional knowledge with scientific knowledge in biodiversity conservation.

3. Imagine you are working with an indigenous community to develop a conservation plan. Outline the steps you would take to ensure their full participation and respect for their traditional knowledge and practices.

Indigenous-led conservation initiatives are essential for preserving biodiversity, protecting ecosystems, and promoting sustainable development. By recognizing the rights and knowledge of indigenous peoples, we can forge stronger partnerships and create a more inclusive and effective approach to conservation. These initiatives serve as inspiring examples of how traditional wisdom and modern science can work hand in hand towards a greener future.

Marine Protected Areas and Biodiversity Conservation

Marine environments cover more than 70% of the Earth's surface and are home to a vast array of plant and animal species. However, these ecosystems are under threat from various human activities such as overfishing, pollution, and habitat destruction. To address these challenges and protect marine biodiversity, the establishment of Marine Protected Areas (MPAs) has become a crucial strategy in conservation biology and wildlife management.

Definition and Importance of Marine Protected Areas

A Marine Protected Area is a designated region in the ocean that is managed to achieve specific conservation objectives. These objectives may include the protection and recovery of vulnerable species, the preservation of important habitats, and the maintenance of ecological processes. MPAs can range in size and complexity, from small coastal reserves to large-scale marine parks covering vast oceanic areas.

MPAs play a crucial role in biodiversity conservation for several reasons. Firstly, they provide a refuge for marine species to reproduce, feed, and grow without human interference. By restricting certain activities such as fishing or mining, MPAs can allow ecosystems to recover and restore their natural functions. This enables the creation of "spillover effects" where organisms and larvae from protected areas migrate to surrounding areas, benefiting local fisheries and enhancing overall ecosystem health.

Secondly, MPAs contribute to the conservation of unique habitats such as coral reefs, seagrass meadows, and mangrove forests. These ecosystems are hotspots of biodiversity, supporting a wide variety of marine species and providing vital services such as carbon sequestration and coastal protection. By safeguarding these habitats, MPAs ensure the long-term survival of endangered and threatened species, protecting the overall stability of marine ecosystems.

Design and Management of Marine Protected Areas

The design and management of MPAs are critical factors in their effectiveness. Several key principles and strategies are employed to ensure the conservation goals of MPAs are achieved:

1. **Zoning and Spatial Planning:** MPAs often employ zoning techniques that establish different usage zones within the protected area. These may include strict no-take zones where all extractive activities are prohibited, buffer zones where limited activities are permitted, and zones for recreational use. Spatial planning

SUCCESS STORIES IN CONSERVATION BIOLOGY

ensures the optimal distribution of different zones to maximize conservation benefits.

2. **Collaborative Governance:** MPAs are typically managed through collaborative processes involving various stakeholders, including government bodies, local communities, scientists, and conservation organizations. This participatory approach ensures the integration of local knowledge and expertise, as well as the inclusion of diverse perspectives for effective decision-making and enforcement.

3. **Monitoring and Enforcement:** Regular monitoring is crucial to assess the effectiveness of MPAs and make informed management decisions. This includes monitoring the health and abundance of target species, as well as tracking the recovery of habitats. Effective enforcement, through patrols, surveillance systems, and legal instruments, is essential to prevent illegal activities within the protected areas.

4. **Education and Awareness:** Public outreach, educational programs, and community engagement are essential components of successful MPA management. By increasing public awareness about the importance of marine biodiversity and the role of MPAs, individuals are more likely to support and actively participate in conservation efforts.

Case Studies: Successful Marine Protected Areas

Several notable examples of successful MPAs demonstrate their effectiveness in conserving marine biodiversity:

1. **Great Barrier Reef Marine Park, Australia:** The Great Barrier Reef Marine Park is one of the largest and most iconic MPAs in the world. It covers an area of approximately 344,400 square kilometers and protects diverse coral reefs, seagrass meadows, and mangroves. The park's zoning system allows for a range of uses while minimizing ecological impacts, resulting in the preservation of critical habitats and the recovery of several threatened species.

2. **Cabrits National Park, Dominica:** Located in the Caribbean, Cabrits National Park encompasses both terrestrial and marine ecosystems. The marine portion of the park includes seagrass beds, coral reefs, and mangroves, serving as important habitats for numerous marine species. The establishment of a designated marine reserve within the park has led to the recovery of fish populations and the protection of vulnerable species such as sea turtles and nurse sharks.

3. **Chagos Marine Protected Area, Indian Ocean:** The Chagos Marine Protected Area is one of the largest fully protected marine reserves in the world.

CHAPTER 3: CONSERVATION BIOLOGY AND WILDLIFE MANAGEMENT

Covering an area of approximately 640,000 square kilometers, it safeguards coral reefs, seagrass beds, and extensive deep-sea ecosystems. The MPA has successfully protected the biodiverse and pristine marine habitats, ensuring the conservation of numerous rare and endangered species, including the critically endangered hawksbill turtle.

Challenges and Future Directions

While MPAs have made significant contributions to marine biodiversity conservation, they still face various challenges and require continuous improvement:

1. **Effectiveness and Connectivity:** Ensuring the effectiveness of MPAs requires addressing issues related to their size, connectivity, and design. Small and isolated MPAs may not provide sufficient protection for wide-ranging species or support the migration and dispersal of larvae. Enhancing the connectivity between MPAs and establishing networks of protected areas is essential for achieving more comprehensive conservation outcomes.

2. **Illegal Fishing and Enforcement:** Illegal, unreported, and unregulated (IUU) fishing remains a major threat to marine biodiversity and undermines the effectiveness of MPAs. Strengthening enforcement capabilities, increasing surveillance technologies, and promoting international cooperation are crucial for combating illegal fishing activities within MPAs.

3. **Climate Change Impacts:** Climate change poses significant challenges to MPAs, including rising sea temperatures, ocean acidification, and sea-level rise. These factors can negatively impact coral reefs, fish populations, and other sensitive marine habitats. Adapting MPA management strategies to address climate change impacts and integrating climate resilience into MPA design and planning are critical for long-term success.

4. **Socioeconomic Considerations:** Balancing conservation objectives with socioeconomic considerations is vital for the sustainability of MPAs. Engaging local communities, securing alternative livelihoods for those dependent on traditional fishing practices, and incorporating traditional ecological knowledge into management plans can help mitigate conflicts and enhance community support for MPAs.

In conclusion, Marine Protected Areas are instrumental in conserving marine biodiversity and ensuring the long-term sustainability of our oceans. By protecting critical habitats, supporting fish populations, and preserving vulnerable species, MPAs play a significant role in mitigating human impacts on marine ecosystems. However, addressing ongoing challenges, such as climate change and illegal fishing,

SUCCESS STORIES IN CONSERVATION BIOLOGY

and promoting effective management strategies will be key to the future success of MPAs and the conservation of marine biodiversity.

Conservation of Endangered Birds in New Zealand

New Zealand is well-known for its rich biodiversity and unique avian species. However, many of these bird species are facing severe threats and are on the verge of extinction. In this section, we will explore the conservation efforts in New Zealand aimed at protecting endangered birds and safeguarding their habitats.

Background

New Zealand is home to a diverse range of bird species, including the iconic Kiwi, Kakapo, and Takahe. These birds evolved in the absence of mammalian predators, leading to their flightlessness and vulnerability to introduced predators such as rats, stoats, and possums. Additionally, habitat loss and fragmentation due to deforestation and land development have further contributed to the decline of these bird populations.

Conservation Principles

To effectively conserve endangered bird species in New Zealand, several conservation principles are employed:

1. **Predator control:** Controlling introduced predators is crucial to protect the breeding grounds and nesting sites of endangered birds. This involves implementing trapping programs, poison bait stations, and predator-proof fencing to create predator-free sanctuaries.

2. **Habitat restoration:** Restoring and creating suitable habitats is essential for the survival of endangered birds. Efforts include reforestation, wetland restoration, and removal of invasive plant species that compete with native plants.

3. **Species reintroduction:** Certain bird species have disappeared from certain regions due to predation or habitat loss. Conservation programs involve relocating and reintroducing these species to areas where predators have been successfully controlled and habitat conditions have improved.

4. **Community engagement:** Engaging local communities, landowners, and iwi (Māori tribes) is crucial for the success of conservation efforts. Collaboration fosters support, shared knowledge, and stewardship of the land.

CHAPTER 3: CONSERVATION BIOLOGY AND WILDLIFE
MANAGEMENT

Case Study: The Takah☐ Conservation

The Takahē (Porphyrio hochstetteri) is a flightless bird endemic to New Zealand. It was believed to be extinct until its rediscovery in the 1940s. Today, the Takahē is one of the most endangered bird species in New Zealand, with a population of fewer than 400 individuals.

The Department of Conservation (DOC) in collaboration with Te Rūnanga o Ngāi Tahu, a Māori tribal organization, has implemented a comprehensive conservation program for the Takahē. The key aspects of the program include:

1. **Predator control:** Intensive predator control measures, such as trapping and aerial pest control, have been implemented in key Takahē habitats, such as Fiordland and Murchison Mountains. These efforts have significantly reduced stoat and rat populations, allowing the Takahē to thrive.

2. **Habitat restoration:** Wetland restoration projects have been initiated to provide suitable breeding and feeding habitats for the Takahē. These projects involve removing invasive plant species and restoring wetland hydrology to create ideal conditions for the growth of native plants.

3. **Breeding and monitoring:** The DOC operates captive-breeding centers to ensure the survival and genetic diversity of the Takahē population. Chicks are hatched in captivity and later transferred to predator-free islands or carefully managed mainland sites. Regular population monitoring is carried out to assess the success of conservation efforts.

Successes and Challenges

The conservation efforts for endangered birds in New Zealand have yielded some notable successes. The efforts to protect the Chatham Island Black Robin (Petroica traversi) from extinction in the 1980s are a prime example of successful conservation. Through captive breeding and predator control, the population of this bird has increased from a mere five individuals to over 200.

However, there are also challenges in conserving endangered birds in New Zealand. Some of the key challenges include:

1. **Lack of funding:** Conservation programs require substantial financial resources to implement and sustain. Limited funding poses a challenge to scaling up conservation efforts and addressing the needs of multiple endangered bird species.

SUCCESS STORIES IN CONSERVATION BIOLOGY

2. **Climate change:** The effects of climate change, such as extreme weather events and altered ecosystems, pose additional challenges to the survival of endangered bird species. Rising sea levels and increased temperatures may impact nesting sites and disrupt food availability.

3. **Genetic diversity:** Small populations of endangered birds can suffer from reduced genetic diversity, making them more vulnerable to diseases and environmental changes. Maintaining genetic diversity through careful breeding programs is crucial for their long-term survival.

Resources and Further Reading

If you are interested in learning more about the conservation of endangered birds in New Zealand, here are some recommended resources:

1. *Birdwatching in New Zealand* by Derek Onley and Paul Scofield

2. *Protecting Paradise: Conservation in New Zealand* by Colin Miskelly

3. *Kakapo: Rescued from the Brink of Extinction* by Alison Ballance

4. *Tētēkura: The Chatham Island Black Robin* by David Butler and Don Merton

5. Department of Conservation website: https://www.doc.govt.nz/

Conclusion

The conservation of endangered birds in New Zealand is a complex and challenging endeavor. Through predator control, habitat restoration, and community engagement, significant progress has been made in protecting these unique bird species. However, ongoing efforts are required to overcome funding limitations, address the impacts of climate change, and ensure the long-term survival of these remarkable birds. By raising awareness and supporting conservation initiatives, we can contribute to the preservation of New Zealand's avian treasures for future generations.

Chapter 5: Renewable Energy and Green Technologies

Renewable energy and green technologies have become increasingly important in the global effort to mitigate climate change and reduce dependence on fossil fuels. In this chapter, we will explore the need for renewable energy, different renewable energy technologies, their successes and challenges, and the future directions in this field.

The Need for Renewable Energy

The burning of fossil fuels, such as coal, oil, and natural gas, for energy production is a major contributor to greenhouse gas emissions and climate change. To tackle this problem, the transition to renewable energy sources is necessary. Renewable energy offers several benefits:

- **Reduced greenhouse gas emissions:** Renewable energy sources produce little to no carbon dioxide or other harmful greenhouse gas emissions when generating electricity. This helps combat climate change and reduce air pollution.

- **Energy security and independence:** Unlike fossil fuels, which are finite resources, renewable energy sources are abundant and widely distributed. Investing in renewable energy reduces reliance on imported fuels and increases energy security.

- **Job creation and economic growth:** The renewable energy sector has the potential to create new employment opportunities and stimulate economic growth. Developing and implementing renewable energy technologies can lead to a green economy.

Renewable Energy Technologies

There are various renewable energy technologies that harness the power of naturally replenishing resources. Let's explore some of the key renewable energy technologies:

1. **Solar Energy and Photovoltaics:** Solar energy is obtained from the sun's radiation. Photovoltaic (PV) systems convert sunlight directly into electricity, utilizing semiconductor materials. Solar thermal systems use the sun's heat to generate electricity or produce hot water for heating.

SUCCESS STORIES IN CONSERVATION BIOLOGY

2. **Wind Power and Wind Turbines:** Wind power harnesses the kinetic energy of the wind to generate electricity. Wind turbines convert the wind's motion into rotational energy, which is then converted into electrical energy using a generator.

3. **Hydroelectric Power and Tidal Energy:** Hydroelectric power utilizes the energy of flowing or falling water to generate electricity. It is obtained through the construction of dams, which create reservoirs of water. Tidal energy, on the other hand, harnesses the power of tidal currents and waves.

4. **Geothermal Energy and Biogas:** Geothermal energy harnesses heat stored beneath the Earth's surface to generate electricity or provide heating and cooling. Biogas is produced from organic waste materials, such as agricultural residues and animal manure, through a process called anaerobic digestion. It can be used for heating, electricity generation, and as a transportation fuel.

5. **Biomass and Bioenergy:** Biomass refers to organic matter derived from plants and animals. It can be used to produce heat, electricity, and transportation fuels through processes like combustion, gasification, and anaerobic digestion.

Case Study: Costa Rica's Renewable Energy Revolution

Costa Rica is renowned for its commitment to renewable energy and sustainable development. In recent years, Costa Rica has achieved significant milestones in the transition to renewable energy sources. Here are some key aspects of their success:

1. **Hydropower Dominance:** Costa Rica heavily relies on hydropower for electricity generation. Its mountainous terrain and abundant rivers provide the perfect conditions for hydropower plants.

2. **Geothermal Power Expansion:** Costa Rica has utilized its volcanic activity to tap into geothermal energy. Geothermal power plants produce a substantial amount of electricity and are considered reliable and renewable.

3. **Investment in Wind and Solar:** Costa Rica has been investing in wind and solar energy projects to diversify its energy mix. Wind farms and solar installations have become increasingly common across the country.

CHAPTER 3: CONSERVATION BIOLOGY AND WILDLIFE MANAGEMENT

4. **Government Policies and Incentives:** The Costa Rican government has implemented favorable policies and incentives to encourage renewable energy development. This includes feed-in tariffs, tax exemptions, and streamlined permitting processes.

Consequently, Costa Rica has achieved impressive feats by relying on renewable energy sources. In some instances, the country has been able to produce 100% of its electricity from renewables for extended periods.

Success Stories in Renewable Energy

In addition to Costa Rica's renewable energy revolution, here are some other notable success stories in renewable energy:

1. **Germany's Energy Transition:** Germany, also known as the "Energiewende," has made significant progress in transitioning from nuclear and fossil fuel-based energy to renewables. The country has implemented robust policies and incentives to support the expansion of wind, solar, and biomass energy.

2. **Solar Energy in India:** India has emerged as one of the leading countries in solar energy adoption. The government's initiatives, such as the Jawaharlal Nehru National Solar Mission, have facilitated the rapid growth of solar power installations, making India one of the top solar energy producers globally.

3. **Offshore Wind Farms in Europe:** Several countries in Europe, including the United Kingdom, Denmark, and the Netherlands, have made significant investments in offshore wind farms. These wind farms utilize the strong and consistent winds at sea to produce clean and renewable electricity.

4. **Hydropower in Brazil:** Brazil has a long-standing history of hydropower development. It has developed large-scale hydropower projects, such as the Itaipu Dam, which is one of the world's largest hydropower plants. Hydropower contributes significantly to Brazil's energy mix.

5. **Biogas Production and Utilization in Sweden:** Sweden has implemented comprehensive biogas production and utilization programs. The country utilizes organic waste materials, such as sewage sludge and food waste, to produce biogas for heating, electricity, and transportation. Sweden's biogas initiatives have achieved significant success and reduced greenhouse gas emissions.

SUCCESS STORIES IN CONSERVATION BIOLOGY   215

Challenges and Future Directions

While renewable energy has experienced remarkable growth in recent years, there are still challenges that need to be addressed:

1. **Energy Storage Technologies and Grid Integration:** The intermittent nature of renewable energy sources, such as solar and wind, requires efficient energy storage technologies to bridge the gaps in energy supply. Moreover, integrating renewable energy into existing power grids poses technical and logistical challenges.

2. **Policy and Regulatory Support:** Governments need to create and implement supportive policies and regulations to encourage renewable energy development. This includes feed-in tariffs, tax incentives, and renewable energy targets.

3. **Addressing Intermittency and Variability:** Renewable energy sources are subject to natural variability and intermittency, depending on weather conditions. Developing advanced forecasting models and technologies to manage variability is essential for the stable integration of renewable energy into the grid.

4. **Renewable Energy for Industrial Processes:** Finding renewable energy alternatives for industrial processes, which often rely on fossil fuels, is a significant challenge. Innovative technologies, such as hydrogen fuel cells and sustainable biofuels, show promise in addressing this challenge.

5. **Advancements in Green Technologies and Innovations:** Continued research and development are crucial to unlocking the full potential of renewable energy technologies. This includes improving the efficiency of solar panels, developing new wind turbine designs, and exploring emerging technologies like wave energy and hydrogen power.

Conclusion

Renewable energy and green technologies are essential for achieving a sustainable and low-carbon future. By transitioning to renewable energy sources, we can mitigate the impacts of climate change, reduce pollution, and foster economic growth. However, addressing the challenges associated with renewable energy requires collective efforts from governments, industry, and individuals. By investing in research, supporting policy frameworks, and embracing technological

Challenges and Future Directions in Conservation Biology

Human-Wildlife Conflicts

Human-wildlife conflicts arise when there is a clash between human activities and the conservation of wildlife. As human populations expand and encroach upon natural habitats, the competition for resources and space between humans and wildlife intensifies. These conflicts can have serious consequences for both humans and wildlife, including economic losses, damage to property, and threats to human safety and wildlife populations. In this section, we will explore the causes and impacts of human-wildlife conflicts, as well as strategies for mitigating these conflicts.

Causes of Human-Wildlife Conflicts

Human-wildlife conflicts can stem from various factors, including habitat loss, resource scarcity, and human activities. The conversion of natural habitats for agriculture, urbanization, and infrastructure development reduces available habitat for wildlife, forcing them to seek food and shelter in human-dominated areas. As wildlife populations decrease, their search for resources also becomes more intense and leads them to come into direct contact with human settlements.

Resource scarcity is another major driver of human-wildlife conflicts. Competition for limited resources such as water, food, and grazing land can lead to conflicts between humans and wildlife, particularly in areas where water and food sources are scarce.

Human activities, such as hunting, poaching, and illegal wildlife trade, can disrupt ecosystems and exacerbate human-wildlife conflicts. For example, the illegal trade of elephant ivory increases the demand for tusks, leading to increased poaching and conflicts between elephants and local communities.

Impacts of Human-Wildlife Conflicts

Human-wildlife conflicts can have significant impacts on both human and wildlife populations. For humans, these conflicts can result in economic losses, especially for farmers whose crops are damaged by wildlife. Livestock predation by large

CHALLENGES AND FUTURE DIRECTIONS IN CONSERVATION BIOLOGY

carnivores, such as lions or wolves, can also lead to significant economic losses for pastoral communities.

In addition to economic losses, human-wildlife conflicts can pose threats to human safety. Attacks by wildlife, especially large predators, can result in injuries or fatalities. This can generate fear and resentment towards wildlife, leading to retaliation and further conflicts.

For wildlife populations, human-wildlife conflicts can have serious implications. Direct killing of wildlife in response to conflicts can lead to the decline of populations, especially for threatened or endangered species. Habitat fragmentation and disruption of animal behaviors, such as migration patterns, can also result from conflicts, further impacting wildlife populations.

Strategies for Mitigating Human-Wildlife Conflicts

Mitigating human-wildlife conflicts requires a multi-faceted approach that considers the needs and perspectives of both humans and wildlife. Here are some strategies that can help reduce conflicts and promote coexistence:

1. **Fencing and barriers:** Physical barriers, such as fences or trenches, can be effective in preventing wildlife from entering human settlements or agricultural areas. Electric fences have been successful in protecting crops and livestock from large herbivores and carnivores.

2. **Land-use planning:** Strategic planning of human activities and wildlife conservation can help minimize conflicts. By designating protected areas and wildlife corridors, human settlements can be separated from crucial wildlife habitats, reducing the likelihood of encounters. Proper zoning and regulation of land use can also help prevent habitat destruction and fragmentation.

3. **Alternative livelihoods:** Providing alternative sources of income for communities living near wildlife habitats can reduce pressure on resources and decrease the need for direct interaction with wildlife. Initiatives such as ecotourism, sustainable agriculture, and community-based conservation projects can provide economic benefits while promoting positive attitudes towards wildlife.

4. **Conflict resolution and compensation programs:** Establishing mechanisms for conflict resolution and compensation can help address conflicts when they occur. This can involve providing financial compensation for crop or livestock

CHAPTER 3: CONSERVATION BIOLOGY AND WILDLIFE MANAGEMENT

losses, as well as implementing protocols for reporting and addressing wildlife attacks.

5. **Education and awareness:** Public education and awareness campaigns can play a crucial role in fostering understanding and tolerance towards wildlife. By raising awareness about the importance of biodiversity and the conservation of wildlife, communities can develop a sense of ownership and responsibility in managing conflicts.

Example: Human-Elephant Conflicts in India

One prominent example of human-wildlife conflict is the ongoing conflicts between humans and elephants in India. As human settlements continue to expand into elephant habitats, encounters between elephants and humans have become increasingly frequent. The destruction of crops by elephants often leads to economic losses for farmers, exacerbating the conflict.

In response to this issue, several initiatives have been implemented to mitigate human-elephant conflicts. One successful strategy involves the use of trained elephants and their mahouts (handlers) to drive away wild elephants from human settlements. This humane approach allows for the safe removal of elephants without causing harm or stress to the animals. In addition, the creation of elephant corridors, which are strips of land connecting fragmented elephant habitats, has helped reduce conflicts by providing safe passage for elephants to roam and access resources.

These efforts highlight the importance of innovative and community-driven solutions in addressing human-wildlife conflicts. By involving local communities and considering their perspectives, conservationists can develop more effective strategies that promote coexistence between humans and wildlife.

Conclusion

Human-wildlife conflicts are complex issues that require a holistic and interdisciplinary approach. By understanding the causes and impacts of these conflicts, as well as implementing strategies for mitigation and coexistence, we can work towards a future where humans and wildlife can thrive together. Through effective land-use planning, alternative livelihood options, and education, we can reduce conflicts, protect wildlife populations, and ensure the conservation of our natural heritage.

CHALLENGES AND FUTURE DIRECTIONS IN CONSERVATION BIOLOGY

Illegal Wildlife Trade

Illegal wildlife trade is a significant environmental issue that poses a grave threat to biodiversity and the survival of many species. It encompasses the illicit buying, selling, and transportation of wildlife and wildlife products, including live animals, animal parts, and derivatives. This underground market is driven by a demand for exotic pets, traditional medicine, fashion accessories, and trophies, among other things. The illegal wildlife trade is a highly profitable criminal enterprise that is estimated to be worth billions of dollars annually.

The Threat to Biodiversity

The illegal wildlife trade is one of the major drivers of species extinction and habitat destruction. It affects a wide range of plants and animals, including mammals, birds, reptiles, amphibians, and marine species. Many of these species are already highly endangered due to habitat loss and climate change, making them especially vulnerable to exploitation.

The trade not only depletes populations of target species but also disrupts ecosystems and threatens the delicate balance of biodiversity. For example, the poaching of keystone species, such as elephants and tigers, can have cascading effects on the entire ecosystem, leading to the loss of other species and destabilizing ecological processes.

Causes and Actors of the Illegal Wildlife Trade

The illegal wildlife trade thrives due to a combination of factors, including poverty, weak law enforcement, corruption, and a lack of awareness and regulations. It involves a network of organized criminal syndicates, local poachers, traders, and middlemen who exploit both national and international markets.

In some cases, the trade is driven by cultural beliefs and traditional practices that assign medicinal or symbolic value to certain animal parts. For example, rhino horns are erroneously believed to have medicinal properties in some cultures, leading to a demand that drives poaching.

Similarly, the demand for exotic pets drives the smuggling of live animals, often resulting in poor welfare conditions and high mortality rates. Reptiles, birds, and primates are among the most commonly trafficked animals in the illegal wildlife trade.

Impacts on Local Communities

The illegal wildlife trade not only threatens biodiversity but also has significant socio-economic impacts on local communities. In many developing countries, poaching and illegal wildlife trade provide a source of income for impoverished communities, who are often exploited by criminal syndicates. However, this income is often short-lived and unsustainable, leading to increased poverty and reliance on unsustainable practices.

Furthermore, the loss of wildlife can disrupt local ecosystems, impacting the availability of resources such as clean water, food, and traditional livelihoods like fishing and agriculture. This, in turn, can exacerbate poverty and lead to social unrest.

Addressing the Illegal Wildlife Trade

Efforts to combat the illegal wildlife trade require a multi-faceted approach that involves international cooperation, improved law enforcement, community engagement, and demand reduction strategies.

Stronger legislation and enforcement, along with forensic techniques and the use of advanced technologies, can help disrupt trafficking networks and bring wildlife criminals to justice. International cooperation and collaboration between countries are crucial to tackling the transnational aspects of the trade.

Engaging local communities in conservation initiatives, providing alternative livelihoods, and promoting sustainable development can help address the root causes of the illegal wildlife trade. Raising awareness and education about the importance of biodiversity and the negative impacts of the trade are also essential for changing consumer behavior and reducing demand.

Success Stories in Combating Illegal Wildlife Trade

There have been several notable success stories in combating the illegal wildlife trade. One example is the conservation efforts to protect African elephants from poaching for their ivory. International bans on ivory trade and increased enforcement have resulted in a significant decline in elephant poaching in recent years.

Another success story is the conservation of the black rhinoceros in Namibia, where community involvement, improved law enforcement, and innovative conservation strategies have helped stabilize the population, despite ongoing challenges.

Furthermore, organizations like the Wildlife Conservation Society and the International Fund for Animal Welfare have played a significant role in raising

CHALLENGES AND FUTURE DIRECTIONS IN CONSERVATION BIOLOGY

awareness, conducting research, and supporting on-the-ground conservation efforts to combat the illegal wildlife trade.

Challenges and Future Directions

Despite these successes, the illegal wildlife trade continues to present significant challenges. The trade continues to evolve, with criminals adapting to new tactics and exploiting regulatory loopholes. Greater cooperation between countries, improved intelligence sharing, and increased investment in law enforcement are required to keep pace with these changes.

Furthermore, addressing the root causes of the trade, such as poverty and weak governance, remains a complex task. Sustainable development initiatives that promote economic opportunities while preserving biodiversity are crucial for long-term success.

Additionally, innovative approaches, such as the use of technology for monitoring and tracking, and engaging local communities as partners in conservation efforts, hold promise for combating the illegal wildlife trade.

In conclusion, the illegal wildlife trade poses a serious threat to biodiversity and ecosystems worldwide. It is driven by various factors, including cultural beliefs, poverty, and weak law enforcement. Addressing this issue requires a comprehensive approach that involves international collaboration, improved legislation, community engagement, and demand reduction strategies. While there have been successes in combating the trade, ongoing challenges and the evolving nature of the problem demand continued efforts and innovative solutions. By working together, we can help protect our wildlife and ensure a sustainable future for generations to come.

CHAPTER 3: CONSERVATION BIOLOGY AND WILDLIFE MANAGEMENT

Habitat Fragmentation and Loss

Habitat fragmentation and loss are two of the greatest challenges facing biodiversity conservation today. As human populations continue to expand and develop, natural habitats are being converted into urban areas, agricultural fields, roads, and other human-dominated landscapes. This process disrupts and fragments ecosystems, leading to the loss of critical habitats for many species.

Causes of Habitat Fragmentation and Loss

There are several driving forces behind habitat fragmentation and loss. One of the primary causes is urbanization, as cities and towns expand, they consume surrounding natural areas. This development often results in the creation of fragmented patches of habitat, separated by roads, buildings, and other infrastructure. Agricultural expansion is another significant driver, as forests and grasslands are cleared to make way for crops and livestock.

Infrastructure development, such as roads, highways, and dams, also plays a role in habitat fragmentation. These structures can act as barriers, physically preventing the movement of species and fragmenting habitats. Additionally, habitat loss can occur due to logging, mining, and other extractive activities that remove or alter natural vegetation.

Impacts of Habitat Fragmentation and Loss

Habitat fragmentation and loss have severe consequences for biodiversity. Smaller, isolated habitat patches support fewer species and are more prone to population declines and local extinctions. Fragmentation disrupts ecological processes by reducing gene flow, altering nutrient cycling, and changing species interactions.

Species that require large continuous habitats, such as large carnivores or migratory birds, are particularly vulnerable to habitat fragmentation. These species often face barriers, such as roads or agricultural fields, which impede their movement and access to resources. As a result, their populations become fragmented, isolated, and more susceptible to the negative effects of small population sizes, such as inbreeding and reduced genetic diversity.

Moreover, habitat loss can lead to the loss of critical resources, such as food and nesting sites, which many species depend on for survival and reproduction. This can have cascading effects on entire ecosystems, as the loss of one species can disrupt the intricate web of interactions between species.

CHALLENGES AND FUTURE DIRECTIONS IN CONSERVATION
BIOLOGY

Mitigation Strategies for Habitat Fragmentation and Loss

Addressing habitat fragmentation and loss requires a multi-faceted approach that combines conservation, land-use planning, and sustainable development practices. Here are some mitigation strategies that can help alleviate the impacts of habitat fragmentation and loss:

1. **Nature reserves and protected areas:** Establishing protected areas can help safeguard large intact habitats and provide refuge for vulnerable species. These areas serve as stepping stones for species to move between fragmented habitats and maintain connectivity.

2. **Habitat corridors:** Creating corridors that connect fragmented patches of habitat allows species to move between them more easily. These corridors can be in the form of vegetated strips along roads, above or below ground passages, or riverine corridors.

3. **Green infrastructure planning:** Incorporating ecological considerations into urban planning and infrastructure development can help minimize habitat fragmentation. Designing green spaces, such as parks and green roofs, can provide habitats and connectivity for urban-dwelling species.

4. **Land-use planning and zoning:** Implementing land-use plans and zoning regulations that prioritize the conservation of natural habitats can help limit further fragmentation. This involves designating areas for protected habitats, agricultural lands, and urban development.

5. **Agroforestry and habitat restoration:** Promoting agroforestry practices, such as planting trees on agricultural lands, can help restore some of the lost habitat connectivity. Habitat restoration efforts, such as reforestation and wetland restoration, can also contribute to reducing habitat fragmentation.

6. **Collaboration and partnerships:** Collaboration between stakeholders, including governments, conservation organizations, private landowners, and local communities, is crucial for effective habitat conservation and restoration. Partnerships can help leverage resources and expertise for implementing conservation actions.

Case Study: Amazon Rainforest Fragmentation

The Amazon rainforest, one of the most biodiverse regions on Earth, has been significantly affected by habitat fragmentation and loss. The expansion of

CHAPTER 3: CONSERVATION BIOLOGY AND WILDLIFE MANAGEMENT

agriculture, primarily for cattle ranching and soybean production, has led to extensive deforestation and fragmentation of the forest. Roads and rivers also fragment the Amazon, creating barriers to wildlife movement.

The impacts of habitat fragmentation in the Amazon are far-reaching. Many large mammal species, such as jaguars and tapirs, require large continuous habitats to thrive. However, their populations have become increasingly isolated, leading to genetic isolation and reduced population viability. Furthermore, the loss of habitat has resulted in the decline of many bird and amphibian species, impacting the functioning of the ecosystem.

To address the issue of habitat fragmentation in the Amazon, conservation organizations and governments are working together to establish protected areas, create biological corridors, and promote sustainable land-use practices. By increasing connectivity between forest fragments and protecting intact habitats, these efforts aim to mitigate the negative impacts of habitat fragmentation on biodiversity.

Future Directions

As the global population continues to grow, the pressures on habitats will only increase. It is essential to prioritize the conservation and restoration of habitats to prevent further fragmentation and loss. Additionally, the incorporation of ecological considerations into land-use planning and infrastructure development will be crucial for maintaining connectivity between habitats.

Advancements in remote sensing technologies, such as satellite imagery and drones, can aid in mapping fragmented habitats and monitoring the effectiveness of conservation efforts. Genetic techniques, such as landscape genomics, can provide insights into the impacts of habitat fragmentation on population connectivity and genetic diversity.

Education and awareness programs can also play a vital role in promoting public engagement and understanding of the importance of healthy and connected habitats. By raising awareness about the impacts of habitat fragmentation and loss, individuals can contribute to conservation efforts and help shape policies and practices that prioritize habitat protection.

In conclusion, habitat fragmentation and loss pose significant challenges to biodiversity conservation. Addressing these issues requires a holistic approach that combines protected areas, habitat corridors, sustainable land-use practices, and collaborations among various stakeholders. By implementing these strategies and investing in the conservation and restoration of habitats, we can work towards

CHALLENGES AND FUTURE DIRECTIONS IN CONSERVATION BIOLOGY

minimizing the impacts of habitat fragmentation and preserving our planet's rich biodiversity.

Climate Change and Species Conservation

Climate change is one of the greatest challenges facing our planet today. It is caused by the increasing concentration of greenhouse gases in the Earth's atmosphere, primarily carbon dioxide (CO_2) from the burning of fossil fuels, deforestation, and other human activities. This rise in greenhouse gas levels leads to global warming, resulting in a range of impacts on ecosystems and species.

Impacts of Climate Change on Species

The effects of climate change are already being observed worldwide, with significant impacts on biodiversity and species conservation. Rising temperatures, altered precipitation patterns, and changing seasonal cycles disrupt ecosystems and challenge the ability of species to survive.

Habitat Loss and Range Shifts As temperatures rise, many species are facing habitat loss and fragmentation. Some habitats, such as polar ice caps and high-altitude areas, are particularly vulnerable to warming temperatures. As a result, species that depend on these habitats, such as polar bears and mountain-dwelling plants, face significant risks to their survival.

Additionally, climate change is causing a shift in the geographical ranges of many species. As suitable habitats and conditions for certain species change, they may need to move to new areas. This can lead to competition with other species, loss of genetic diversity, and reduced access to food and resources.

Altered Ecosystem Interactions Climate change also disrupts the intricate web of interactions between species within ecosystems. For example, changes in temperature and precipitation can impact the timing of flowering and plant-pollinator interactions. If the flowering period of a plant shifts out of sync with the arrival of its pollinators, it can result in reduced pollination rates and decreased reproductive success.

Similarly, changes in ocean temperatures and acidity levels can disrupt marine food webs, affecting the availability of prey for predators. This can have cascading effects throughout the ecosystem, impacting the abundance and distribution of species at multiple trophic levels.

CHAPTER 3: CONSERVATION BIOLOGY AND WILDLIFE MANAGEMENT

Increased Risk of Extinction   The combined impacts of habitat loss, range shifts, and disrupted ecosystem interactions make many species more vulnerable to extinction. Climate change acts as an additional stressor that can push already threatened species over the edge.

For example, coral reefs are highly sensitive to even slight increases in water temperatures. Coral bleaching, a phenomenon caused by warmer oceans, can lead to the death of coral colonies and the subsequent loss of habitat for countless marine species. Similarly, species that rely on specific temperatures for breeding or food availability, such as migratory birds or insects, may struggle to adapt to rapidly changing conditions.

Mitigation and Adaptation Strategies

To address the challenges posed by climate change on species conservation, a combination of mitigation and adaptation strategies is needed.

Mitigation   Mitigation involves reducing greenhouse gas emissions to limit the magnitude of climate change. This can be achieved through various measures, including:

- Transitioning to renewable energy sources, such as solar and wind power, to reduce reliance on fossil fuels.

- Improving energy efficiency in transportation, buildings, and industries to reduce carbon emissions.

- Protecting and restoring forests, which act as carbon sinks by absorbing CO_2 from the atmosphere.

By mitigating climate change, we can help reduce the severity of its impacts on species and ecosystems.

Adaptation   Adaptation strategies aim to assist species and ecosystems in adjusting to the changing climate. This may involve:

- Conserving and restoring critical habitats to ensure the long-term survival of species.

- Creating corridors and connectivity between fragmented habitats to allow for the movement of species.

CHALLENGES AND FUTURE DIRECTIONS IN CONSERVATION BIOLOGY

- Assisting species with limited mobility, such as plants or sedentary animals, by establishing protected areas or translocation programs to more suitable habitats.

- Implementing managed relocation, also known as assisted migration, for species that are unable to adapt quickly enough to changing conditions.

- Promoting genetic diversity within species to enhance their resilience to changing environments.

By implementing both mitigation and adaptation strategies, we can reduce the risks faced by species and ecosystems and improve their chances of survival in a changing climate.

Real-World Examples

Numerous conservation initiatives around the world are addressing the impacts of climate change on species. Here are a few real-world examples of successful projects:

Polar Bear Conservation Polar bears are directly impacted by the loss of Arctic sea ice caused by warming temperatures. To protect this iconic species, conservation efforts focus on reducing greenhouse gas emissions through increased energy efficiency and renewable energy sources. Additionally, initiatives are in place to protect key polar bear habitats and regulate activities that may further disrupt their way of life.

Assisted Migration of Mountain Species In some cases, conservationists have resorted to managed relocation to assist species that cannot migrate on their own to more suitable habitats. For example, in the mountains of Europe, where temperatures have risen rapidly, the Alpine ibex has been successfully translocated to higher altitudes to adapt to changing climatic conditions.

Coral Reef Restoration Efforts to restore coral reefs involve interventions to improve their resilience to climate change. This may include techniques such as coral gardening, where fragments of resilient coral species are grown and transplanted onto damaged reefs. Additionally, reducing local stressors like pollution and unsustainable fishing practices helps preserve the health of coral reefs and increase their chances of survival in a changing ocean environment.

Challenges and Future Directions

Despite the progress made in addressing climate change's impacts on species conservation, numerous challenges persist. These challenges include:

Lack of Political Will Climate change is a global issue that requires international cooperation and commitment. The lack of political will and global consensus on climate action poses a significant challenge to comprehensive and effective responses.

Uncertainties in Predicting Species Responses The complex interactions between climate and species make it challenging to predict how individual species will respond to climate change. Uncertainties in modeling future scenarios and assessing the resilience of species can hinder informed decision-making.

Limited Financial Resources Implementing mitigation and adaptation strategies requires significant financial resources. Limited funding can impede the scaling up of conservation efforts and hinder the protection of vulnerable species.

Socioeconomic Implications Addressing climate change and species conservation can have significant socioeconomic implications. Balancing the needs of local communities, economic development, and conservation objectives requires careful planning and stakeholder engagement.

To overcome these challenges and advance species conservation in the face of climate change, future directions should focus on:

- Increasing public awareness and education about the impacts of climate change on species and the need for conservation action.

- Strengthening international collaboration and agreements to ensure effective climate change mitigation and adaptation measures.

- Expanding research efforts to better understand the interactions between climate change and species, improving predictive models, and identifying effective conservation strategies.

- Integrating climate considerations into existing conservation plans and policies at regional and national levels.

- Engaging local communities and indigenous peoples in conservation decision-making processes, recognizing their traditional knowledge and practices as valuable contributions to climate change adaptation.

CHALLENGES AND FUTURE DIRECTIONS IN CONSERVATION BIOLOGY 229

By addressing these challenges and pursuing these future directions, we can better protect and conserve species in the face of climate change, ensuring their long-term survival in a rapidly changing world.

Exercises

1. Research a species that is being impacted by climate change. Describe its specific vulnerabilities and the conservation efforts in place to protect it.

2. Imagine you are developing an adaptation strategy for a species at risk due to climate change. Outline the key steps you would take to ensure its survival in a changing environment.

3. Debate the ethical implications of assisted migration as a conservation strategy to address climate change impacts on species. Consider the potential risks and benefits associated with this approach.

4. Investigate the role of citizen science in monitoring and understanding the impacts of climate change on species. Describe a citizen science project focused on species conservation in the context of climate change.

5. Research and discuss a case study where local communities have successfully participated in climate change adaptation and species conservation efforts. Highlight the key factors that contributed to the success of the project.

Further Reading

- "Climate Change and Biodiversity" by Thomas E. Lovejoy and Lee Hannah
- "Biodiversity and Climate Change: Transforming the Biosphere" by Thomas E. Lovejoy and Lee Hannah
- "The Sixth Extinction: An Unnatural History" by Elizabeth Kolbert
- "Climate Change Biology" by Lee Hannah

Engaging Local Communities in Conservation Efforts

Engaging local communities in conservation efforts is crucial for achieving effective and sustainable results in conservation biology. Local communities often have a direct impact on the natural resources and wildlife within their areas, and their active involvement can lead to better understanding, support, and long-term success in conservation initiatives. In this section, we will explore various strategies and approaches to engage local communities in conservation efforts.

CHAPTER 3: CONSERVATION BIOLOGY AND WILDLIFE MANAGEMENT

Importance of Local Community Engagement

Conservation efforts that fail to involve local communities often face challenges and limitations. Local communities possess valuable traditional knowledge about the ecosystems and species in their areas. They are also directly affected by the consequences of environmental degradation, such as the loss of ecosystem services and decreased livelihood opportunities.

Engaging local communities in conservation efforts can:

+ Increase awareness and understanding of the value of biodiversity and the importance of conservation.

+ Foster a sense of ownership and responsibility for natural resources and wildlife.

+ Enhance the chances of successful implementation by integrating local knowledge and practices.

+ Improve conservation outcomes by establishing mutually beneficial partnerships and collaborations.

+ Promote sustainable and alternative livelihood options that alleviate poverty and reduce pressure on natural resources.

Approaches to Engaging Local Communities

There are several approaches that can be employed to engage local communities in conservation efforts:

1. **Community-based conservation:** This approach involves actively involving local communities in decision-making processes and empowering them to take ownership of conservation initiatives. It recognizes the rights and interests of local communities and promotes collaborative management of natural resources. This can be achieved through community-based organizations, participatory mapping exercises, and the establishment of community-managed protected areas.

2. **Environmental education and awareness:** Education and awareness programs play a critical role in engaging local communities. These programs can be conducted through schools, community centers, and workshops, and aim to enhance knowledge and understanding of environmental issues, conservation practices, and sustainable development. They can also

CHALLENGES AND FUTURE DIRECTIONS IN CONSERVATION BIOLOGY

highlight the economic and social benefits that conservation efforts can bring to local communities.

3. **Capacity building and training:** Building the capacity of local communities enables them to actively participate and contribute to conservation efforts. This can involve training programs on sustainable resource management, wildlife monitoring techniques, sustainable agriculture practices, and eco-tourism. Capacity building programs empower local communities with the skills and knowledge necessary to actively engage in conservation activities.

4. **Partnerships and collaborations:** Collaboration between local communities, governmental and non-governmental organizations, and researchers is essential for successful conservation. By forming partnerships, different stakeholders can pool their resources, knowledge, and expertise to develop effective conservation strategies. These partnerships can involve joint research projects, sustainable development initiatives, and sharing of best practices.

5. **Incentives and alternative livelihoods:** To encourage local community engagement, it is important to provide incentives and alternative livelihood options that are sustainable and economically viable. This can include eco-tourism ventures, sustainable agriculture programs, handicraft production, and cultural tourism initiatives. By providing economic benefits, local communities are more likely to actively participate in conservation efforts and become champions of the environment.

Case Studies

Let's explore two case studies that highlight the successful engagement of local communities in conservation efforts:

Case Study 1: Mahatma Gandhi Marine National Park, Andaman and Nicobar Islands, India

The Mahatma Gandhi Marine National Park is located in the Andaman and Nicobar Islands, a biodiversity hotspot in India. The park is home to numerous coral reefs, mangroves, and marine species. To protect this fragile ecosystem, the local communities living around the park actively participate in its conservation.

The park authorities, in collaboration with local communities, have implemented various conservation initiatives, including:

- Creating awareness programs in local schools and community centers to educate children and adults about the importance of marine biodiversity.

- Involving local fishermen in monitoring and surveillance activities to prevent illegal fishing practices and protect sensitive marine habitats.

- Supporting alternative livelihood opportunities, such as eco-tourism and sustainable fisheries, to reduce dependency on destructive fishing practices.

- Establishing community-based patrolling systems to monitor and prevent illegal activities within the park.

As a result of these efforts, the local communities have become staunch advocates for conserving the marine park. They actively participate in conservation activities and play a crucial role in protecting the rich biodiversity of the area.

Case Study 2: Maasai Mara Conservancies, Kenya

The Maasai Mara in Kenya is one of the most iconic wildlife reserves in Africa, known for its abundant wildlife, including the Great Migration of wildebeest. To address the challenges of habitat loss and human-wildlife conflicts, local Maasai communities have been involved in the creation of conservancies.

Through the establishment of conservancies, the Maasai communities:

- Have been granted land rights and ownership, allowing them to manage and benefit directly from conservation efforts.

- Have formed partnerships with tourism operators, ensuring that a portion of the revenue generated from tourism activities goes back to the local communities.

- Have implemented sustainable grazing practices to prevent overgrazing and maintain healthy rangelands for wildlife.

- Have been trained as local guides and rangers, providing employment opportunities and reducing poaching incidents.

The Maasai communities' involvement in the management and conservation of the Maasai Mara has led to reduced human-wildlife conflicts, sustainable tourism practices, and improved livelihoods for the local communities.

CHALLENGES AND FUTURE DIRECTIONS IN CONSERVATION BIOLOGY

Challenges and Future Directions

While engaging local communities in conservation efforts is crucial, it also comes with challenges. Some of the common challenges include:

- **Lack of awareness and understanding:** Many local communities may lack awareness and understanding of conservation issues. Effective communication and education programs are essential to address this challenge.

- **Limited resources and capacity:** Local communities often struggle with limited resources and capacity to actively engage in conservation activities. Providing training, technical support, and access to resources can help overcome this challenge.

- **Conflicts of interest:** Conflicts of interest between conservation goals and local communities' immediate needs can hinder engagement. Balancing conservation objectives with the socio-economic needs of local communities requires careful consideration and negotiation.

- **Lack of local empowerment:** In some cases, decision-making power remains centralized with external stakeholders, limiting the meaningful involvement of local communities. Empowering local communities through participatory processes is crucial for their active engagement.

To address these challenges and further enhance the engagement of local communities in conservation efforts, some future directions include:

- Strengthening partnerships and collaboration between local communities, researchers, governments, and non-governmental organizations.

- Supporting initiatives that promote the recognition and respect of indigenous rights and traditional knowledge.

- Incorporating the principles of social justice and equity in conservation policies and practices.

- Investing in long-term monitoring and evaluation of community-led conservation initiatives to assess their effectiveness.

- Promoting participatory approaches that involve local communities in decision-making processes and planning.

CHAPTER 3: CONSERVATION BIOLOGY AND WILDLIFE MANAGEMENT

Engaging local communities in conservation efforts is not only ethically and morally important but also provides a pathway to more effective and sustainable conservation outcomes. By recognizing the importance of local knowledge and involving local communities as key stakeholders, we can achieve a better understanding of the interconnectedness between humans and nature, and work towards a more harmonious and resilient future.

Chapter 4: Sustainable Agriculture and Food Systems

Chapter 4: Sustainable Agriculture and Food Systems

Chapter 4: Sustainable Agriculture and Food Systems

In this chapter, we will explore the concept of sustainable agriculture and its importance in ensuring food security and environmental conservation. We will discuss various principles and practices of sustainable agriculture, as well as innovative technologies that are revolutionizing the way we produce and consume food. By the end of this chapter, you will gain a deeper understanding of how sustainable agriculture can contribute to a greener and more secure future.

Definition and Principles of Sustainable Agriculture

Sustainable agriculture is an approach to farming that aims to meet the needs of the present generation without compromising the ability of future generations to meet their own needs. It places emphasis on the ecological, economic, and social aspects of agricultural systems. The following are some key principles of sustainable agriculture:

1. Conservation of Resources: Sustainable agriculture seeks to minimize the use of non-renewable resources such as fossil fuels and synthetic fertilizers. It promotes the efficient use of energy, water, and nutrients, as well as the recycling of organic materials.

2. Biodiversity Enhancement: Sustainable agriculture recognizes the importance of biodiversity in maintaining ecosystem resilience and productivity. It encourages the preservation of natural habitats, the use of crop rotations and cover crops, and the promotion of beneficial insects and wildlife.

236 CHAPTER 4: SUSTAINABLE AGRICULTURE AND FOOD SYSTEMS

3. Soil Health and Fertility: Sustainable agriculture promotes practices that improve soil health and fertility, such as organic matter addition, crop diversification, and reduced soil tillage. These practices enhance nutrient cycling, prevent soil erosion, and increase the soil's capacity to retain water.

4. Water Conservation: Sustainable agriculture emphasizes efficient water use through measures such as drip irrigation, precision farming, and water harvesting techniques. It aims to minimize water pollution and protect the quality of water resources.

5. Social Equity: Sustainable agriculture recognizes the importance of ensuring fair wages and working conditions for farmers and farmworkers. It promotes local food systems, encourages community engagement, and supports small-scale farmers.

Innovations in Agricultural Technologies

In recent years, advancements in agricultural technologies have provided promising solutions to the challenges faced by traditional farming practices. These innovations have the potential to improve productivity, reduce environmental impact, and enhance the resilience of food systems. Here are some notable examples:

1. Precision Agriculture and Smart Farming: Precision agriculture utilizes technologies such as GPS, sensors, and data analytics to optimize decision-making in farming practices. It enables farmers to apply inputs, such as fertilizers and pesticides, precisely where and when they are needed, minimizing waste and reducing environmental pollution.

2. Hydroponics and Vertical Farming: Hydroponics is a soilless cultivation technique that involves growing plants in nutrient-rich water. Vertical farming takes this concept a step further by stacking multiple layers of plants in vertically-oriented systems. These innovative approaches allow for year-round production, reduced water consumption, and higher crop yields per unit area.

3. Biotechnology in Agriculture: Biotechnology has revolutionized agriculture through the development of genetically modified organisms (GMOs) and gene editing techniques. GMOs can improve crop resistance to pests, diseases, and environmental stress, while gene editing offers precise modifications to plant genomes. These technologies have the potential to enhance crop productivity and reduce the need for chemical inputs.

CHAPTER 4: SUSTAINABLE AGRICULTURE AND FOOD SYSTEMS 237

Case Studies in Sustainable Agriculture

Let's examine some real-world examples of sustainable agriculture practices and their positive impacts:

1. Community-supported Agriculture (CSA): CSA is a model in which consumers directly support local farmers by purchasing shares of the seasonal harvest in advance. This approach ensures a stable income for farmers and provides consumers with fresh, locally-grown produce. CSA also fosters a sense of community and strengthens the connection between producers and consumers.

2. Regenerative Farming Practices: Regenerative farming goes beyond sustainable agriculture by actively restoring degraded land and ecosystems. Practices such as rotational grazing, agroforestry, and soil regeneration techniques increase soil organic matter, improve soil structure, and promote carbon sequestration. These practices contribute to climate change mitigation and enhance long-term soil fertility.

3. Rooftop and Urban Farming: In densely populated urban areas, rooftop and urban farming offer opportunities to produce fresh food locally. By utilizing rooftops, balconies, and unused urban spaces, these initiatives reduce food transportation distances and enhance food security in cities. They also provide educational and community-building opportunities.

Success Stories in Sustainable Agriculture

The success stories in sustainable agriculture showcase the positive impact of adopting environmentally friendly farming practices. Here are some notable examples:

1. Organic Farming Movement: Organic farming has gained global recognition for its emphasis on ecological sustainability and ethical practices. Organic farms avoid synthetic pesticides and fertilizers, prioritize soil health, and promote biodiversity. The growth of the organic farming movement has led to increased availability of organic products and improved environmental stewardship.

2. Agroecology in Cuba: Due to limited access to chemical inputs during the economic crisis in the 1990s, Cuba transitioned to agroecological farming practices. The country prioritized sustainable agriculture, including organic farming, small-scale farming, and urban agriculture. This shift resulted in increased food security, reduced dependence on imports, and a resurgence of local food systems.

3. Integrated Pest Management (IPM): IPM is an approach that utilizes a combination of biological, cultural, and chemical control methods to manage pests effectively. It reduces reliance on synthetic pesticides while maintaining crop

238 *CHAPTER 4: SUSTAINABLE AGRICULTURE AND FOOD SYSTEMS*

productivity. IPM has been successfully applied in various cropping systems, leading to reduced pesticide use, improved ecosystem health, and reduced risk to human health.

Challenges and Future Directions in Sustainable Agriculture

While sustainable agriculture has shown great potential, several challenges must be addressed to achieve widespread adoption and transformation of our food systems. Some key challenges and future directions in sustainable agriculture include:

1. Ensuring Food Security and Eliminating Hunger: Sustainable agriculture should prioritize the goal of achieving global food security while minimizing negative environmental impacts. It requires addressing the root causes of hunger, such as poverty, inequality, and inadequate access to resources.

2. Sustainable Farming Policy and Government Support: Governments need to implement policies that promote sustainable agricultural practices and provide financial and technical support to farmers. This includes incentivizing sustainable farming practices, supporting research and development, and fostering collaborations between farmers and researchers.

3. Reducing Food Waste and Loss: A significant challenge in achieving sustainable agriculture is the staggering amount of food waste and loss throughout the supply chain. Efforts should be made to reduce post-harvest losses, improve storage and transportation systems, and educate consumers about the importance of minimizing food waste.

4. Climate Change Resilience in Agriculture: Climate change poses significant threats to agricultural productivity and stability. Adaptation strategies, such as developing drought-tolerant crop varieties, implementing water management techniques, and enhancing soil carbon sequestration, are essential for mitigating climate change impacts and building resilience in agriculture.

5. Advancing Agroecological Research and Education: Agroecology, as a holistic approach to sustainable agriculture, requires ongoing research and education. Investments in agroecological research can contribute to the development of innovative practices and technologies that improve the sustainability and productivity of farming systems.

In conclusion, sustainable agriculture holds immense potential in addressing the challenges of feeding a growing population while safeguarding the environment. By embracing the principles of sustainable agriculture and harnessing innovative technologies, we can create a future in which agriculture is truly sustainable, resilient, and equitable. Through various case studies and success stories, we have seen the positive impacts of sustainable agriculture around the

PRINCIPLES OF SUSTAINABLE AGRICULTURE

world. However, addressing the challenges and pursuing future research and education are essential for realizing the full potential of sustainable agriculture. It is up to us to embrace and support these practices to create a greener and more sustainable future for food systems worldwide.

Principles of Sustainable Agriculture

Organic Farming and Permaculture

Organic farming and permaculture are two sustainable and environmentally-friendly approaches to agriculture that have gained significant attention in recent years. These methods prioritize the use of natural resources, minimize the use of synthetic inputs, and promote the creation of self-sustaining ecosystems. In this section, we will explore the principles, techniques, and benefits of organic farming and permaculture.

Principles of Organic Farming

Organic farming is based on several key principles that guide its practices. These principles include:

- **Soil health:** Organic farmers prioritize the health of the soil by using natural fertilizers, such as compost and manure, to enrich the soil with organic matter and nutrients. They also cultivate soil-friendly practices, such as crop rotation and cover cropping, to prevent soil erosion and maintain soil fertility.

- **Biodiversity:** Organic farming aims to promote biodiversity by creating a diverse and balanced ecosystem. This includes intercropping, where different crops are grown together, and incorporating natural predators to control pests instead of relying on chemical pesticides.

- **No synthetic chemicals:** Organic agriculture prohibits the use of synthetic chemicals, including synthetic pesticides, herbicides, and fertilizers. Instead, natural alternatives are used to manage pests and weeds, such as crop rotation, companion planting, and the use of herbal extracts.

- **Animal welfare:** Organic farming emphasizes the ethical treatment of animals. Livestock raised on organic farms must have access to outdoor areas and be provided with appropriate living conditions and natural feed.

CHAPTER 4: SUSTAINABLE AGRICULTURE AND FOOD SYSTEMS

+ **Sustainability:** Organic agriculture aims to minimize its environmental impact and promote long-term sustainability. This includes conserving water resources, reducing energy consumption, and implementing efficient waste management practices.

Principles of Permaculture

Permaculture, short for "permanent agriculture," is a design approach that seeks to create sustainable and self-sufficient systems that mimic natural ecosystems. It integrates various elements, such as plants, animals, and structures, to create productive and harmonious environments. The principles of permaculture include:

+ **Observation and interaction:** Permaculture begins with careful observation of the natural environment and its patterns. By understanding the interactions between different elements, permaculturists can design systems that work in harmony with nature.

+ **Catch and store energy:** Permaculture focuses on capturing and storing energy from natural resources such as sunlight, wind, and water. Techniques like rainwater harvesting, passive solar design, and wind turbines are used to harness and utilize these energy sources.

+ **Use renewable resources:** Permaculture emphasizes the use of renewable resources instead of relying on non-renewable ones. This includes utilizing organic waste as compost, using renewable energy sources, and growing perennial crops that require minimal inputs.

+ **Integrate functions:** Permaculture systems aim to maximize efficiency by integrating multiple functions within a single element. For example, a pond can provide water for irrigation, habitat for fish, and a cooling effect for surrounding areas.

+ **Produce no waste:** Permaculture encourages the creation of closed-loop systems where waste from one element becomes a resource for another. Composting, vermiculture, and recycling are used to minimize waste generation and promote resource efficiency.

Techniques in Organic Farming and Permaculture

Both organic farming and permaculture employ a variety of techniques to achieve their goals of sustainability and self-sufficiency. Some common techniques include:

PRINCIPLES OF SUSTAINABLE AGRICULTURE

- **Composting:** Composting is the process of decomposing organic materials, such as food scraps, yard waste, and animal manure, to create nutrient-rich compost. This compost is then used as a natural fertilizer to enhance soil fertility and improve plant health.

- **Crop rotation:** Crop rotation is the practice of growing different crops in a specific sequence on the same piece of land. This technique helps prevent the buildup of pests and diseases, improves soil nutrition, and reduces the need for synthetic pesticides and fertilizers.

- **Polyculture:** Polyculture involves growing multiple crops together in the same area. This promotes biodiversity, enhances pest control, and improves soil fertility through the exchange of nutrients among different plants.

- **Agroforestry:** Agroforestry integrates trees, crops, and livestock in a single system. The trees provide shade, prevent soil erosion, and attract beneficial insects, while the crops and livestock contribute to the overall productivity of the system.

- **Water management:** Organic farming and permaculture prioritize efficient water use through techniques such as rainwater harvesting, drip irrigation, and contour planting. These methods help conserve water resources and reduce reliance on external sources of irrigation.

Benefits of Organic Farming and Permaculture

Organic farming and permaculture offer several benefits for both the environment and society, including:

- **Environmental sustainability:** By minimizing the use of synthetic chemicals, organic farming and permaculture help protect soil and water quality, preserve biodiversity, and reduce pollution. These practices also contribute to climate change mitigation by sequestering carbon in the soil.

- **Improved food quality:** Organic farming and permaculture prioritize the production of nutritious and chemical-free food. These methods enhance the taste, flavor, and nutritional value of crops, promoting better health outcomes for consumers.

- **Enhanced resilience:** Organic farming and permaculture systems are more resilient to climate variability and extreme weather events. Their diverse and

242 CHAPTER 4: SUSTAINABLE AGRICULTURE AND FOOD SYSTEMS

interconnected nature can adapt to changing conditions and minimize the risk of crop failure.

+ **Local and community development:** Organic farming and permaculture systems often emphasize local food production and distribution, which strengthens local economies and reduces dependence on global food supply chains. They also foster community engagement and education through shared knowledge and resources.

+ **Cost-effective:** While organic farming and permaculture may require more labor-intensive practices initially, they can be economically viable in the long run. By reducing the dependence on expensive inputs and improving soil fertility, these methods can lower production costs and increase profitability for farmers.

Example: Organic Farming in Action

To illustrate the principles and techniques of organic farming, let's consider the example of an organic vegetable farm. The farmer employs practices such as crop rotation, composting, and natural pest control methods to maintain the health of the soil and plants. Instead of using synthetic pesticides, the farmer encourages beneficial insects, such as ladybugs and lacewings, to control pests.

The farm also practices polyculture, growing a variety of crops together. For example, the farmer may plant beans, corn, and squash together—a classic combination known as the "Three Sisters." The corn provides support for the beans to climb, while the beans fix nitrogen in the soil and the squash acts as a natural mulch, suppressing weeds and retaining soil moisture.

To improve water management, the farmer implements drip irrigation, which delivers water directly to the plants' root zones, minimizing water loss due to evaporation. Additionally, rain barrels are used to collect and store rainwater for irrigation during dry periods.

By following these organic farming techniques, the farmer produces high-quality, nutrient-rich vegetables while minimizing the use of synthetic inputs and reducing environmental impact.

Conclusion

Organic farming and permaculture are sustainable and ecologically sound approaches to agriculture. They prioritize soil health, biodiversity, and resource conservation while producing nutritious food and supporting local communities.

PRINCIPLES OF SUSTAINABLE AGRICULTURE

By embracing these principles and techniques, farmers can contribute to a greener and more sustainable future.

Agroecology and Agroforestry

Agroecology and agroforestry are two important concepts in sustainable agriculture that promote the use of ecological principles and practices to enhance agricultural productivity while minimizing negative environmental impacts. These approaches prioritize the conservation of natural resources, biodiversity, and soil fertility, and aim to create resilient and self-sustaining agricultural systems. In this section, we will explore the principles and benefits of agroecology and agroforestry, as well as some successful case studies.

Agroecology

Agroecology is a scientific discipline that studies the interactions between plants, animals, humans, and the environment within agricultural systems. It recognizes that agricultural practices are deeply interconnected with ecological processes and aims to enhance the sustainability and productivity of farming systems through the application of ecological principles.

Principles of Agroecology Agroecology is guided by several key principles:

1. **Biodiversity:** Agroecology recognizes the importance of biodiversity in maintaining ecological balance and enhancing ecosystem resilience. By promoting the use of diverse crop varieties, intercropping, and crop rotation, agroecological practices help reduce the reliance on monocultures and enhance pest management and soil fertility.

2. **Soil Health:** Agroecology emphasizes the importance of maintaining and enhancing soil health. This involves practices such as organic matter addition, cover cropping, and reduced tillage to promote soil structure, water retention, and nutrient cycling.

3. **Water Management:** Agroecology promotes efficient water use by minimizing water losses through techniques like drip irrigation, rainwater harvesting, and contour plowing. By optimizing water use, agroecological systems reduce the pressures on freshwater resources and increase drought resilience.

244 CHAPTER 4: SUSTAINABLE AGRICULTURE AND FOOD SYSTEMS

4. **Nutrient Cycling:** Agroecology emphasizes the recycling of nutrients within agricultural systems. This involves using organic fertilizers, incorporating leguminous crops to fix nitrogen, and implementing nutrient management strategies to minimize nutrient losses.

5. **Biological Pest Control:** Agroecology promotes the use of natural enemies and biological control agents to manage pests and diseases. By enhancing the presence of beneficial insects, birds, and microorganisms, agroecological systems reduce the need for synthetic pesticides and foster ecological balance.

6. **Social Equity:** Agroecology recognizes the importance of social and economic equity in agricultural systems. It emphasizes the involvement and empowerment of farmers, local communities, and stakeholders in decision-making processes, promoting food sovereignty and social justice.

Benefits of Agroecology Agroecology offers numerous benefits for farmers, ecosystems, and society as a whole. Some of the key benefits include:

- **Increased Resilience:** Agroecological systems are more resilient to climate change impacts, such as extreme weather events and changing rainfall patterns. By diversifying crops and adopting conservation practices, farmers can adapt to changing conditions and minimize crop failures.

- **Enhanced Biodiversity:** Agroecological practices create habitats and provide resources for beneficial insects, birds, and other wildlife. This promotes biodiversity conservation and contributes to the preservation of ecosystem services, such as pollination and biological control.

- **Improved Soil Fertility:** Agroecological practices, such as the use of organic inputs and cover cropping, improve soil health and fertility. Healthy soils support higher crop yields, reduce erosion, and contribute to carbon sequestration, mitigating climate change.

- **Reduced Environmental Impacts:** Agroecology minimizes the use of synthetic inputs, such as fertilizers and pesticides, which can have negative impacts on water quality, biodiversity, and human health. By adopting agroecological practices, farmers can reduce pollution and contribute to cleaner ecosystems.

- **Enhanced Food Security:** Agroecological systems are designed to produce diverse and nutritious crops while maintaining long-term productivity. By

PRINCIPLES OF SUSTAINABLE AGRICULTURE

prioritizing local food production, agroecology can contribute to improved food security and reduced dependence on global food supply chains.

Agroforestry

Agroforestry is a land-use management system that integrates trees or shrubs with crops and/or livestock in agricultural landscapes. This approach combines the benefits of agriculture and forestry to create diverse and sustainable production systems.

Types of Agroforestry There are several different types of agroforestry systems, each with its own unique characteristics and benefits:

- **Silvopasture:** Silvopasture integrates trees, forage crops, and livestock. Trees provide shade and shelter for animals, while their fallen leaves contribute to soil fertility. This system can improve animal welfare, reduce heat stress, and increase overall productivity.

- **Alley Cropping:** Alley cropping involves planting rows of trees or shrubs in between rows of crops. This provides multiple benefits, such as erosion control, windbreaks, and the potential to harvest timber or fruits from the trees. The space between trees allows for crop production and reduces competition.

- **Forest Farming:** Forest farming refers to the cultivation of high-value specialty crops, such as medicinal herbs, mushrooms, or nuts, under the shade of a forest canopy. This system takes advantage of existing forests to produce non-timber forest products sustainably.

- **Windbreaks:** Windbreaks are rows of trees or shrubs planted along field edges or as hedgerows within agricultural landscapes. They provide protection from wind erosion, prevent soil degradation, and create microclimates that enhance crop production.

Benefits of Agroforestry Agroforestry systems offer a range of ecological, economic, and social benefits:

- **Ecosystem Services:** Agroforestry systems provide a variety of ecosystem services, including carbon sequestration, water filtration, and habitat for wildlife. Trees in agroforestry systems contribute to biodiversity conservation and enhance the overall ecological balance.

246 CHAPTER 4: SUSTAINABLE AGRICULTURE AND FOOD SYSTEMS

+ **Financial Returns:** Agroforestry can provide additional sources of income for farmers through the sale of timber, fruits, or non-timber forest products. By diversifying their production, farmers can reduce risks associated with a single crop and improve their financial resilience.

+ **Soil Conservation:** Agroforestry helps prevent soil erosion by intercepting rainfall, reducing runoff, and improving soil structure. The presence of trees and their deep root systems contribute to soil stability and fertility, reducing the need for synthetic fertilizers.

+ **Climate Change Mitigation:** Agroforestry systems sequester carbon dioxide from the atmosphere and store it in trees and soils. This helps mitigate climate change by reducing greenhouse gas emissions and enhancing carbon sinks.

+ **Livelihood Support:** Agroforestry systems provide opportunities for rural development and poverty alleviation. By diversifying income sources and promoting sustainable land management, agroforestry can contribute to the well-being of local communities.

Case Studies in Agroecology and Agroforestry   There are several successful case studies that highlight the benefits and potential of agroecology and agroforestry:

+ **The Land Institute - Kernza®:** The Land Institute, based in Kansas, USA, has been working on developing perennial grain crops, such as Kernza®. Kernza® is a perennial wheatgrass that has the potential to reduce soil erosion, enhance soil health, and provide sustainable grain production.

+ **Agriculture and Forestry Research Center - Vietnam:** The Agriculture and Forestry Research Center in Vietnam has been promoting agroforestry systems that combine fruit trees with annual crops. This integrated approach has led to increased income for farmers, reduced dependence on synthetic inputs, and improved biodiversity.

+ **Finca Las Piedras - Costa Rica:** Finca Las Piedras is a farm in Costa Rica that practices agroforestry by integrating timber trees with cattle ranching. This system provides shade for livestock, enhances soil fertility, and produces high-value timber, contributing to both economic and environmental sustainability.

PRINCIPLES OF SUSTAINABLE AGRICULTURE

Challenges and Future Directions   Despite the numerous benefits of agroecology and agroforestry, there are still challenges that need to be addressed. Some of these challenges include:

- **Knowledge and Awareness:** There is a need to raise awareness about the benefits of agroecology and agroforestry among farmers, policymakers, and consumers. Education and capacity building programs can help promote the adoption of these sustainable practices.

- **Policy Support:** Government policies and regulations need to support agroecological and agroforestry practices. This includes providing financial incentives, technical assistance, and infrastructure for farmers to transition to sustainable production systems.

- **Market Access:** Access to markets can be a challenge for farmers practicing agroecology and agroforestry. Efforts should be made to connect farmers with local markets, support value-added processing, and create certification schemes that recognize the sustainability of these practices.

- **Research and Innovation:** Continued research and innovation are crucial to further develop and improve agroecological and agroforestry systems. This includes studying the interactions between crops, trees, and livestock, as well as developing new technologies and techniques for sustainable farming.

In conclusion, agroecology and agroforestry offer sustainable and holistic approaches to agriculture that prioritize environmental conservation, biodiversity, and food security. These systems provide multiple benefits, ranging from increased resilience to climate change to enhanced livelihoods for farmers. By adopting and supporting agroecological and agroforestry practices, we can create a more sustainable and resilient agricultural future.

Sustainable Soil Management

Soil is a vital resource for agriculture and plays a crucial role in supporting plant growth and providing nutrients to sustain life. However, unsustainable soil management practices have led to soil degradation, erosion, and loss of fertility. In this section, we will explore the principles and techniques of sustainable soil management, which aim to maintain or improve soil health while minimizing negative impacts on the environment.

248 *CHAPTER 4: SUSTAINABLE AGRICULTURE AND FOOD SYSTEMS*

Understanding Soil Health

Before delving into sustainable soil management practices, it is important to understand the concept of soil health. Soil health refers to the capacity of soil to function as a living ecosystem, sustaining plants, animals, and humans. A healthy soil system is characterized by adequate nutrient content, good soil structure, balanced pH levels, and high microbial activity.

Soil health is crucial because it affects plant productivity, water quality, and carbon sequestration. Healthy soils retain water better, prevent runoff and erosion, and support diverse microbial communities that play a vital role in nutrient cycling and disease suppression.

Principles of Sustainable Soil Management

Sustainable soil management is guided by several key principles aimed at protecting and enhancing soil health. These principles include:

1. **Conservation of soil organic matter:** Soil organic matter (SOM) is vital for maintaining soil health and fertility. It improves soil structure, promotes water retention, and provides nutrients to plants. Sustainable soil management practices aim to increase or maintain SOM through organic amendments (e.g., compost, cover crops) and reduced tillage.

2. **Minimizing soil erosion:** Soil erosion is a major concern that leads to the loss of topsoil and essential nutrients. Sustainable soil management practices employ techniques to control erosion, such as contour plowing, terracing, and the use of windbreaks and cover crops.

3. **Managing nutrient balance:** Sustainable soil management involves optimizing nutrient availability while minimizing nutrient losses. This is achieved through practices like nutrient recycling, precision fertilization, and the use of organic amendments. Soil testing and analysis play a crucial role in determining nutrient requirements for different crops.

4. **Reducing chemical inputs:** Sustainable soil management aims to minimize the use of synthetic pesticides and fertilizers. Instead, it promotes integrated pest management (IPM) strategies that rely on biological control agents, crop rotation, and resistant cultivars. Precision farming techniques, such as variable rate application, can help optimize the use of inputs.

PRINCIPLES OF SUSTAINABLE AGRICULTURE

249

5. **Managing soil compaction:** Soil compaction can negatively impact root growth, water infiltration, and soil aeration. Sustainable soil management involves practices like reduced tillage, use of cover crops, and proper timing of field operations to minimize compaction.

6. **Promoting crop diversity and rotation:** Monocropping depletes soil nutrients and increases susceptibility to pests and diseases. Sustainable soil management emphasizes crop rotation and diversification to enhance soil fertility, suppress pests, and reduce the need for synthetic inputs.

Techniques for Sustainable Soil Management

There are several techniques and practices that can be employed to achieve sustainable soil management. Let's explore some of the commonly used techniques:

1. **Conservation tillage:** Conservation tillage practices, such as no-till or reduced tillage, minimize soil disturbance and help retain soil organic matter. By leaving crop residue on the soil surface, these practices reduce erosion, improve water infiltration, and promote soil biodiversity.

2. **Cover cropping:** Cover crops are non-commercial crops planted between cash crops to protect the soil from erosion, improve soil structure, and enhance nutrient cycling. They also provide additional organic matter when incorporated into the soil.

3. **Crop rotation:** Crop rotation involves systematically changing crops grown in a specific field over time. This practice helps break pest and disease cycles, improves soil fertility, and reduces the reliance on chemical inputs.

4. **Composting:** Composting is the process of converting organic waste into nutrient-rich humus. The application of compost improves soil structure, water-holding capacity, and nutrient content. It also reduces landfill waste and decreases the need for synthetic fertilizers.

5. **Precision farming:** Precision farming utilizes advanced technologies like Global Positioning Systems (GPS), remote sensing, and data analytics to optimize resource use in agriculture. By precisely applying inputs such as water and fertilizers according to the specific needs of different soil zones, it reduces waste and enhances efficiency.

250 CHAPTER 4: SUSTAINABLE AGRICULTURE AND FOOD SYSTEMS

6. **Agroforestry:** Agroforestry combines agricultural crops with trees or shrubs, providing multiple benefits such as improved soil structure, enhanced nutrient cycling, and increased biodiversity. Agroforestry systems can also sequester carbon, mitigate climate change, and provide supplemental income through timber or fruit production.

Challenges and Future Directions

While sustainable soil management practices have shown promise in improving soil health and productivity, there are still several challenges to overcome. These challenges include:

1. **Knowledge and awareness gaps:** Adoption of sustainable soil management practices requires awareness, education, and access to information. Farmers need to be aware of the benefits of sustainable practices and have access to training and technical support.

2. **Economic viability:** Transitioning to sustainable soil management practices may require upfront investments and changes in farming systems. Farmers need financial incentives and supportive policies to make these transitions economically viable.

3. **Scaling up adoption:** While there are successful case studies of sustainable soil management, widespread adoption of these practices is still limited. More efforts are needed to encourage farmers to adopt sustainable practices and overcome barriers like infrastructure limitations and resistance to change.

4. **Climate change resilience:** Climate change poses additional challenges to soil management. Extreme weather events, such as floods and droughts, can damage soil structure and nutrient availability. Developing practices that enhance soil resilience to climate change is essential.

To address these challenges and advance sustainable soil management, ongoing research and innovation are crucial. Governments, research institutions, and agricultural organizations need to collaborate closely to develop and disseminate best practices, facilitate knowledge exchange, and support farmers in transitioning to sustainable soil management systems.

Summary

Sustainable soil management is an essential component of promoting environmentally friendly agriculture and ensuring food security in the face of increasing global challenges. By conserving soil organic matter, minimizing erosion, optimizing nutrient balance, reducing chemical inputs, combating soil compaction, and promoting diversity and rotation, we can improve soil health, enhance crop productivity, and reduce environmental impacts.

Through the adoption of techniques such as conservation tillage, cover cropping, crop rotation, composting, precision farming, and agroforestry, farmers can contribute to sustainable soil management while maintaining or improving their economic viability. However, addressing the challenges of knowledge gaps, economic viability, scaling up adoption, and climate change resilience is crucial for the widespread implementation of sustainable soil management practices.

By valuing and protecting our soil resources, we can contribute to a greener future and ensure the long-term sustainability of agricultural systems. It is our collective responsibility to support and promote sustainable soil management for the benefit of current and future generations.

Innovations in Agricultural Technologies

Precision Agriculture and Smart Farming

Precision Agriculture and Smart Farming are innovative approaches that utilize advanced technologies to optimize agricultural practices. These methods aim to improve crop yield, reduce resource wastage, and enhance overall sustainability in the agriculture industry. In this section, we will explore the principles and techniques behind precision agriculture and smart farming, discuss their benefits and challenges, and provide real-world examples of their successful implementation.

Principles of Precision Agriculture

Precision Agriculture is based on the principle of site-specific management, which involves tailoring agricultural practices to the specific needs of different areas within a field. This approach recognizes that within a given field, soil conditions, nutrient levels, and crop health can vary significantly. By collecting and analyzing data on these variations, farmers can make well-informed decisions and apply

252 CHAPTER 4: SUSTAINABLE AGRICULTURE AND FOOD SYSTEMS

inputs such as fertilizers, pesticides, and water strategically, optimizing crop growth and minimizing waste.

The key principles of Precision Agriculture include:

+ **Data Collection:** Gathering accurate and detailed information about soil properties, crop growth, weather conditions, and other relevant factors is crucial for making informed decisions. This data can be collected using remote sensing technologies, satellite imagery, drones, and ground-based sensors.

+ **Data Analysis and Decision-making:** Once the data is collected, it needs to be processed and analyzed to extract meaningful insights. This involves the use of advanced analytics tools and algorithms that can identify patterns, detect anomalies, and generate actionable recommendations.

+ **Variable Rate Technology (VRT):** VRT allows farmers to apply inputs, such as fertilizers and irrigation, at variable rates across different areas of a field. This technique ensures that resources are used efficiently, minimizing environmental impacts and reducing costs.

+ **Remote Monitoring and Automation:** Precision Agriculture relies on remote monitoring systems that continuously track crop health, soil moisture, and other relevant parameters. Automated systems can deliver real-time alerts and enable farmers to respond promptly to changing conditions.

+ **Integration of Equipment and Systems:** Precision Agriculture involves integrating various technologies, such as GPS (Global Positioning System), GIS (Geographical Information System), and farm management software. This integration facilitates data sharing, improves operational efficiency, and enables seamless decision-making.

Techniques in Precision Agriculture

Precision Agriculture employs a range of techniques to optimize agricultural practices. Some of the key techniques include:

1. **Soil Mapping and Analysis:** Soil mapping techniques, such as electromagnetic conductivity measurement and soil sampling, help identify variations in soil characteristics. By analyzing these variations, farmers can

INNOVATIONS IN AGRICULTURAL TECHNOLOGIES

make targeted decisions regarding nutrient application and land management strategies.

2. **Variable Rate Application (VRA):** VRA technology allows farmers to apply inputs, such as fertilizers, herbicides, and pesticides, at precise rates across different sections of a field. This technique ensures that resources are used efficiently and reduces the risk of over-application, minimizing environmental impacts.

3. **Remote Sensing and Imagery:** Remote sensing technologies, including satellite imagery, aerial photography, and drones, enable farmers to monitor crop health, identify pest infestations, and assess soil moisture levels. This information helps farmers make timely interventions and optimize resource allocation.

4. **Precision Irrigation:** Precision irrigation techniques, such as drip irrigation and soil moisture sensors, allow farmers to deliver the right amount of water to crops based on their specific needs. This approach reduces water wastage, improves water use efficiency, and prevents soil erosion.

5. **Crop Yield Monitoring:** Precision Agriculture utilizes yield monitoring systems to track crop performance throughout the growing season. These systems provide valuable information about crop growth, yield variations, and factors affecting productivity. Farmers can use this data to fine-tune their management practices and maximize overall yield.

Benefits and Challenges of Precision Agriculture

Precision Agriculture and Smart Farming offer numerous benefits to farmers, the environment, and society as a whole. Some of the key advantages include:

- **Increased Efficiency:** Precision Agriculture enables farmers to optimize the use of inputs such as water, fertilizer, and pesticides, reducing waste and improving resource efficiency.

- **Higher Crop Yield and Quality:** By tailoring management practices to the specific needs of crops, Precision Agriculture can enhance productivity and improve the quality of agricultural produce.

- **Environmental Sustainability:** Precision Agriculture minimizes the impact of agriculture on the environment by reducing the use of agrochemicals, preventing soil degradation, and conserving water resources.

254 CHAPTER 4: SUSTAINABLE AGRICULTURE AND FOOD SYSTEMS

- **Cost Savings:** Optimized resource utilization and improved efficiency result in cost savings for farmers, making agriculture more economically viable.

- **Data-driven Decision-making:** Precision Agriculture empowers farmers with data and insights that enable informed decision-making and proactive management.

Despite the numerous benefits, Precision Agriculture also faces certain challenges that need to be addressed for widespread adoption:

- **Cost and Accessibility:** The initial investment in precision agriculture technologies, such as sensors and equipment, can be expensive. Additionally, access to these technologies may be limited in certain regions or for small-scale farmers.

- **Data Management and Analysis:** The collection and analysis of data require specialized skills and resources. Farmers need support and training to effectively manage and interpret the vast amounts of data generated by precision agriculture technologies.

- **Technological Integration:** Precision Agriculture relies on the integration of various technologies and systems. Ensuring compatibility and interoperability between different software and hardware components can be a challenge.

- **Data Security and Privacy:** As precision agriculture involves the collection and analysis of sensitive data, ensuring data security and privacy is crucial. Farmers must be aware of potential risks and implement appropriate measures to protect their data.

- **Knowledge and Education:** A successful transition to Precision Agriculture requires awareness and knowledge among farmers. Access to training programs and educational resources that provide guidance on implementing precision agriculture techniques is essential.

Real-world Examples

Precision Agriculture and Smart Farming have been successfully implemented in various parts of the world. Here are two notable examples:

INNOVATIONS IN AGRICULTURAL TECHNOLOGIES

1. **The Netherlands:** The Netherlands is known for its innovative farming practices, including Precision Agriculture. Farmers in the Netherlands have embraced technologies such as GPS-guided machinery, remote sensing, and data-driven decision-making. These techniques have helped optimize resource utilization, reduce environmental impacts, and increase crop yields.

2. **Brazil:** In Brazil, precision agriculture techniques have been employed extensively in large-scale commercial farming operations. Farmers use satellite imagery and drones to monitor crop health, implement precision irrigation systems, and apply inputs at variable rates. These practices have led to improved efficiency, reduced costs, and increased profitability in the agriculture sector.

These examples demonstrate the potential of Precision Agriculture and Smart Farming to transform conventional farming practices and contribute to a more sustainable and productive agricultural system.

Conclusion

Precision Agriculture and Smart Farming offer a promising solution to address the challenges facing modern agriculture. By leveraging advanced technologies and data-driven approaches, farmers can optimize resource utilization, increase productivity, and minimize environmental impacts. However, the successful adoption of precision agriculture requires overcoming certain challenges, including cost, accessibility, data management, and technological integration. With continued advancements in technology and increased awareness among farmers, Precision Agriculture has the potential to revolutionize the way we produce food and contribute to a more sustainable future.

256 CHAPTER 4: SUSTAINABLE AGRICULTURE AND FOOD SYSTEMS

Exercises

1. Suppose you are a farmer who wants to implement precision agriculture techniques on your farm. Identify two specific areas where precision agriculture can be beneficial and explain the potential advantages.

2. What are some potential risks or challenges farmers might face when adopting precision agriculture practices? Provide examples and discuss possible ways to mitigate these challenges.

3. Research and analyze a case study that highlights the successful implementation of precision agriculture techniques in a specific region or farming operation. Present your findings, including the key strategies employed and the outcomes achieved.

Additional Resources

- Precision Agriculture: Technology and Economic Perspectives - *Pierre R. Crosson, Keith O. Fuglie*

- Precision Agriculture: A Practitioner's Guide to Remote Sensing and GIS-based Precision Agriculture Management - *David W. Archer, Lawrie G. C. Richey, Richard A. Cooke*

- Precision Agriculture for Sustainability and Environmental Protection - *Xinping Chen, Huasong Huang*

- Precision Agriculture Videos - *John Deere* (Available at: https://www.deere.com/en/videos/)

Note: The exercises and additional resources are intended to encourage further exploration and understanding of the topic. It is recommended to actively engage with the exercises and explore the additional resources to enhance your knowledge of Precision Agriculture and Smart Farming.

Fun Fact

In Japan, farmers are experimenting with robot bees called "pollinator drones." These unmanned aerial vehicles are equipped with a special gel to collect and transfer pollen, helping to pollinate crops in areas where natural pollinators are scarce. The pollinator drones mimic the behavior of bees, contributing to increased crop yield and food production.

INNOVATIONS IN AGRICULTURAL TECHNOLOGIES

Hydroponics and Vertical Farming

Hydroponics and vertical farming are innovative and sustainable agricultural practices that are gaining significant attention in recent years. In this section, we will explore the principles, techniques, and benefits of hydroponics and vertical farming, as well as their potential to revolutionize the way we produce food.

Principles of Hydroponics

Hydroponics is a soilless method of growing plants that relies on a nutrient-rich water solution instead of traditional soil. The principles of hydroponics are based on providing plants with the essential nutrients they need to grow, while also optimizing water and resource efficiency.

In traditional soil-based agriculture, plants obtain nutrients from the soil through their root systems. However, in hydroponics, the nutrients are dissolved in water and delivered directly to the roots. This allows for better control over nutrient uptake, resulting in increased plant growth and productivity.

There are different types of hydroponic systems, including nutrient film technique (NFT), deep water culture (DWC), and drip irrigation systems. Each system has its own set of advantages and considerations, but they all share the common principle of providing plants with a carefully balanced nutrient solution.

Principles of Vertical Farming

Vertical farming takes the concept of hydroponics a step further by utilizing vertical space to maximize crop production. In vertical farms, plants are grown in stacked layers or vertically inclined surfaces, such as shelves or walls, using artificial lighting systems.

The principles of vertical farming are rooted in optimizing space utilization, reducing land requirements, and minimizing the use of resources. With vertical farming, multiple layers of crops can be grown in a controlled environment, allowing for year-round production and increased crop yields.

Vertical farming systems often incorporate advanced technologies, such as LED lights for optimized plant growth, climate control systems for temperature and humidity regulation, and automation for efficient operation. These technologies enable precise control over environmental parameters, resulting in optimal growing conditions for plants.

258 CHAPTER 4: SUSTAINABLE AGRICULTURE AND FOOD SYSTEMS

Benefits of Hydroponics and Vertical Farming

Hydroponics and vertical farming offer numerous benefits compared to traditional agriculture methods:

1. Water Efficiency: Hydroponics and vertical farming use a fraction of the water required in traditional agriculture. The closed-loop systems in hydroponics minimize water loss through evaporation and ensure that water is efficiently recycled and reused.

2. Land Conservation: Vertical farming allows for high-density crop production in urban areas or limited land spaces. By utilizing vertical space, the need for extensive land is minimized, conserving valuable agricultural land for other purposes.

3. Controlled Environment: Hydroponic and vertical farming systems provide an ideal environment for plant growth. The controlled conditions eliminate the reliance on weather conditions and reduce the risk of pests and diseases, resulting in consistent crop yields throughout the year.

4. Increased Crop Yields: The controlled environment, optimized nutrient delivery, and efficient use of resources in hydroponics and vertical farming result in higher crop yields compared to traditional farming methods. This increased productivity can help meet the growing demand for food in an ever-expanding global population.

5. Reduced Environmental Impact: Hydroponics and vertical farming have a lower environmental impact compared to conventional agriculture. The absence of chemical fertilizers and pesticides, reduced water usage, and efficient resource management contribute to sustainable food production and minimize ecological degradation.

Challenges and Future Directions

While hydroponics and vertical farming offer promising solutions for sustainable agriculture, there are also challenges that need to be addressed:

1. Energy Consumption: The indoor and artificial lighting systems used in vertical farming require significant energy inputs. Finding energy-efficient solutions and integrating renewable energy sources will be crucial to reduce the environmental footprint of these systems.

2. Cost considerations: The initial setup costs of hydroponic and vertical farming systems can be high, making it less accessible for small-scale farmers. Scaling up production and developing cost-effective technologies will play a crucial role in making these practices more financially viable.

INNOVATIONS IN AGRICULTURAL TECHNOLOGIES

3. Nutrient Management: Ensuring a well-balanced nutrient solution is critical for the success of hydroponics. Monitoring nutrient levels and preventing nutrient imbalances or deficiencies require careful attention and expertise.

4. Public Perception and Acceptance: Educating the public about the benefits and safety of hydroponics and vertical farming is essential for wider adoption. Overcoming skepticism and promoting awareness can help create a supportive environment for these innovative agricultural practices.

The future of hydroponics and vertical farming lies in continuous innovation and technological advancements. As researchers and practitioners explore new techniques, materials, and approaches, these methods have the potential to enhance food security, promote sustainability, and contribute to a greener future.

Resources

For further reading and resources on hydroponics and vertical farming in sustainable agriculture, the following references are recommended:

1. "The Vertical Farm: Feeding the World in the 21st Century" by Dr. Dickson Despommier 2. "Hydroponics: A Comprehensive Guide to Hydroponics at Home" by Bruce Walsh 3. "Vertical Farming: A Beginner's Guide to Vertical Farming" by Jason Green 4. "Hydroponic Food Production: A Definitive Guidebook for the Advanced Home Gardener and the Commercial Hydroponic Grower" by Howard M. Resh 5. Websites: The Association for Vertical Farming (https://vertical-farming.net/), The Hydroponics Association (https://www.hydroponicsassociation.org/)

Exercise

In a hydroponic system, the concentration of nutrient solution needs to be carefully controlled to ensure optimal plant growth. It is desired to prepare a nutrient solution with a targeted concentration of 800 ppm (parts per million) of nitrogen (N). The commercially available nitrogen stock solution has a concentration of 1500 ppm. How much nitrogen stock solution (in mL) should be added to 1 liter of water to prepare the desired nutrient solution? (Assume no dilution of the stock solution is required.)

Solution

The desired concentration of nitrogen (N) in the nutrient solution is 800 ppm. The stock solution has a concentration of 1500 ppm. Let's assume V mL of the stock solution is added to 1 liter of water.

The concentration of nitrogen in the final solution can be calculated using the formula:

$$C_f = \frac{C_i \times V_i}{V_f}$$

where: C_f = final concentration of nitrogen in the nutrient solution (800 ppm) C_i = initial concentration of nitrogen in the stock solution (1500 ppm) V_i = volume of the stock solution added (unknown) V_f = final volume of the nutrient solution (1000 mL)

Rearranging the formula to solve for V_i, we have:

$$V_i = \frac{C_f \times V_f}{C_i}$$

Plugging in the values, we get:

$$V_i = \frac{800 \times 1000}{1500} = 533.33 \text{ mL}$$

Therefore, approximately 533.33 mL of the nitrogen stock solution should be added to 1 liter of water to prepare the desired nutrient solution with a concentration of 800 ppm of nitrogen.

Note: It is important to handle and measure the stock solution accurately to achieve the desired nutrient concentration.

Biotechnology in Agriculture

Biotechnology, a field at the intersection of biology and technology, has revolutionized many areas of human life, and agriculture is no exception. By harnessing the power of biotechnology, scientists have developed innovative techniques to enhance crop productivity, improve plant resistance to pests and diseases, and reduce the environmental impact of agriculture. In this section, we will explore the principles, applications, and challenges of biotechnology in agriculture.

Principles of Biotechnology

Biotechnology in agriculture utilizes the principles of genetic engineering, molecular biology, and cell culture techniques to manipulate the genetic material of plants. The key principles involved in biotechnology include:

INNOVATIONS IN AGRICULTURAL TECHNOLOGIES

1. **Genetic modification:** Genetic modification involves the transfer of specific genes from one organism to another, resulting in the expression of desired traits in the recipient organism. This process is achieved through techniques such as gene cloning, recombinant DNA technology, and gene editing.

2. **DNA sequencing and analysis:** DNA sequencing allows scientists to determine the precise sequence of nucleotides within genes. This information is crucial for identifying genes responsible for specific traits and designing genetic modifications.

3. **Gene expression regulation:** Gene expression refers to the process by which genetic information is used to synthesize functional products such as proteins. Biotechnology techniques enable scientists to control gene expression, allowing the production of desirable proteins or suppression of unwanted proteins.

4. **Cell and tissue culture:** Cell and tissue culture techniques involve growing plant cells or tissues in vitro under controlled conditions. This allows for the production of genetically identical plants on a large scale, known as micropropagation, and the regeneration of whole plants from small tissue samples.

Applications of Biotechnology in Agriculture

Biotechnology has a wide range of applications in agriculture, contributing to the development of crops with improved traits, efficient plant breeding techniques, and sustainable farming practices. Some key applications include:

1. **Genetically modified (GM) crops:** GM crops have been developed to exhibit traits such as increased yield, improved nutritional content, pest and disease resistance, and tolerance to environmental stresses. For example, Bt cotton, a genetically modified cotton variety, produces a naturally occurring insecticide that protects the plant against bollworms.

2. **Crop improvement through genetic markers:** Genetic markers are specific sequences of DNA associated with desirable traits. By identifying and selecting plants with these markers, breeders can accelerate the process of developing new crop varieties with improved traits, such as drought tolerance or disease resistance.

3. **Molecular diagnostics for plant diseases:** Biotechnology techniques, such as polymerase chain reaction (PCR), allow for the rapid and accurate detection of plant pathogens. This helps farmers identify and manage diseases early, preventing widespread crop damage and reducing the need for chemical pesticides.

4. **Plant tissue culture for propagation:** Plant tissue culture techniques enable the production of a large number of genetically identical plants from a small piece of tissue. This is particularly useful for propagating plants with desirable traits, such as high-yielding or disease-resistant varieties, and for preserving rare or endangered plant species.

5. **Phytoremediation:** Biotechnology is used in phytoremediation, which involves using plants to clean up contaminated soils or water. Certain plants have the ability to accumulate pollutants, such as heavy metals, and degrade or immobilize toxic compounds, thus helping to restore polluted environments.

Challenges in Biotechnology

While biotechnology offers immense potential for improving agricultural practices, it also raises several ethical, environmental, and regulatory challenges that need to be addressed. Some of the key challenges are:

1. **Regulation and public acceptance:** Biotechnology, especially genetically modified organisms (GMOs), has faced significant regulatory scrutiny and public skepticism. Striking a balance between ensuring safety and fostering innovation is essential to harness the benefits of biotechnology in agriculture.

2. **Resistance and unintended effects:** Pests and diseases can develop resistance to genetically modified traits over time, rendering them less effective. Furthermore, unintended effects, such as off-target gene modifications, can occur during genetic engineering and need to be carefully monitored and managed.

3. **Intellectual property and access:** The patenting of genetically modified crops and biotechnological processes can limit access for small-scale farmers and hinder the development of region-specific, context-appropriate solutions. Ensuring fair and equitable access to biotechnological innovations is crucial for sustainable agricultural development.

CASE STUDIES IN SUSTAINABLE AGRICULTURE 263

4. **Environmental impact:** Biotechnology should be harnessed in a way that minimizes potential negative environmental impacts. For example, the cultivation of genetically modified crops should be coupled with appropriate management strategies to prevent the development of resistant weeds or the disruption of non-target organisms.

Case Study: Golden Rice

Golden Rice is a genetically modified rice variety that has been engineered to produce beta-carotene, a precursor of vitamin A. Vitamin A deficiency is a leading cause of childhood blindness and other health issues in developing countries. Golden Rice was developed to address this issue by providing a dietary source of vitamin A, particularly in regions where rice is a staple food.

The development and deployment of Golden Rice faced significant challenges, including regulatory barriers, intellectual property concerns, and opposition from anti-GMO activists. However, efforts have been underway to address these challenges and make Golden Rice available to those who can benefit from it. The case of Golden Rice highlights the potential of biotechnology to address important nutritional challenges, while also emphasizing the need for an evidence-based and inclusive approach to agricultural innovation.

Conclusion

Biotechnology has transformed the field of agriculture, offering innovative solutions to enhance crop productivity, improve plant resilience, and promote sustainable farming practices. By utilizing genetic engineering, molecular biology, and other biotechnological tools, scientists have developed genetically modified crops, efficient breeding techniques, and environmentally friendly farming methods. However, the successful integration of biotechnology into agriculture requires addressing ethical, environmental, and regulatory challenges. Through careful and responsible application, biotechnology can contribute to a greener and more secure future for agriculture.

Case Studies in Sustainable Agriculture

Community-supported Agriculture

Community-supported agriculture (CSA) is an alternative model of food production and distribution that directly connects farmers and consumers. In a

264 CHAPTER 4: SUSTAINABLE AGRICULTURE AND FOOD SYSTEMS

CSA system, individuals or families become members by purchasing a share or subscription in advance, typically at the beginning of a growing season. In return, members receive a regular supply of fresh, locally grown produce throughout the season.

Principles of Community-supported Agriculture

CSA is based on several key principles that set it apart from conventional agricultural practices. These principles include:

1. Local and sustainable: CSAs prioritize the production of locally grown food, reducing transportation emissions and promoting food sovereignty. By using sustainable farming practices, such as organic or agroecological methods, CSAs aim to minimize the environmental impact of food production.

2. Direct farmer-consumer relationship: CSAs foster direct connections between farmers and consumers, allowing for transparency and trust in the food system. Through on-farm visits, newsletters, and other forms of communication, members have the opportunity to learn about the farming practices and challenges faced by their local farmers.

3. Shared risk and reward: CSAs operate on a shared risk and reward system. Members acknowledge that farming is influenced by various factors beyond the control of farmers, such as weather conditions and pests. By sharing both the risks and potential rewards, CSAs create a sense of community and mutual support.

4. Seasonality and diversity: CSAs embrace the natural cycle of seasonal produce, providing members with a diverse range of fruits, vegetables, and sometimes other products like eggs, honey, or meat. This encourages members to try new foods, appreciate the uniqueness of each season, and reconnect with the rhythms of nature.

Benefits of Community-supported Agriculture

CSAs offer several benefits to both farmers and consumers:

1. Fresh and nutritious produce: CSA members enjoy access to fresh, locally grown food that is harvested at its peak ripeness, ensuring maximum flavor and nutritional value. The shorter time between harvest and consumption also reduces nutrient loss.

2. Support for local farmers: By participating in a CSA, members directly support local farmers and their livelihoods. This financial stability enables farmers to invest in sustainable and regenerative farming practices, ensuring the long-term health of the land.

CASE STUDIES IN SUSTAINABLE AGRICULTURE

3. Strengthened community connections: CSAs create a sense of community by bringing farmers and consumers together. Members often have the opportunity to visit the farm, participate in farm events, and interact with other members. This strengthens social ties and fosters a shared commitment to local food systems.

4. Environmental sustainability: CSAs prioritize sustainable farming practices such as organic or agroecological methods. These practices promote soil health, biodiversity, and reduce the use of synthetic pesticides and fertilizers. By supporting CSAs, consumers contribute to the preservation of the environment.

Challenges and Future Directions of Community-supported Agriculture

While community-supported agriculture has many benefits, it also faces several challenges:

1. Financial constraints: Starting and maintaining a CSA requires significant financial investment. Farmers need resources to purchase seeds, equipment, and infrastructure, as well as cover labor and operational costs. Establishing a stable customer base through a CSA membership model helps address these challenges.

2. Seasonal limitations: Since CSA farms prioritize seasonal and local produce, there may be limitations on the variety of crops available throughout the year. However, some CSAs offer extended or year-round memberships by partnering with other local producers or incorporating preservation techniques like canning or drying.

3. Distribution logistics: Efficient distribution of CSA shares can be a logistical challenge. Farmers need to coordinate harvest schedules, packing, and delivery to ensure that members receive their shares in a timely manner. This requires careful planning and organization.

4. Education and awareness: Increasing awareness and educating the public about the benefits of CSAs is crucial for their long-term success. By promoting the value of local and sustainable food systems, CSAs can attract more members and build stronger community support.

The future directions of CSAs involve embracing technological advancements, such as online platforms for ordering and tracking shares, to enhance convenience and streamline communication between farmers and members. Additionally, exploring partnerships with other local food initiatives, such as farmers' markets or restaurants, can help expand the reach and impact of CSAs.

Overall, community-supported agriculture plays a vital role in promoting sustainable food systems, supporting local economies, and fostering vibrant communities. By actively participating in a CSA, individuals can become active agents of change and contribute to a more resilient and equitable food future.

266 CHAPTER 4: SUSTAINABLE AGRICULTURE AND FOOD SYSTEMS

Regenerative Farming Practices

Regenerative farming practices are a set of agricultural techniques that aim to restore and improve the health of soils while reducing the negative impacts of conventional farming on the environment. These practices focus on building soil organic matter, increasing biodiversity, enhancing water retention, and sequestering carbon.

Importance of Regenerative Farming

Conventional farming methods often involve heavy use of synthetic fertilizers, pesticides, and intensive tillage, which can degrade soil fertility, decrease biodiversity, contribute to water pollution, and release greenhouse gases into the atmosphere. In contrast, regenerative farming practices promote sustainable and resilient agricultural systems by working with natural processes rather than against them.

Principles of Regenerative Farming

Regenerative farming practices are guided by several key principles:

1. **Minimize Soil Disturbance:** Instead of extensive tilling, regenerative farmers use minimal or no-till practices to preserve soil structure and prevent erosion. This helps maintain soil health and encourages the growth of beneficial microorganisms.

2. **Maximize Soil Cover:** Cover crops, crop residues, and organic mulches are used to protect the soil from erosion, suppress weed growth, and improve soil moisture retention. This practice also enhances soil organic matter content and promotes nutrient cycling.

3. **Diversify Crop Rotation:** Planting a variety of crops in a rotation pattern breaks pest and disease cycles, increases nutrient availability, and improves soil fertility. Cover crops, legumes, and nitrogen-fixing plants are often included in crop rotations to enhance soil nitrogen levels naturally.

4. **Integrate Livestock:** Integrating livestock into cropping systems allows for the recycling of nutrients through managed grazing and manure application. This practice helps improve soil fertility, enhances nutrient cycling, and reduces the need for synthetic fertilizers.

5. **Promote Biodiversity:** Creating diverse ecosystems on and around farmland encourages the presence of beneficial insects, birds, and other wildlife that can

CASE STUDIES IN SUSTAINABLE AGRICULTURE 267

help control pests naturally. Planting hedgerows, windbreaks, and wildflower strips provides habitat for these beneficial organisms.

Examples of Regenerative Farming Techniques

Several regenerative farming techniques are being adopted around the world with promising results. Let's explore a few examples:

1. **Agroforestry Systems:** Agroforestry combines the cultivation of trees with agricultural crops or livestock. This technique enhances biodiversity, improves soil structure, provides shade and wind protection, and offers additional income streams from timber, fruits, or nuts.

2. **Keyline Design:** Keyline design is a water management technique that aims to capture and distribute rainfall evenly across a farm. By creating contour lines on the landscape, water runoff is minimized, groundwater recharge is increased, and soil erosion is reduced.

3. **Holistic Planned Grazing:** This technique involves managed grazing of livestock, mimicking the natural movement and behavior of herds. Proper grazing rotation ensures that the land is not overgrazed, promotes the growth of diverse grass species, and improves soil health.

4. **Aquaponics:** Aquaponics combines aquaculture (fish farming) with hydroponics (cultivating plants in water) in a symbiotic system. The waste produced by fish is used as nutrients for plants, while the plants filter the water for the fish. This closed-loop system minimizes water use and allows for year-round crop production.

Benefits and Challenges

Regenerative farming practices offer several benefits:

- **Improved Soil Health:** These practices enhance soil fertility, biodiversity, and water retention, resulting in healthier and more productive soils.

- **Carbon Sequestration:** Regenerative techniques promote the capture and storage of atmospheric carbon dioxide in the soil, mitigating climate change.

- **Reduced Chemical Use:** By minimizing synthetic fertilizer and pesticide application, regenerative farming reduces the negative impacts on human health and the environment.

- **Better Water Management:** Cover crops and reduced tillage help conserve water, prevent soil erosion, and improve water quality by reducing runoff of pollutants.

- **Resilient Farming Systems:** Regenerative practices increase the resilience of farms to extreme weather events, pest outbreaks, and market fluctuations.

However, there are also challenges to implementing regenerative farming practices:

- **Knowledge and Education:** Farmers need access to training, technical assistance, and research to adopt and apply regenerative practices effectively.

- **Economic Viability:** Transitioning to regenerative systems may require upfront investments and changes in production methods, which can be financially challenging for some farmers.

- **Policy and Market Support:** Adequate policies and incentives are needed to promote and reward farmers practicing regenerative agriculture. Consumer demand for sustainably produced food also plays a crucial role.

- **Scaling Up:** While there are numerous success stories of regenerative farming, scaling up these practices to a broader agricultural landscape requires collaboration, knowledge sharing, and supportive infrastructure.

Conclusion

Regenerative farming practices offer a promising approach to address sustainability challenges in agriculture. By focusing on soil health, biodiversity, and ecosystem resilience, these practices can contribute to a more sustainable and ecologically sound food system. As more farmers adopt regenerative techniques and policymakers provide support, we have the potential to transform the agricultural sector and create a greener future.

Further Reading

- Gliessman, S.R. (2018). *Agroecology: The Ecology of Sustainable Food Systems.* CRC Press.

- Montgomery, D.R. (2017). *Growing a Revolution: Bringing Our Soil Back to Life.* W.W. Norton & Company.

CASE STUDIES IN SUSTAINABLE AGRICULTURE 269

- Altieri, M.A. (2018). *Agroecology: Science and Politics.* Agroecology Program, UC Berkeley.

- Kiss, A. (2020). *Soil Regeneration: A Practical Guide to Regenerative Agriculture.* Chelsea Green Publishing.

Rooftop and Urban Farming

Rooftop and urban farming is an innovative and sustainable approach to agriculture that involves growing crops and raising animals in urban areas, especially on rooftops. This practice has gained popularity in recent years due to its numerous environmental, social, and economic benefits. In this section, we will explore the principles and techniques of rooftop and urban farming, as well as examine some successful case studies and the challenges associated with this practice.

Principles of Rooftop and Urban Farming

Rooftop and urban farming is rooted in the principles of sustainable agriculture and aims to maximize the use of urban spaces for food production. The following principles guide the practice of rooftop and urban farming:

1. **Land conservation:** With the growing demand for food and the decreasing availability of arable land, utilizing rooftops and other urban spaces for farming helps conserve valuable land resources.

2. **Food security:** Urban farming can contribute to local food security by reducing dependence on long-distance transportation and increasing access to fresh, nutritious produce within cities.

3. **Environmental sustainability:** By bringing agriculture into urban areas, rooftop farming reduces the carbon footprint associated with transportation and distribution of food. It also promotes biodiversity and reduces the urban heat island effect by providing more green spaces.

4. **Resource efficiency:** Rooftop farming can be designed to optimize resource use, such as water and energy. Conservation techniques like rainwater harvesting and recycling of organic waste can be integrated into rooftop farming systems.

270 CHAPTER 4: SUSTAINABLE AGRICULTURE AND FOOD SYSTEMS

5. **Community engagement:** Rooftop farming has the potential to bring communities together, fostering social connections and creating opportunities for education, skill-building, and employment.

Techniques of Rooftop and Urban Farming

The success of rooftop and urban farming relies on the adoption of suitable techniques that maximize productivity in limited space. Some common techniques employed in rooftop farming include:

1. **Container gardening:** Utilizing containers, such as pots, raised beds, or vertical structures, allows plants to be grown even on small rooftop spaces. This technique enables better control over soil quality, drainage, and plant spacing.

2. **Hydroponics:** Hydroponic systems, which involve growing plants in a nutrient-rich water solution without soil, are well-suited for rooftop farming. These systems use less water and space compared to traditional soil-based cultivation. Nutrients are supplied directly to the roots, leading to faster plant growth and higher yields.

3. **Green roofs:** Green roofs combine vegetation with a waterproofing membrane, providing additional insulation and stormwater management benefits. By incorporating edible plants into green roofs, rooftop farming can be integrated seamlessly into building design.

4. **Vertical farming:** Vertical farming utilizes vertical space to maximize the number of crops grown. Systems such as vertical hydroponics or vertical aeroponics allow plants to be stacked or placed in towers, maximizing space utilization while minimizing the footprint.

5. **Aquaponics:** Aquaponics is a symbiotic system that combines hydroponics with fish farming. Nutrient-rich fish wastewater is used as a fertilizer for plants, while the plants filter and purify the water for the fish. This closed-loop system reduces the need for external inputs and increases overall resource efficiency.

Case Studies of Rooftop and Urban Farming

Numerous successful rooftop and urban farming projects around the world are changing the way cities produce food. Let's explore a few noteworthy examples:

CASE STUDIES IN SUSTAINABLE AGRICULTURE

1. **Brooklyn Grange, New York City, USA:** Brooklyn Grange operates the world's largest rooftop soil farm, spanning over two and a half acres. They utilize extensive green roofs to grow a wide variety of vegetables and herbs. The farm not only supplies fresh produce to local markets and restaurants but also offers educational programs and event spaces.

2. **Lufa Farms, Montreal, Canada:** Lufa Farms operates commercial rooftop greenhouses across Montreal, producing over 40 different types of vegetables year-round. Their innovative model incorporates hydroponics and advanced environmental control systems to optimize plant growth and maximize resource efficiency.

3. **Sky Greens, Singapore:** Singapore, known for its limited land availability, has embraced urban farming on a vertical scale. Sky Greens operates vertical hydroponic farms using rotating towers, allowing them to grow vegetables at different heights. These farms provide fresh greens to local communities and reduce reliance on food imports.

4. **Rooftop Republic, Hong Kong:** Rooftop Republic converts underutilized rooftops into productive urban farming spaces, providing fresh produce to local residents. They collaborate with building owners and offer workshops and training to promote sustainable agriculture and community engagement.

Challenges and Future Directions

While rooftop and urban farming offer exciting opportunities, several challenges need to be addressed for its widespread adoption:

1. **Limited space and structural constraints:** Rooftop farming requires careful consideration of weight limitations, access, and structural stability. Developing strategies to address these constraints is essential for successful implementation.

2. **Water and energy requirements:** Urban environments often face constraints in water availability and energy resources. Developing efficient irrigation systems and exploring renewable energy sources can enhance the sustainability of rooftop farming.

3. **Policies and regulations:** Clear guidelines and policies are needed to facilitate the integration of rooftop farming into urban planning. This includes addressing issues related to building codes, land use, and food safety regulations.

272 CHAPTER 4: SUSTAINABLE AGRICULTURE AND FOOD SYSTEMS

4. **Education and awareness:** Creating awareness among urban dwellers about the benefits of rooftop farming and providing resources for education and training can help build public support and participation.

In conclusion, rooftop and urban farming hold immense potential to transform our cities into sustainable food-producing hubs. By leveraging underutilized spaces, employing innovative techniques, and fostering community engagement, we can contribute to food security, environmental sustainability, and the well-being of urban communities. With the right strategies and support, rooftop and urban farming can play a significant role in creating a greener and more resilient future.

Exercise:

Think about a building in your city that has suitable rooftop space for farming. Develop a plan to convert the rooftop into a productive urban farm. Consider the techniques, resources, and potential challenges you might face. Discuss your plan with a group and gather feedback to improve your proposal.

Agroforestry Systems

Agroforestry is a sustainable land use system that combines trees and shrubs with agricultural crops and/or livestock production. It is an ancient practice that has gained significant attention in recent years due to its numerous environmental and economic benefits. Agroforestry systems are designed to mimic natural ecosystems, where different species coexist and interact synergistically. This section will explore the principles and advantages of agroforestry, discuss various agroforestry techniques, and provide examples of successful agroforestry projects.

Principles of Agroforestry

Agroforestry is based on the following principles:

1. **Diversity:** Agroforestry systems promote biodiversity by integrating multiple species of trees, shrubs, and crops. This diversity enhances ecosystem resilience and provides habitat for a wide range of organisms.

2. **Complementary Interactions:** Agroforestry combines plants with symbiotic relationships, such as nitrogen-fixing trees that enhance soil fertility, or shade trees that provide protection and microclimates for understory crops.

3. **Productivity and Stability:** Agroforestry systems aim to increase overall farm productivity and provide stable yields over time. By diversifying

CASE STUDIES IN SUSTAINABLE AGRICULTURE

income sources, farmers are less vulnerable to market fluctuations and climate variability.

4. **Sustainable Resource Management:** Agroforestry promotes the efficient use of natural resources such as water, nutrients, and sunlight. By optimizing resource allocation, agroforestry reduces environmental degradation and improves ecosystem services.

5. **Resilience to Climate Change:** Agroforestry systems are inherently resilient to climate change. The diversity and structural complexity of these systems provide protection against extreme weather events and help mitigate the impacts of climate change.

Types of Agroforestry Systems

There are various types of agroforestry systems, each with its own unique combination of trees, crops, and livestock. Some common types include:

1. **Alley Cropping:** In this system, rows of trees or shrubs are planted in between rows of crops. The trees provide shade, windbreak, and contribute organic matter to the soil, while the crops benefit from reduced soil erosion and increased moisture retention.

2. **Silvopasture:** Silvopasture combines trees, forage crops, and livestock grazing. The trees offer shade and shelter for animals, while the animals help to control weeds and provide natural fertilizers. This system enhances livestock productivity and improves pasture quality.

3. **Forest Gardens:** Forest gardens mimic the structure and functions of natural forests. They consist of layers of fruit and nut trees, shrubs, herbs, and ground covers. This system maximizes productivity and biodiversity, providing a sustainable source of food and income.

4. **Windbreaks and Shelterbelts:** Windbreaks are rows of trees or shrubs planted to provide protection from strong winds. They reduce soil erosion, create microclimates, and improve crop yields. Shelterbelts serve a similar purpose but are designed to protect livestock or farm infrastructure.

5. **Riparian Agroforestry:** Riparian areas, such as riverbanks or wetlands, are planted with trees and shrubs to enhance water quality, reduce erosion, and provide habitat for wildlife. This system improves stream health and provides ecological functions in addition to agricultural benefits.

274 CHAPTER 4: SUSTAINABLE AGRICULTURE AND FOOD SYSTEMS

Success Stories in Agroforestry

Agroforestry has been successfully implemented in various parts of the world, bringing multiple benefits to farmers and the environment. Here are a few examples:

1. **The Shamba System in East Africa:** The Shamba system combines fruit and timber trees with agricultural crops such as corn, beans, and vegetables. It has proven to be highly productive, providing food security, improved soil fertility, and increased income for smallholder farmers. The system also contributes to climate change mitigation by sequestering carbon in trees.

2. **The Coffee Agroforestry System in Central and South America:** Coffee agroforestry is an example of shade-grown coffee production, where coffee plants are grown under a canopy of shade trees. This system enhances biodiversity by providing habitat for birds and other animals. It also improves coffee quality and resilience to climate change by providing a more stable microclimate.

3. **Taungya System in Southeast Asia:** The Taungya system combines tree planting with shifting cultivation. Farmers plant trees on fallow lands as they prepare for the next crop cycle. This system helps to restore degraded forests, provides additional income from timber sales, and improves soil fertility.

Challenges and Future Directions

While agroforestry offers many benefits, there are challenges that need to be addressed to promote widespread adoption. Some of these challenges include:

1. **Knowledge and Awareness:** Many farmers are unaware of the potential benefits of agroforestry and lack the necessary knowledge and skills to implement these systems effectively. Providing education and training programs can help overcome this barrier.

2. **Access to Resources:** Farmers may face challenges in accessing quality seeds, technical support, and financing for initial investments. Policies and initiatives that support agroforestry practices and provide resources to farmers are needed.

CASE STUDIES IN SUSTAINABLE AGRICULTURE

3. **Policy and Land Tenure:** Agroforestry often requires long-term land tenure to provide incentives for farmers to invest in tree planting. Policymakers need to create an enabling environment that recognizes the value of agroforestry and supports land tenure security.

4. **Scaling Up:** While there are successful agroforestry projects, scaling up these systems to a landscape or regional level remains a challenge. Collaboration between governments, NGOs, and research institutions is crucial to promote large-scale adoption.

5. **Climate Change Adaptation and Resilience:** As climate change continues to pose challenges, agroforestry systems need to be adapted to changing conditions. Research efforts should focus on identifying tree species and management practices that are resilient to climate change.

In conclusion, agroforestry systems offer a sustainable and resilient approach to agriculture that can address multiple environmental and economic challenges. By integrating trees, crops, and livestock, agroforestry promotes biodiversity, enhances soil fertility, and provides multiple income sources for farmers. While challenges exist, successful agroforestry projects around the world demonstrate the potential of these systems. With increased knowledge, awareness, and support, agroforestry has the potential to contribute to a more sustainable and resilient food system.

Indigenous Farming Practices and Traditional Knowledge

Indigenous farming practices and traditional knowledge play a crucial role in sustainable agriculture. These practices have been developed over centuries by indigenous communities around the world, who have successfully cultivated crops while preserving biodiversity and maintaining environmental balance. In this section, we will explore the principles behind indigenous farming practices, their importance in sustainable agriculture, and some success stories in employing these practices.

Principles of Indigenous Farming Practices

Indigenous farming practices are rooted in a deep understanding of the natural environment and its cycles. They are based on principles that promote harmony between humans and nature, ensuring the long-term sustainability of agricultural systems. Some key principles of indigenous farming practices include:

276 CHAPTER 4: SUSTAINABLE AGRICULTURE AND FOOD SYSTEMS

1. Agrobiodiversity: Indigenous communities recognize the importance of growing a diverse range of crops. They cultivate a variety of traditional and local crop varieties that are adapted to their specific climatic conditions and have evolved over generations. This agrobiodiversity not only ensures food security but also enhances resilience against pests, diseases, and climatic variations.

2. Organic and low-input farming: Indigenous farmers adopt organic and low-input farming techniques, avoiding the use of synthetic fertilizers, pesticides, and herbicides. Instead, they rely on natural methods such as composting, crop rotation, intercropping, and the use of natural pest control measures. This ensures that the soil remains fertile, conserves biodiversity, and minimizes harm to the environment.

3. Traditional knowledge and cultural practices: Indigenous farming practices are deeply rooted in traditional knowledge passed down through generations. This knowledge encompasses a deep understanding of local ecosystems, including soil types, water cycles, and plant interactions. Indigenous farmers employ traditional techniques such as seed saving, traditional breeding, and land-use practices, which have been refined over centuries to optimize agricultural production sustainably.

Importance in Sustainable Agriculture

Indigenous farming practices hold immense value in achieving sustainable agricultural systems. They offer several benefits that contribute to environmental conservation, food security, and the preservation of cultural heritage:

1. Biodiversity conservation: Indigenous farming practices are closely linked to the conservation of agrobiodiversity. By growing a wide range of crop varieties, indigenous farmers contribute to the preservation of traditional crops that might otherwise be lost. This biodiversity not only provides a diverse and nutritious diet but also ensures the availability of genetic resources for future crop improvement efforts.

2. Climate resilience: Indigenous farming practices are often adapted to local climatic conditions. Traditional knowledge allows farmers to predict weather patterns, select suitable crops, and employ specific cultivation techniques that enhance resilience against climate change impacts such as droughts, floods, and temperature fluctuations. These practices can play a vital role in ensuring food security in the face of climate uncertainties.

3. Cultural heritage preservation: Indigenous farming practices are closely intertwined with cultural traditions and identities. They form an integral part of indigenous communities' cultural heritage and social fabric. By promoting and preserving these practices, not only are we conserving agricultural knowledge, but

CASE STUDIES IN SUSTAINABLE AGRICULTURE

we are also acknowledging and respecting the rights and contributions of indigenous communities.

Success Stories in Indigenous Farming Practices

Many success stories demonstrate the efficacy of indigenous farming practices in achieving sustainable agriculture. Here are a few notable examples:

1. **Milpa Agriculture in Mexico:** Milpa is a traditional farming system practiced by indigenous communities in Mexico. It involves the intercropping of maize, beans, and squash, which complement each other in terms of nutrient requirements and growth patterns. This agricultural system not only provides a balanced diet but also promotes the efficient use of resources and enhances soil fertility.

2. **Terraced Farming in the Philippines:** The Ifugao people in the Philippines have created an intricate system of terraced rice fields on steep slopes. These terraces prevent soil erosion, conserve water, and maximize land utilization. This sustainable farming practice has been recognized as a UNESCO World Heritage site and has sustained local communities for more than 2,000 years.

3. **Permaculture in Australia:** Permaculture, a system that integrates agriculture with design principles from natural ecosystems, has been widely adopted by indigenous communities in Australia. By mimicking the patterns and processes of nature, such as soil regeneration, water harvesting, and companion planting, permaculture offers sustainable and productive farming systems.

4. **Traditional Rice Cultivation in Bali:** In Bali, Indonesia, the traditional rice cultivation system known as Subak has been practiced for centuries by Balinese farmers. This cooperative water management system ensures equitable distribution of water resources among farmers while respecting ecological balances. The conservation of water sources and the integration of temple rituals into agricultural practices make Subak a successful example of sustainable and culturally significant farming.

Challenges and Future Directions

Despite the proven benefits of indigenous farming practices, a range of challenges threatens their continuation and mainstream adoption. Some key challenges include:

1. Land rights and tenure: Indigenous communities often face issues of land tenure, as their traditional lands are at risk of being taken over by commercial agriculture or development projects. Protecting indigenous land rights is crucial for maintaining traditional farming practices.

278 CHAPTER 4: SUSTAINABLE AGRICULTURE AND FOOD SYSTEMS

2. Knowledge transmission: The transmission of traditional ecological knowledge from one generation to another is declining due to various factors, including urbanization, globalization, and the erosion of cultural connections. Efforts should be made to document and revive traditional knowledge systems.

3. Institutional support: Indigenous farmers often lack access to technical support, financial resources, and markets for their produce. Strengthening support networks and institutions that encourage indigenous farming practices is essential for their long-term sustainability.

4. Recognition and empowerment: Indigenous knowledge and practices are often undervalued and marginalized. Recognizing the contributions of indigenous communities in sustainable agriculture and empowering them to make decisions about their lands and resources are crucial steps towards promoting indigenous farming practices.

The future direction for indigenous farming practices lies in fostering collaborations between indigenous communities, scientists, policymakers, and other stakeholders. By combining traditional knowledge with modern scientific approaches, we can develop innovative and context-specific solutions for sustainable agriculture. Furthermore, integrating indigenous farming practices into educational curricula and research programs will help ensure the preservation and advancement of this invaluable knowledge.

In conclusion, indigenous farming practices and traditional knowledge offer valuable insights and contributions to sustainable agriculture. Harnessing their principles and wisdom can enhance agrobiodiversity, climate resilience, and cultural heritage preservation. Recognizing the importance of indigenous farming and addressing the challenges they face will help pave the way for a more sustainable and inclusive agricultural future.

Success Stories in Sustainable Agriculture

Organic Farming Movement

The organic farming movement has gained significant momentum in recent years as consumers become more concerned about the environmental and health impacts of conventional agricultural practices. Organic farming is an approach that emphasizes sustainable and natural methods of crop and livestock production while minimizing the use of synthetic chemicals and genetic modifications. In this section, we will explore the principles of organic farming, its benefits, and some successful examples of its implementation.

SUCCESS STORIES IN SUSTAINABLE AGRICULTURE

Principles of Organic Farming

Organic farming is based on several key principles that guide its practices:

1. **Soil Health:** Organic farmers prioritize the health of the soil, recognizing it as a living system that provides the foundation for plant growth. They focus on building and maintaining soil fertility through practices such as crop rotation, cover cropping, and the use of organic matter like compost and manure.

2. **Biodiversity:** Organic farmers promote biodiversity by creating and preserving habitats for a variety of plant and animal species. They avoid the use of synthetic pesticides and genetically modified seeds, which can harm beneficial organisms and reduce biodiversity.

3. **Ecological Balance:** Organic farming aims to maintain ecological balance by using natural methods to control pests and diseases. This includes techniques like crop rotation, biological pest control, and the use of natural predators.

4. **No Synthetic Chemicals:** Organic farmers avoid the use of synthetic chemicals, including synthetic fertilizers, pesticides, and herbicides. Instead, they rely on natural alternatives, such as compost, crop rotation, and beneficial insects, to manage pests and maintain the health of their crops.

5. **Animal Welfare:** Organic farming places a strong emphasis on the humane treatment and welfare of animals. Livestock are raised in conditions that allow them to express their natural behaviors, and the use of antibiotics and growth hormones is prohibited.

Benefits of Organic Farming

The organic farming movement has gained popularity due to its numerous benefits, which include:

1. **Environmental Sustainability:** Organic farming practices help protect the environment by reducing pollution, conserving water and energy, and preserving biodiversity. Organic farmers work in harmony with nature, promoting the health of ecosystems and the conservation of natural resources.

2. **Improved Soil Quality:** Organic farming methods prioritize soil health, leading to improved soil structure, nutrient cycling, and water-holding capacity. This, in turn, supports better plant growth, enhances crop resilience to climate change, and reduces soil erosion.

3. **Safer Food:** Organic farming eliminates the use of synthetic pesticides, herbicides, and genetically modified organisms, resulting in food that is free from potentially harmful residues. Organic produce is also often richer in essential nutrients and antioxidants.

280 CHAPTER 4: SUSTAINABLE AGRICULTURE AND FOOD SYSTEMS

4. **Support for Small-Scale Farmers:** Organic farming provides opportunities for small-scale farmers to thrive by offering premium prices for their products. It allows them to differentiate themselves in the market and connect directly with environmentally conscious consumers.

Successful Examples of Organic Farming

The success of organic farming can be seen in various parts of the world. Here are a few notable examples:

1. **Sikkim, India:** In 2016, the Indian state of Sikkim became the world's first fully organic state. Through government initiatives and support for farmers, Sikkim transitioned its entire agricultural land to organic methods, resulting in improved soil health, increased biodiversity, and reduced chemical pollution.

2. **Château Maris, France:** Château Maris, a winery located in the Languedoc-Roussillon region of France, has gained recognition for its organic and biodynamic practices. By employing sustainable viticulture techniques, such as cover cropping and the use of natural predators, they have produced high-quality, environmentally friendly wines.

3. **Rodale Institute, USA:** The Rodale Institute, a nonprofit research institution in Pennsylvania, has been a pioneer in organic farming since the 1940s. Their long-term Farming Systems Trial has demonstrated the environmental and agronomic benefits of organic farming, including improved soil health, higher yields, and reduced greenhouse gas emissions.

Challenges and Future Directions

While the organic farming movement has witnessed significant growth, it still faces certain challenges and opportunities for improvement. Some of these include:

1. **Scaling Up:** Although organic farming has expanded, it still represents a small fraction of overall agricultural production. Scaling up organic farming practices while ensuring economic viability for farmers remains a challenge.

2. **Access to Organic Food:** Organic food can be more expensive than conventionally produced food, making it less accessible to lower-income populations. Increasing access to organic food through initiatives like community-supported agriculture (CSA) and farmers' markets is essential.

3. **Research and Innovation:** Continued research and innovation are needed to address the technical and agronomic challenges of organic farming, such as improving yields, developing pest-resistant varieties, and optimizing nutrient management.

SUCCESS STORIES IN SUSTAINABLE AGRICULTURE

4. **Policy Support:** Governments play a crucial role in supporting the organic farming movement through policies that provide incentives for farmers, promote access to organic food, and ensure transparent labeling and certification standards.

As the demand for sustainable and healthy food continues to grow, the organic farming movement is likely to evolve and expand. By addressing the challenges and harnessing the opportunities, organic farming can contribute to a more sustainable and resilient agricultural system.

Did you know?

In addition to producing food, organic farming has other applications such as organic cotton farming to promote sustainable and ethical practices in the textile industry. By avoiding the use of synthetic pesticides and fertilizers, organic cotton farming reduces the environmental impact of conventional cotton production and promotes the health of both farmers and consumers.

Agroecology in Cuba

Agroecology is a sustainable and holistic approach to agriculture that promotes the harmonious integration of plants, animals, and humans within ecological systems. It emphasizes the use of traditional and indigenous knowledge, as well as modern scientific principles, to design and manage agricultural systems that are both productive and environmentally friendly. One country that has successfully implemented agroecology principles on a national scale is Cuba.

Background

Cuba's journey towards agroecology began in the early 1990s, following the collapse of the Soviet Union and the loss of its primary trading partner. This event led to a severe economic crisis in Cuba, which severely affected its agriculture sector. The country faced food shortages, as well as a decline in agricultural inputs such as fertilizers, pesticides, and machinery.

In response to these challenges, the Cuban government launched the "Special Period in Peacetime" program, which aimed to decentralize agriculture and promote sustainable practices. This program placed a strong emphasis on agroecology and organic farming techniques as a means of ensuring food security and self-sufficiency.

Principles of Agroecology in Cuba

The success of agroecology in Cuba can be attributed to several key principles and strategies. These include:

282 CHAPTER 4: SUSTAINABLE AGRICULTURE AND FOOD SYSTEMS

- **Diversification of crops and farms:** Agroecology promotes the cultivation of a wide variety of crops and the integration of different agricultural activities within a farm. This diversification helps to improve soil fertility, prevent pests and diseases, and increase resilience to climate change.

- **Crop rotation and intercropping:** In Cuba, farmers practice crop rotation and intercropping to optimize resource utilization and reduce the risk of crop failure. These techniques also promote natural pest control and improve soil health by enhancing nutrient cycling.

- **Use of organic fertilizers and biological pest control:** Cuba's agroecological systems rely on organic fertilizers such as compost, green manure, and biofertilizers. Similarly, biological pest control methods, such as the use of beneficial insects and biopesticides, are preferred over synthetic pesticides.

- **Conservation of water resources:** Given the limited availability of water in Cuba, agroecology emphasizes water conservation practices such as rainwater harvesting, drip irrigation, and mulching. These practices help to minimize water wastage and improve the efficiency of irrigation.

- **Promotion of local food systems:** Agroecology in Cuba aims to establish local and decentralized food systems to reduce dependence on imports. This involves supporting small-scale farmers, promoting farmers' markets, and fostering community participation in food production.

Successes of Agroecology in Cuba

The implementation of agroecology in Cuba has yielded impressive results in terms of food production, environmental sustainability, and social equity. Some of the notable successes include:

- **Increase in organic farming:** Cuba has one of the highest percentages of organic farmland in the world, with over 400,000 hectares dedicated to organic agriculture. This shift towards organic farming has reduced the use of synthetic fertilizers and pesticides, resulting in healthier ecosystems and reduced environmental pollution.

- **Self-sufficiency in food production:** Despite facing economic challenges and limited access to external resources, Cuba has been able to achieve near self-sufficiency in food production. Agroecology has played a crucial role in ensuring food security by empowering small-scale farmers and promoting local food systems.

SUCCESS STORIES IN SUSTAINABLE AGRICULTURE

+ **Conservation of biodiversity:** Agroecological practices in Cuba have contributed to the preservation of biodiversity on agricultural lands. By promoting the use of diverse crop varieties and preserving traditional farming practices, agroecology has helped to conserve native plant species and protect habitats for wildlife.

+ **Empowerment of rural communities:** The adoption of agroecology has empowered rural communities in Cuba by providing them with opportunities for self-employment, income generation, and food sovereignty. Small-scale farmers are supported through training programs, access to land, and credit facilities, enabling them to sustain their livelihoods.

Challenges and Future Directions

While agroecology in Cuba has achieved significant successes, it also faces several challenges and future directions for further improvement. These include:

+ **Limited access to inputs and resources:** The economic constraints faced by Cuba pose challenges in terms of accessing necessary inputs and resources for agroecological practices. Investments are needed to improve infrastructure, provide appropriate technologies, and enhance research and development in agroecology.

+ **Climate change adaptation:** Cuba, like many other countries, is vulnerable to the impacts of climate change. The agroecological systems in Cuba need to continually adapt to changing climatic conditions to ensure their resilience and sustainability. This requires ongoing research, monitoring, and the development of appropriate adaptation strategies.

+ **Scaling up and knowledge transfer:** While agroecology has been successful at the local and regional levels in Cuba, there is a need to scale up these practices to achieve wider impact. This involves knowledge transfer, capacity building, and the dissemination of best practices to farmers across the country.

+ **Integration of agroecology into policy frameworks:** To ensure the long-term success of agroecology in Cuba, it is essential to integrate agroecological principles into national agricultural policies and regulations. This includes providing incentives and support for agroecological practices, as well as addressing any barriers or conflicts with existing policies.

284 *CHAPTER 4: SUSTAINABLE AGRICULTURE AND FOOD SYSTEMS*

In conclusion, agroecology has played a crucial role in Cuba's agricultural transformation, contributing to increased food security, environmental sustainability, and rural empowerment. The success of agroecology in Cuba highlights the potential of this approach in addressing the challenges faced by modern agriculture. By embracing agroecology, other countries can learn valuable lessons from Cuba's experiences and work towards building more sustainable and resilient food systems.

Integrated Pest Management

Integrated Pest Management (IPM) is a holistic approach to pest control that aims to minimize the impact of pests on agriculture while reducing reliance on chemical pesticides. IPM incorporates a combination of strategies such as biological control, cultural practices, and chemical control, with a focus on sustainable and environmentally friendly methods. This section will explore the principles and techniques of IPM, highlight successful case studies, and discuss the challenges and future directions in implementing IPM.

Principles of Integrated Pest Management

The principles of IPM revolve around the integration of multiple strategies to manage pests effectively. These strategies include:

1. **Monitoring and identification of pests:** Regular monitoring of pest populations and accurate identification of pests are fundamental to IPM. This allows farmers to understand the type and extent of pest problems, enabling them to make informed decisions regarding pest control strategies.

2. **Prevention and cultural practices:** IPM emphasizes the importance of preventive measures and cultural practices that discourage pest infestation. This includes crop rotation, planting resistant varieties, maintaining proper sanitation, and optimizing irrigation and nutrient management. By creating unfavorable conditions for pests, farmers can reduce their reliance on pesticides.

3. **Biological control:** Biological control involves using natural enemies of pests to regulate their populations. These natural enemies can be predators, parasitoids, or pathogens that prey upon or infect pests. IPM promotes the conservation and augmentation of these beneficial organisms to control pests effectively. For example, ladybugs are introduced to control aphids, and Bacillus thuringiensis (Bt) is used to combat certain caterpillar pests.

4. **Chemical control as a last resort:** While chemical pesticides are part of IPM, they are considered a last resort and are used only when other strategies are

SUCCESS STORIES IN SUSTAINABLE AGRICULTURE

inadequate. When chemical control is necessary, selective and target-specific pesticides are preferred, minimizing non-target impacts. Additionally, pesticide application timing and dosage should be optimized to maximize effectiveness and minimize environmental risks.

5. **Continuous evaluation and adaptation:** IPM is a dynamic process that requires continuous evaluation and adaptation. Regular monitoring of pest populations, efficacy of control methods, and environmental impacts helps farmers fine-tune their pest management strategies. By learning from previous experiences and adopting new technologies, farmers can improve their overall pest control practices.

Case Studies in Integrated Pest Management

1. Integrated Weed Management in Organic Farming

Weeds are a significant challenge in organic farming, where synthetic herbicides cannot be used. To address this, integrated weed management practices have been implemented. These include the use of cover crops, mulching, mechanical cultivation, and hand weeding. Cover crops suppress weed growth through competition, mulching acts as a physical barrier, and mechanical cultivation removes weed seedlings. By combining these practices, farmers can effectively manage weed populations while maintaining organic certification.

2. Integrated Pest Management in Rice Cultivation

Rice is a staple crop that is susceptible to the attacks of pests and diseases. In traditional rice cultivation, the excessive use of chemical pesticides posed environmental and health risks. However, with the adoption of IPM, the pest management approach has shifted towards a more sustainable and integrated approach. Farmers now rely on pheromone traps for monitoring and mass trapping of insect pests, biological control agents such as spiders and predatory insects, and resistant rice varieties. This integrated approach has reduced pesticide use, minimized pest damage, and improved the overall sustainability of rice production.

3. Integrated Pest Management in Greenhouse Production

Greenhouse production faces unique challenges due to the controlled environment that can favor pest outbreaks. In this context, IPM has proven to be effective in managing pests while minimizing pesticide use. Integrated strategies include the use of insect screens to prevent pest entry, deployment of beneficial insects to control pests, and the implementation of cultural practices such as crop rotation and sanitation. This combination of techniques ensures that the greenhouse environment remains pest-free and minimizes the impact on the surrounding ecosystem.

286 CHAPTER 4: SUSTAINABLE AGRICULTURE AND FOOD SYSTEMS

Challenges and Future Directions in Integrated Pest Management

Despite the successes of IPM, several challenges need to be addressed for its wider adoption and implementation. These challenges include:

1. **Knowledge and awareness:** Many farmers, especially small-scale farmers, may not have access to comprehensive knowledge about IPM practices and their benefits. Training programs, extension services, and knowledge-sharing platforms need to be developed to increase awareness and understanding of IPM principles and techniques.

2. **Financial limitations:** Implementing IPM requires an initial investment in infrastructure, equipment, and training, which may be a barrier for resource-constrained farmers. Government support, subsidies, and financial incentives are crucial to enable farmers to adopt and implement IPM on a larger scale.

3. **Limited research and development:** Continued research and development are essential to improving existing IPM strategies and developing new, innovative approaches. Investment in research to identify and develop effective biological control agents, optimize cultural practices, and enhance monitoring techniques is necessary to overcome the challenges faced in managing pests sustainably.

4. **Socioeconomic and cultural factors:** Socioeconomic factors, such as market demand and consumer preferences, can influence the adoption of IPM practices. Farmers need to be assured of price premiums or market access for adopting IPM methods. Additionally, cultural factors, traditions, and perceptions of pest control also play a role in the adoption and implementation of IPM. Addressing these factors is essential for the successful integration of IPM in diverse agricultural systems.

5. **Climate change impacts:** Climate change can influence pest dynamics, with changes in temperature, rainfall patterns, and extreme weather events affecting the distribution and abundance of pests. Adaptation of IPM strategies to these changing conditions, such as developing pest forecasting models and promoting climate-resilient cultural practices, will be crucial in managing pests effectively in a changing climate.

In conclusion, Integrated Pest Management is a sustainable and environmentally friendly approach to pest control. By integrating various strategies and minimizing chemical pesticide use, IPM offers a holistic and effective solution to managing pests in agriculture. Successful case studies in various agricultural systems demonstrate the potential of IPM to reduce the reliance on chemical pesticides while maintaining crop productivity and minimizing environmental impacts. However, challenges related to knowledge dissemination, financial limitations, research and development, socioeconomic factors, and climate change

SUCCESS STORIES IN SUSTAINABLE AGRICULTURE

impacts need to be addressed for the broader adoption and implementation of IPM.

Smallholder Farmers' Cooperatives

In this section, we will explore the concept of smallholder farmers' cooperatives and their role in sustainable agriculture. Smallholder farmers, who typically own or lease small plots of land, face numerous challenges in today's agricultural industry. These challenges include limited access to resources, lack of market power, and vulnerability to climate change. Smallholder farmers' cooperatives aim to address these issues by promoting collective action, resource sharing, and improved market opportunities.

The Importance of Smallholder Farmers

Smallholder farmers play a crucial role in global food production and ensuring food security, particularly in developing countries. Despite owning relatively small plots of land, they collectively account for a significant portion of agricultural output. Smallholder farming systems often prioritize diversified and sustainable farming practices, such as agroecology, organic farming, and the preservation of traditional knowledge and crop diversity. Additionally, smallholder farmers are often deeply rooted in their communities and have a strong connection to the land, maintaining cultural heritage and fostering rural development.

Definition and Principles of Cooperatives

A cooperative is an autonomous association of individuals or organizations voluntarily united to meet their common economic, social, and cultural needs through a jointly owned and democratically controlled enterprise. Smallholder farmers' cooperatives bring together farmers in a specific geographic area to collectively address their common challenges and improve their livelihoods. They operate based on the following principles:

1. Voluntary and Open Membership: Membership in a smallholder farmers' cooperative is open to all farmers in the designated area who share the common goals and objectives.

2. Democratic Control: Cooperatives are autonomous organizations controlled by their members, who actively participate in decision-making processes.

288 CHAPTER 4: SUSTAINABLE AGRICULTURE AND FOOD SYSTEMS

3. Economic Participation: Members contribute equitably to, and democratically control, the capital of the cooperative. Surplus generated is allocated in a manner agreed upon by the members.

4. Cooperation Among Cooperatives: Cooperatives work together to strengthen their collective power and achieve common goals through mutual support and collaboration.

5. Education, Training, and Information: Cooperatives provide education and training opportunities to their members to enhance their knowledge and skills in various aspects of agriculture and cooperative management.

6. Concern for the Community: Cooperatives work for the sustainable development of their communities by actively engaging in activities that address social and environmental needs.

Benefits of Smallholder Farmers' Cooperatives

Smallholder farmers' cooperatives offer numerous benefits to their members and the broader agricultural sector:

1. Access to Resources Cooperatives pool resources, such as land, capital, machinery, and inputs like seeds and fertilizers, to overcome individual limitations. This collective approach enables smallholder farmers to access resources and technologies that would otherwise be financially out of reach for individual farmers.

2. Market Power By joining forces, smallholder farmers' cooperatives can negotiate better prices for their produce, increase their bargaining power with buyers, and access larger markets. Cooperatives can also provide storage facilities, transportation services, and marketing support to help farmers connect with consumers directly or through value-added processing.

3. Knowledge Sharing and Technical Assistance Cooperatives facilitate the exchange of knowledge and best practices among members. Farmers can learn from each other's experiences, adopt innovative farming techniques, and receive technical assistance from agricultural experts, extension services, and research institutions.

SUCCESS STORIES IN SUSTAINABLE AGRICULTURE

4. Risk Mitigation Smallholder farmers are often vulnerable to climate change, pests, diseases, and volatile market conditions. Cooperatives can help farmers manage risks by collectively investing in climate-smart practices, crop insurance, diversification strategies, and market information systems.

5. Empowerment and Social Cohesion Cooperatives strengthen farmers' collective decision-making power and enhance their participation in local governance. By working together, farmers build social networks, foster solidarity, and promote gender equality in agriculture.

Case Study: Smallholder Coffee Farmers' Cooperative in Colombia

One inspiring example of a successful smallholder farmers' cooperative is the Coffee Growers' Cooperative in Colombia. With over 500,000 small coffee farmers as members, this cooperative has played a pivotal role in transforming Colombia into one of the world's leading coffee producers.

The cooperative facilitates knowledge sharing among members to improve farming practices and increase the quality and productivity of coffee production. By collectively investing in infrastructure and processing facilities, the cooperative has enabled farmers to add value to their coffee beans and access specialty markets. Additionally, the cooperative provides technical assistance, financial services, and fair trade certification to its members.

Through the cooperative, smallholder coffee farmers in Colombia have gained greater control over the production and marketing of their coffee. They have improved their livelihoods and reduced their vulnerability to international coffee price fluctuations. The cooperative has also empowered farmers to invest in local community development projects, such as schools, healthcare facilities, and clean water initiatives.

Challenges and Solutions

While smallholder farmers' cooperatives offer numerous benefits, they also face various challenges:

1. Financing and Capital Accessing affordable credit and capital is often a significant challenge for smallholder farmers' cooperatives. Lack of collateral and financial literacy among farmers can hinder their ability to secure loans. Solutions include establishing microfinance initiatives tailored to the needs of smallholder

290 CHAPTER 4: SUSTAINABLE AGRICULTURE AND FOOD SYSTEMS

farmers, advocating for favorable cooperative financing policies, and developing innovative financing models such as community-based revolving funds.

2. Governance and Management Cooperatives require effective governance structures and skilled management to ensure transparency, accountability, and efficient operations. Training programs on cooperative management, leadership, and financial management are essential to build the capacity of cooperative members and staff.

3. Market Access and Value Chain Integration Integrating smallholder farmers' cooperatives into mainstream agricultural value chains can be challenging. Cooperatives need to establish partnerships with other actors in the value chain, invest in processing and quality assurance, and comply with market standards and certifications. Capacity-building programs focusing on market knowledge, negotiation skills, and product diversification can enhance market access for cooperative members.

4. Climate Change Resilience Smallholder farmers are particularly vulnerable to the impacts of climate change. Cooperatives can help farmers adapt to these challenges by promoting climate-smart agricultural practices, facilitating access to climate information and technologies, and supporting diversification strategies that increase resilience against climate-related risks.

Conclusion

Smallholder farmers' cooperatives are an important tool for promoting sustainable agriculture and improving the livelihoods of small-scale farmers. By fostering collective action, resource sharing, and market access, cooperatives empower farmers to overcome challenges and achieve greater economic and social well-being. Governments, development agencies, and civil society organizations must continue to support the establishment and strengthening of smallholder farmers' cooperatives to create a fair and sustainable agricultural system for all stakeholders involved.

Sustainable Fisheries and Aquaculture

In this section, we will explore the concept of sustainable fisheries and aquaculture and discuss the importance of these practices in ensuring the long-term viability of

SUCCESS STORIES IN SUSTAINABLE AGRICULTURE

seafood production. We will also delve into the challenges faced by the fishing industry and highlight success stories in sustainable fisheries management.

Understanding Sustainable Fisheries

Sustainable fisheries refer to the responsible management of fish populations and their habitats in order to maintain their productivity over time. This involves practices that aim to prevent overfishing, minimize bycatch, protect critical marine habitats, and promote the overall health of aquatic ecosystems.

The Importance of Sustainable Fisheries

Maintaining sustainable fisheries is crucial for several reasons. Firstly, it helps to ensure the long-term availability of seafood for consumption. With the global population increasing, the demand for seafood is also rising. Unsustainable fishing practices can deplete fish stocks and disrupt the delicate balance of marine ecosystems, leading to a decline in the availability of seafood.

Secondly, sustainable fisheries protect biodiversity and the overall health of marine ecosystems. Fish play a vital role in maintaining the balance of aquatic ecosystems. When fish populations are overexploited, it can have ripple effects throughout the food web, affecting other marine organisms, including coral reefs, seabirds, and marine mammals.

Furthermore, sustainable fisheries contribute to the economic well-being of coastal communities. Many communities around the world rely on fishing as a primary source of income and livelihood. By implementing sustainable practices, we can ensure the long-term viability of these communities and support their economic development.

Challenges in Fisheries Management

Despite the importance of sustainable fisheries, the fishing industry faces several challenges:

1. Overfishing: Overfishing occurs when fish stocks are harvested at an unsustainable rate, leading to the depletion of populations. This is often driven by high demand and inadequate fisheries management.

2. Bycatch: Bycatch refers to the unintentional capture of non-target species, such as dolphins, sea turtles, and seabirds, during fishing operations. Bycatch can have significant negative impacts on these species and contribute to overall ecosystem degradation.

292 CHAPTER 4: SUSTAINABLE AGRICULTURE AND FOOD SYSTEMS

3. Illegal, unreported, and unregulated (IUU) fishing: IUU fishing undermines conservation and management efforts by disregarding fishing regulations and quotas. It not only depletes fish populations but also threatens the livelihoods of legal fishers and the sustainability of the fishing industry.

Success Stories in Sustainable Fisheries Management

In recent years, several success stories have emerged in the realm of sustainable fisheries management. These examples demonstrate that with effective management measures, it is possible to restore fish populations and promote sustainable practices:

1. The Icelandic Cod Fishery: In the 1980s, the cod population in Icelandic waters was severely depleted due to overfishing. However, through the implementation of science-based quotas, fishing gear modifications, and strict enforcement, the cod stocks have made a remarkable recovery. This success story highlights the importance of effective fisheries management and the potential for recovery even in highly depleted populations.

2. The MSC Certification Program: The Marine Stewardship Council (MSC) is an independent, nonprofit organization that sets standards for sustainable fishing practices. Fisheries that meet these standards are awarded MSC certification, which indicates that their seafood comes from a certified sustainable source. The MSC certification program has helped to raise consumer awareness about sustainable seafood and incentivize fishing practices that prioritize long-term sustainability.

3. Community-Based Fisheries Management: In some regions, local communities have taken the lead in managing their fisheries sustainably. By involving fishermen, scientists, and other stakeholders in decision-making processes, community-based fisheries management has shown promising results. This approach ensures that fishing practices align with the needs and priorities of the communities while maintaining ecological integrity.

Ensuring a Sustainable Future for Aquaculture

Aquaculture, the farming of aquatic organisms, has gained significant importance in meeting global seafood demand. However, ensuring the sustainability of aquaculture practices is essential to avoid negative environmental impacts and maintain long-term productivity.

Key principles for sustainable aquaculture include:

CHALLENGES AND FUTURE DIRECTIONS IN SUSTAINABLE AGRICULTURE

- Site selection: Choosing suitable locations for aquaculture facilities is crucial to minimize environmental impacts. Factors such as water quality, habitat suitability, and carrying capacity should be considered.

- Disease management: Disease outbreaks can have devastating effects on farmed fish populations. Implementing effective disease prevention and control measures, such as proper farm hygiene and monitoring, is vital to ensure the health and welfare of farmed fish.

- Feed sustainability: The use of sustainable and environmentally-friendly feeds is essential in reducing the reliance on wild-caught fish for fishmeal and oil production. Developing alternative feeds based on plant-based ingredients and reducing reliance on wild fish stocks is a key goal.

- Genetic improvement: Selective breeding and genetic improvement programs can enhance the productivity, disease resistance, and overall sustainability of farmed fish populations.

Conclusion

Sustainable fisheries and aquaculture play a crucial role in meeting the world's increasing demand for seafood while protecting marine ecosystems. By implementing effective fisheries management measures, combating overfishing, and promoting sustainable aquaculture practices, we can ensure the future availability of seafood, protect biodiversity, and support the livelihoods of coastal communities. It is our responsibility to make informed choices as consumers and support the transition towards a more sustainable and resilient seafood industry.

Challenges and Future Directions in Sustainable Agriculture

Ensuring Food Security and Eliminating Hunger

Food security is a fundamental concern in today's world, as millions of people around the globe are still suffering from hunger and malnutrition. It is crucial to address this issue and work towards eliminating hunger by adopting sustainable agricultural practices, improving access to nutritious food, and enhancing food distribution systems. This section will explore the principles, challenges, and strategies involved in ensuring food security and reducing hunger.

Background and Principles

Food security can be defined as the condition in which all individuals have physical, social, and economic access to sufficient, safe, and nutritious food that meets their dietary needs and preferences for an active and healthy life. It encompasses four key dimensions: availability, accessibility, utilization, and stability. The principles underlying food security include:

1. **Sustainable agriculture:** To achieve food security, it is essential to adopt sustainable agricultural practices that promote productivity, minimize environmental degradation, and conserve natural resources. This includes practices such as organic farming, permaculture, and agroforestry.

2. **Diversification of food sources:** Ensuring a variety of food sources is crucial to address nutritional deficiencies and improve food access. Promoting diverse crop cultivation, including both traditional and underutilized crops, can enhance food security by increasing diet diversity.

3. **Efficient resource management:** Optimizing the use of resources such as land, water, and energy in agriculture is necessary to minimize waste and increase productivity. Implementing sustainable soil management practices, precision agriculture, and efficient irrigation techniques can improve resource utilization.

4. **Climate resilience:** Building resilience in agricultural systems to climate change is vital for ensuring food security. Developing drought-tolerant crop varieties, implementing climate-smart agricultural practices, and promoting agroecology are essential steps towards climate-resilient agriculture.

5. **Social equity and empowerment:** Addressing food security requires tackling social inequalities and empowering vulnerable populations. Enhancing access to land, credit, technology, and training for smallholder farmers and marginalized communities is crucial for achieving food security and eliminating hunger.

6. **Policy support and governance:** Governments play a critical role in ensuring food security through the formulation and implementation of policies and programs. Effective governance, supportive institutional frameworks, and investment in agricultural research and development are key components of a comprehensive food security strategy.

CHALLENGES AND FUTURE DIRECTIONS IN SUSTAINABLE AGRICULTURE

Challenges in Ensuring Food Security

Achieving food security and eliminating hunger face several challenges, including:

1. **Population growth:** The global population is projected to reach almost 10 billion by 2050, placing increased pressure on food production systems. Feeding the growing population while maintaining sustainability is a major challenge.

2. **Limited resources and land degradation:** Land degradation, water scarcity, and depletion of natural resources pose significant constraints on agricultural productivity. Finding innovative and sustainable solutions to maximize resource use is crucial.

3. **Climate change impacts:** Climate change disrupts agricultural production through extreme weather events, altered precipitation patterns, and shifting pest and disease dynamics. Adapting agriculture to changing climatic conditions is a major challenge for food security.

4. **Income inequality and poverty:** Inadequate access to resources and opportunities, coupled with income inequality, contribute to food insecurity. Addressing poverty and enhancing social protection systems are critical for improving food access.

5. **Food waste and loss:** A significant amount of food produced is wasted or lost throughout the supply chain, from farm to consumer. Minimizing food waste and improving post-harvest handling and storage can enhance food security.

6. **Conflicts and political instability:** Many regions affected by conflicts and political instability face severe food insecurity. Ending conflicts and promoting stability are essential for ensuring access to food.

Strategies for Ensuring Food Security

To address the challenges of food security and eliminate hunger, several strategies are essential:

1. **Investing in sustainable agriculture:** Promoting sustainable agricultural practices, such as organic farming, agroecology, and precision agriculture, can increase productivity, conserve resources, and improve resilience to climate change.

2. **Improving access to credit and resources:** Ensuring smallholder farmers and marginalized communities have access to credit, land, technology, and training is vital for enhancing agricultural productivity and reducing poverty.

3. **Implementing nutrition-sensitive agriculture:** Integrating nutrition considerations into agricultural practices can improve diet diversity and address malnutrition. Promoting the cultivation of nutrient-rich crops and supporting home gardening and livestock rearing can enhance food security.

4. **Enhancing market infrastructure and value chains:** Strengthening market infrastructure, supporting farmer cooperatives, and developing efficient value chains can improve access to markets and enhance farmers' income.

5. **Investing in research and development:** Improving agricultural productivity, developing resilient crop varieties, and adopting innovative technologies require investment in agricultural research and development. Supporting research institutions and promoting knowledge exchange are essential for achieving food security.

6. **Promoting policy coherence and governance:** Coordinating policies across different sectors, such as agriculture, health, and trade, is crucial for addressing food security comprehensively. Strengthening governance, increasing public and private investments, and ensuring policy support are vital.

7. **Enhancing social protection and safety nets:** Ensuring social safety nets, such as cash transfer programs, school feeding initiatives, and maternal and child nutrition programs, can alleviate immediate hunger and improve food access among vulnerable populations.

Case Study: Zero Hunger Initiative in Brazil

Brazil's Zero Hunger Initiative, launched in 2003, is an example of a successful effort to address food security and eliminate hunger. The initiative aimed to integrate several programs and strategies to tackle the root causes of hunger, such as poverty, inequality, and limited access to productive resources.

The Zero Hunger Initiative focused on expanding social protection programs, enhancing small-scale agriculture, promoting agroecology, and improving access to education and healthcare. It encouraged the participation of local communities and prioritized the needs of women, children, and indigenous populations.

CHALLENGES AND FUTURE DIRECTIONS IN SUSTAINABLE AGRICULTURE

Through the Zero Hunger Initiative, Brazil significantly reduced hunger and malnutrition rates, improved agricultural productivity, and empowered marginalized communities. The initiative demonstrated the importance of a comprehensive approach to food security, addressing social, economic, and environmental dimensions.

Conclusion

Ensuring food security and eliminating hunger is a complex challenge that requires the adoption of sustainable agricultural practices, investing in research and development, improving access to resources, and promoting inclusive governance. By integrating principles of sustainable agriculture, diversifying food sources, and addressing social inequalities, it is possible to achieve a world where everyone has access to safe, nutritious, and sufficient food. Governments, policymakers, and individuals all have a role to play in building a sustainable food system that can eliminate hunger and nourish our planet.

Sustainable Farming Policy and Government Support

Sustainable agriculture plays a crucial role in addressing global challenges such as food security, climate change, and environmental degradation. To promote sustainable farming practices, governments around the world have implemented policies and provided support to farmers. In this section, we will explore the principles of sustainable farming, discuss the role of government in supporting sustainable agriculture, and highlight some success stories in sustainable farming policy.

Principles of Sustainable Agriculture

Sustainable agriculture is an approach that aims to meet the needs of the present generation while preserving resources for future generations. It integrates three main principles: environmental stewardship, economic viability, and social equity.

1. **Environmental Stewardship:** Sustainable farming practices prioritize the conservation and protection of natural resources. This includes reducing the use of synthetic fertilizers and pesticides, managing soil erosion, preserving biodiversity, and promoting water and energy efficiency.

2. **Economic Viability:** Sustainable farming should be economically viable for farmers. It involves diversifying income sources, improving market access, reducing production costs, and enhancing productivity and profitability. By adopting

298 CHAPTER 4: SUSTAINABLE AGRICULTURE AND FOOD SYSTEMS

sustainable practices, farmers can reduce input dependency and enhance the resilience of their farming systems.

3. **Social Equity:** Sustainable agriculture emphasizes the well-being of farmers, farmworkers, and local communities. It promotes fair trade, supports family farmers, ensures safe working conditions, and respects the rights of indigenous communities. Social equity also includes providing access to resources, training, and education for all stakeholders in the agricultural sector.

Government Support for Sustainable Agriculture

Governments play a vital role in supporting sustainable farming practices through policies, programs, and financial incentives. Some key ways in which governments provide support include:

1. **Research and Development (R&D) Funding:** Governments allocate funds for agricultural R&D to develop and promote sustainable farming practices. This includes researching innovative technologies, improving crop varieties, and finding sustainable solutions to pest and disease management.

2. **Agricultural Extension Services:** Governments often provide agricultural extension services to disseminate information and knowledge about sustainable farming practices. These services help farmers adopt new technologies, improve their skills, and access training and technical support.

3. **Financial Incentives and Subsidies:** Governments offer financial incentives, grants, and subsidies to encourage farmers to transition to sustainable practices. These incentives may include financial support for organic certification, conservation practices, and renewable energy systems, among others.

4. **Market Support and Market Access:** Governments facilitate market access for sustainable agricultural products by establishing certification schemes, labeling programs, and promoting sustainable agriculture in public procurement. They may also support the development of local food markets and promote sustainable agriculture in trade agreements.

5. **Policy Frameworks and Regulations:** Governments develop policies and regulations to support sustainable farming practices. These may include environmental regulations, land-use policies, agroecological zoning, and regulations on the use of pesticides and genetically modified organisms (GMOs).

Success Stories in Sustainable Farming Policy

Several countries have implemented successful sustainable farming policies that have led to positive environmental, economic, and social outcomes. Let's explore some

CHALLENGES AND FUTURE DIRECTIONS IN SUSTAINABLE
AGRICULTURE

notable examples:

1. **European Union (EU):** The Common Agricultural Policy (CAP) of the EU has evolved over the years to prioritize sustainability. It includes measures to protect the environment, support organic farming, promote agroecological practices, and provide financial incentives for farmers to adopt sustainable practices. The EU has also established the European Innovation Partnership for Agricultural Productivity and Sustainability, which fosters collaboration and innovation in sustainable agriculture.

2. **Denmark:** Denmark has been a pioneer in sustainable agriculture, with a strong focus on organic farming and environmental stewardship. The government has set ambitious targets to increase organic farming and reduce chemical pesticide use. Through subsidies and technical support, Denmark has successfully transitioned a significant portion of its agriculture to organic and sustainable practices.

3. **United States:** The United States Department of Agriculture (USDA) has implemented various programs to support sustainable agriculture. The Conservation Stewardship Program provides financial incentives to farmers for implementing conservation practices on their land. The USDA also supports organic farming through the National Organic Program and provides grants for research and development of sustainable farming practices.

4. **Brazil:** Brazil's government has implemented policies to promote sustainable agriculture in the Amazon rainforest region. Through the Low-Carbon Agriculture Program, farmers receive financial incentives for adopting sustainable practices that reduce deforestation and greenhouse gas emissions. Additionally, the government has implemented strict regulations on land use and illegal deforestation.

Challenges and Future Directions in Sustainable Agriculture

While progress has been made in promoting sustainable farming, several challenges remain. Some of the key challenges and future directions for sustainable agriculture include:

1. **Scaling Up Adoption:** One of the main challenges is scaling up the adoption of sustainable farming practices. Government support should focus on providing technical assistance, training, and financial incentives to encourage more farmers to transition to sustainable practices.

2. **Knowledge and Information Gap:** Farmers often lack access to information and knowledge about sustainable farming practices. Governments should invest in agricultural extension services, farmer-to-farmer knowledge exchange programs,

300 CHAPTER 4: SUSTAINABLE AGRICULTURE AND FOOD SYSTEMS

and digital platforms to bridge the knowledge gap and promote sustainable agriculture.

3. **Policy Coherence:** There is a need for policy coherence across different sectors to promote sustainable agriculture effectively. Governments should ensure that policies related to agriculture, environment, trade, and finance are aligned to support sustainable farming practices.

4. **Climate Change Resilience:** Agriculture is highly vulnerable to climate change impacts. Governments should promote climate-smart agricultural practices that enhance resilience and adaptive capacity. This includes investing in climate-resilient crop varieties, water management systems, and sustainable soil management practices.

5. **Engaging Local Communities:** Effective engagement and participation of local communities are crucial for the success of sustainable agriculture. Governments should involve farmers, indigenous communities, and other stakeholders in the decision-making process and ensure their voices are heard.

In conclusion, sustainable farming policy and government support play a vital role in promoting sustainable agriculture. Governments around the world have implemented policies, provided financial incentives, and supported research and development to encourage farmers to adopt sustainable practices. While challenges remain, successful examples of sustainable farming policies provide valuable lessons and insights for the future of sustainable agriculture. By prioritizing environmental stewardship, economic viability, and social equity, governments can contribute to a greener and more sustainable future in the agricultural sector.

Resources:

1. Food and Agriculture Organization of the United Nations. (2019). *Sustainable Impact Assessment in Agriculture.* Retrieved from [https://www.fao.org/sustainable-assessment/en/](https://www.fao.org/sustainable-

2. European Commission. (2021). *Common Agricultural Policy.* Retrieved from [https://ec.europa.eu/agriculture/cap_overview/index_en.htm](https://ec.europa.eu

3. United States Department of Agriculture. (2021). *Sustainable Agriculture.* Retrieved from [https://www.usda.gov/topics/sustainable-agriculture](https://www.usda.gov/topics

4. Brazilian Agriculture and Livestock Research Corporation. (2021). *Low-Carbon Agriculture Program.* Retrieved from [https://www.embrapa.br/en/tema/agricultura-de-baixo-carbono](https://www.emb

CHALLENGES AND FUTURE DIRECTIONS IN SUSTAINABLE AGRICULTURE

Reducing Food Waste and Loss

Food waste and loss have become significant global challenges that not only contribute to the loss of valuable resources but also exacerbate food insecurity and environmental degradation. According to the Food and Agriculture Organization (FAO) of the United Nations, approximately one-third of all food produced for human consumption is wasted or lost globally, amounting to around 1.3 billion metric tons per year. This wastage occurs throughout the entire food supply chain, from production and transportation to storage and consumption.

Understanding Food Waste and Loss

Food waste refers to the discarding or loss of food that is still safe and suitable for human consumption at some point along the supply chain. It can occur during production, post-harvest handling and storage, processing, distribution, and even at the household level. On the other hand, food loss refers specifically to the decrease in the quantity or quality of food resulting from inefficiencies in the supply chain, including spoilage during transportation, inadequate infrastructure, poor packaging, and inadequate market access.

Reducing food waste and loss is crucial for a sustainable food system and achieving the goal of zero hunger. It not only has economic and social benefits but also helps in conserving natural resources and reducing greenhouse gas emissions.

Causes of Food Waste and Loss

There are various factors that contribute to food waste and loss, including:

+ **Poor agricultural practices:** Inefficient farming practices, such as improper harvesting techniques and lack of proper storage facilities, can lead to significant post-harvest losses.

+ **Inadequate infrastructure:** Limited access to proper storage, transportation, and cooling facilities in many regions results in spoilage and deterioration of perishable food items before they can reach the consumers.

+ **Lack of market opportunities:** Inadequate market linkages and limited demand for certain agricultural products lead to their rejection or discarding by farmers.

+ **Consumer behavior:** Overbuying, improper storage, and expiration of food products at the household level contribute significantly to overall food waste.

302 *CHAPTER 4: SUSTAINABLE AGRICULTURE AND FOOD SYSTEMS*

+ **Food supply chain inefficiencies:** Inefficient logistics, poor inventory management, and improper handling during transportation and storage can result in food losses.

Strategies to Reduce Food Waste and Loss

Addressing the issue of food waste and loss requires a multi-faceted approach involving various stakeholders, including farmers, food processors, distributors, retailers, consumers, and policy-makers. Here are some effective strategies to reduce food waste and loss:

1. **Improving agricultural practices:** Implementing better farming techniques, such as precision agriculture, drip irrigation, and integrated pest management, can help minimize crop losses and improve overall productivity. Additionally, training farmers on proper harvesting, storage, and handling techniques can help reduce post-harvest losses.

2. **Enhancing infrastructure and transportation:** Investment in quality cold storage facilities, transportation infrastructure, and efficient supply chain logistics is necessary to minimize spoilage and reduce food losses during transportation and storage. This includes the development of proper packaging materials to extend the shelf life of perishable food items.

3. **Strengthening market linkages:** Improving market access and providing better market information to farmers can help reduce food losses caused by inadequate market opportunities. Supporting farmers' cooperatives and establishing direct links between producers and consumers, such as community-supported agriculture programs, can also help reduce waste.

4. **Educating consumers:** Raising awareness among consumers about the impacts of food waste and providing information on proper food storage, meal planning, and portion control can significantly reduce waste at the household level. Encouraging food donation and composting can also help divert food waste from landfills.

5. **Implementing policy measures:** Governments and regulatory bodies play a crucial role in reducing food waste and loss through the implementation of supportive policies and regulations. These may include setting targets to reduce food waste, promoting food recovery and redistribution programs, incentivizing

CHALLENGES AND FUTURE DIRECTIONS IN SUSTAINABLE AGRICULTURE

food producers and retailers to donate unsold or surplus food, and implementing stricter food labeling standards.

Case Studies and Success Stories

Several initiatives and success stories around the world have demonstrated the effectiveness of reducing food waste and loss:

1. **The Save Food Initiative:** Launched by FAO and the United Nations Environment Programme (UNEP), this global campaign engages stakeholders from all sectors to reduce food loss and waste. It promotes knowledge exchange and best practices, as well as provides technical support to countries in developing and implementing effective strategies.

2. **The Love Food Hate Waste Campaign:** Initiated by the Waste and Resources Action Programme (WRAP) in the United Kingdom, this campaign aims to raise awareness among consumers about the importance of reducing food waste. It provides practical tips and advice on meal planning, portion control, and creative recipes to utilize leftovers.

3. **The Food Recovery Network:** Operating across various college campuses in the United States, this student-led organization collects surplus food from dining halls and events and distributes it to local food banks and shelters. It not only reduces food waste but also helps address food insecurity in communities.

4. **The Gleaning Network:** Gleaning involves the collection of excess produce from farmers' fields after harvest, which would otherwise go to waste. Organizations like The Gleaning Network in the United Kingdom coordinate volunteers to harvest and distribute fresh produce to local charities and food banks.

5. **The Too Good To Go App:** This mobile application connects consumers with nearby food establishments that have surplus or unsold food at reduced prices towards the end of the day. By allowing users to purchase these food items, the app helps reduce food waste while providing affordable meals to consumers.

Challenges and Future Directions

While progress has been made in reducing food waste and loss, several challenges remain:

304 CHAPTER 4: SUSTAINABLE AGRICULTURE AND FOOD SYSTEMS

- Lack of awareness and coordination: Many consumers are unaware of the significant impacts of food waste and lack the knowledge or resources to minimize waste at the individual level. Coordinating efforts among various stakeholders along the supply chain can be challenging due to fragmented decision-making and lack of collaboration.

- Unequal distribution of resources: Developing countries often face challenges in investing in the necessary infrastructure and technologies to reduce food waste and loss. Adequate financial and technical support is needed to ensure equal access to solutions.

- Changing consumer behavior: Encouraging consumers to adopt behaviors that reduce food waste, such as proper meal planning and portion control, can be challenging. More research and educational initiatives are needed to influence consumer choices and promote sustainable food consumption.

- Policy and regulatory gaps: While policy measures and regulations have been implemented in some countries, there is still a need for stronger and more consistent policies at the national and international levels to encourage waste reduction and facilitate recovery and redistribution efforts.

- Continuous innovation: Advancements in technology and the development of innovative solutions, such as smart packaging, intelligent inventory management systems, and improved tracking and tracing technologies, can further assist in reducing food waste and loss.

Reducing food waste and loss is a complex but critical component of building a sustainable and resilient food system. By implementing a combination of effective strategies, raising awareness among consumers, promoting collaboration among stakeholders, and fostering supportive policies, we can contribute to a more sustainable future with less waste, increased food security, and improved environmental outcomes. Let's make every effort to prevent food waste and loss and ensure that no good food goes to waste.

Climate Change Resilience in Agriculture

Climate change is one of the most pressing challenges that the world is facing today. Its impacts are already being felt in various sectors, including agriculture. As temperatures rise, weather patterns become more unpredictable, and extreme weather events become more frequent, the agricultural sector is being severely affected. Crop yields are declining, pests and diseases are spreading, and water

CHALLENGES AND FUTURE DIRECTIONS IN SUSTAINABLE AGRICULTURE

availability is becoming increasingly unpredictable. In order to ensure food security for a growing population, it is crucial to enhance the resilience of agriculture to climate change.

Understanding Climate Change Impacts on Agriculture

Agriculture is highly dependent on climatic conditions, including temperature, precipitation, and sunlight. Any changes in these factors can significantly affect crop growth, livestock productivity, and overall agricultural productivity. Climate change can lead to the following impacts on agriculture:

1. Changes in temperature: Rising global temperatures can lead to increased evaporation, leading to water stress in crops. Temperature extremes can also affect the growth and development of crops, as different crops have specific temperature ranges within which they thrive.

2. Changes in precipitation patterns: Climate change can result in changes in rainfall patterns, leading to more frequent and intense droughts, as well as heavy rainfall events. Both droughts and excessive rainfall can negatively impact crop productivity and increase the risk of soil erosion.

3. Increased frequency of extreme weather events: Climate change is increasing the frequency and intensity of extreme weather events such as hurricanes, floods, and heatwaves. These events can cause significant damage to crops, livestock, and agricultural infrastructure.

4. Spread of pests and diseases: Changing climatic conditions can create more favorable environments for pests and diseases, leading to increased infestations and outbreaks. This can result in crop losses and the need for increased pesticide use.

5. Changes in water availability: Climate change can affect the availability of water for agriculture, both in terms of water scarcity and an increase in water variability. Changes in precipitation patterns can lead to water shortages, while increased variability can make irrigation planning and management more challenging.

Strategies for Climate Change Resilience in Agriculture

To enhance the resilience of agriculture to climate change, various strategies can be implemented. These strategies aim to minimize the negative impacts of climate change on agricultural systems and maximize their ability to adapt to changing conditions. Some key strategies include:

1. Improved crop selection and breeding: Selecting and breeding crop varieties that are more tolerant to heat, drought, and other climate-related stressors can help

306 CHAPTER 4: SUSTAINABLE AGRICULTURE AND FOOD SYSTEMS

ensure crop productivity in the face of changing climatic conditions. This can involve genetic modifications or traditional breeding techniques.

2. Diversification of crops and cropping systems: Diversifying crops and cropping systems can help reduce the risk of crop failure due to climate-related events. By growing a variety of crops, farmers can increase their resilience to pest and disease outbreaks and improve soil health.

3. Improved water management: Implementing efficient irrigation techniques, such as drip irrigation and precision irrigation, can help conserve water and ensure its optimal use. Water harvesting techniques, such as building reservoirs and using rainwater harvesting systems, can also help enhance water availability during periods of water scarcity.

4. Conservation and restoration of ecosystems: Conserving and restoring natural ecosystems, such as forests, wetlands, and grasslands, can help mitigate climate change by sequestering carbon dioxide and regulating local climatic conditions. Additionally, these ecosystems provide various ecosystem services that are essential for agricultural productivity, including pollination and natural pest control.

5. Adoption of agroecological practices: Agroecological practices, such as organic farming and regenerative agriculture, promote the use of natural processes and ecological principles to enhance soil health, biodiversity, and ecosystem functioning. These practices can improve the resilience of agricultural systems to climate change and reduce their environmental impact.

Example: Climate-Smart Agriculture

One example of a climate change resilience strategy in agriculture is climate-smart agriculture (CSA). CSA aims to sustainably increase agricultural productivity, adapt and build resilience to climate change, and reduce greenhouse gas emissions. It involves the integration of three main pillars: increasing productivity and incomes, building resilience and adapting to climate change, and reducing greenhouse gas emissions.

Implementing CSA practices can include:

1. Conservation agriculture: This involves minimal soil disturbance, diverse crop rotations, and the use of cover crops to improve soil health, reduce erosion, and conserve water.

2. Agroforestry: Combining trees with crops or livestock can provide multiple benefits, such as increased carbon sequestration, improved soil fertility, and enhanced biodiversity.

CHALLENGES AND FUTURE DIRECTIONS IN SUSTAINABLE AGRICULTURE

3. Water management strategies: Implementing efficient irrigation techniques, such as drip irrigation or rainwater harvesting, can help conserve water and ensure its optimal use.

4. Climate-smart livestock management: Introducing improved livestock management practices, such as rotational grazing and feed management, can reduce greenhouse gas emissions from the livestock sector and increase productivity.

5. Integrated pest management: Using a combination of cultural, biological, and chemical control methods can reduce reliance on pesticides and promote natural pest control.

CSA practices can help farmers adapt to changing climatic conditions, increase their resilience to extreme weather events, and reduce their contribution to climate change.

Challenges and Future Directions

Despite the importance of climate change resilience in agriculture, there are several challenges and barriers to its implementation. Some of the key challenges include:

1. Limited access to information and resources: Many farmers, especially those in developing countries, lack access to information, technologies, and financial resources needed to implement climate change resilience strategies.

2. Policy and institutional constraints: Inadequate policy frameworks and institutional support can hinder the adoption of climate change resilience practices. Government support and incentives are necessary to encourage farmers to implement these strategies.

3. Socio-economic factors: Factors such as poverty, land tenure issues, access to markets, and social dynamics can influence the ability of farmers to adopt climate change resilience practices.

4. Knowledge gaps: There are still many knowledge gaps regarding the impacts of climate change on agriculture and the most effective strategies for enhancing resilience. Research and knowledge sharing are crucial for addressing these gaps.

Moving forward, it is essential to prioritize climate change resilience in agriculture and enhance the capacity of farmers to implement climate-smart practices. This requires collaboration between governments, research institutions, NGOs, and farmers themselves. By investing in climate change resilience, agriculture can become more sustainable, productive, and resilient to the challenges posed by climate change.

308 CHAPTER 4: SUSTAINABLE AGRICULTURE AND FOOD SYSTEMS

Advancing Agroecological Research and Education

Agroecology is an interdisciplinary field that combines the principles of ecology, agriculture, and social sciences to develop sustainable and environmentally friendly farming practices. It seeks to optimize agricultural production while minimizing negative impacts on the environment. Advancing agroecological research and education is crucial for promoting sustainable agriculture and addressing the challenges we face in the 21st century.

Importance of Agroecology

Agroecology offers several benefits over conventional agricultural practices. Firstly, it promotes biodiversity and ecosystem resilience by utilizing natural ecological processes. By enhancing the diversity of crops and integrating livestock and plants, agroecological systems can better withstand pests, diseases, and climate variability. This reduces the reliance on synthetic inputs such as pesticides and fertilizers, which can have detrimental effects on the environment and human health.

Secondly, agroecology prioritizes soil health and fertility. It emphasizes the use of organic matter, cover crops, and crop rotation to improve soil structure, nutrient cycling, and water infiltration. Healthy soils not only support plant growth but also sequester carbon, mitigating climate change.

Furthermore, agroecology promotes community resilience and social equity. It encourages small-scale farming and local food systems, reducing dependence on multinational corporations and fostering food sovereignty. Agroecological practices also often require higher levels of knowledge and skills, creating opportunities for farmers to become expert practitioners and researchers.

Current Research Areas in Agroecology

Advancing agroecological research requires addressing key challenges and exploring innovative solutions. Here are some of the current research areas in agroecology:

1. **Climate Change Resilience:** Agroecological systems need to adapt to the changing climate by developing strategies to cope with increased temperatures, altered precipitation patterns, and more frequent extreme weather events. Research focuses on identifying climate-resilient crop varieties, implementing agroforestry practices, and optimizing water management techniques.

2. **Soil Health:** Soil degradation due to erosion, compaction, and nutrient depletion is a significant concern in modern agriculture. Agroecological research

CHALLENGES AND FUTURE DIRECTIONS IN SUSTAINABLE AGRICULTURE

aims to improve soil health through practices such as agroforestry, conservation tillage, and the use of organic amendments. Soil microbiology and biodiversity are also important areas of study to understand their role in nutrient cycling and soil fertility.

3. Integrated Pest Management: Agroecological approaches emphasize natural pest control mechanisms to reduce reliance on chemical pesticides. Current research focuses on developing strategies to attract beneficial insects, use trap crops, implement crop rotation, and incorporate agroecological landscapes that support biological pest control.

4. Agroecosystem Modeling: Mathematical and computer models are essential tools for predicting the long-term impacts of different agricultural practices on productivity, sustainability, and resilience. Research in agroecosystem modeling seeks to improve our understanding of complex ecological processes and inform decision-making for farmers and policymakers.

5. Farmer Education and Knowledge Exchange: Sharing knowledge and best practices is crucial for the widespread adoption of agroecological approaches. Research in farmer education and knowledge exchange explores effective methods of training, capacity building, and participatory research, ensuring that farmers have the necessary skills and information to implement agroecological practices.

Real-World Examples

Several real-world examples showcase the success of agroecology in addressing agricultural and environmental challenges:

1. The System of Rice Intensification (SRI): SRI is an agroecological method for rice cultivation that promotes soil health and reduces water and chemical inputs. By modifying planting techniques, transplanting younger seedlings, and improving soil conditions, SRI has led to increased rice yields and reduced environmental impact in countries such as Madagascar, India, and Cambodia.

2. Organic Agriculture in Denmark: Denmark has made significant progress in advancing organic agriculture through government support and research. The conversion of conventional farms to organic systems has not only reduced pesticide use but also improved soil quality and contributed to water protection. The success

310 CHAPTER 4: SUSTAINABLE AGRICULTURE AND FOOD SYSTEMS

of organic agriculture in Denmark demonstrates the viability of agroecological practices on a larger scale.

3. Agroecological Landscapes in Brazil: The Pontal do Paranapanema region in Brazil has implemented agroecological landscapes as a strategy to restore degraded ecosystems, conserve biodiversity, and improve livelihoods. By integrating native forests, agroforestry systems, and organic farming, the region has seen positive ecological and socio-economic outcomes.

4. Participatory Research in India: Numerous participatory research projects in India have empowered small-scale farmers to experiment with agroecological practices and adapt them to local conditions. These projects have resulted in increased crop yields, reduced input costs, and enhanced farmer resilience.

5. Agroecology in Cuba: Following the collapse of the Soviet Union, Cuba faced severe food shortages due to the loss of external inputs. The country turned to agroecological practices, such as urban farming, organic agriculture, and diversified cropping systems, to address the crisis. Today, agroecology plays a central role in Cuba's food security and sustainable development.

Challenges and Future Directions

While agroecology offers promising solutions, several challenges must be addressed to advance research and education in the field:

1. Funding and Resources: Agroecological research often relies on public funding, which may be limited compared to conventional agricultural research. Increased investment in agroecology is necessary to support long-term studies, farmer training programs, and the development of sustainable farming systems.

2.   Policy Support and Integration: Agroecological practices need to be integrated into national agricultural policies and development plans. Governments should provide incentives and regulatory frameworks that promote sustainable farming practices, research collaboration, and knowledge exchange among stakeholders.

CHALLENGES AND FUTURE DIRECTIONS IN SUSTAINABLE AGRICULTURE

3. Education and Training: Agroecology requires a new generation of farmers, researchers, and policymakers who understand the complexity of ecological farming systems. Enhanced education and training programs at universities and agricultural institutions are essential for building capacity and promoting agroecological thinking.

4. Systemic Change and Paradigm Shift: Transitioning from conventional to agroecological practices requires a systemic change within the agriculture sector. Agroecology challenges the dominance of industrial agriculture and requires a shift in mindset among farmers, consumers, and policymakers. Awareness campaigns, public outreach, and advocacy efforts are crucial for creating the necessary societal change.

5. Scaling Up and Upscaling: Agroecological practices need to be scaled up to have a significant impact on global food systems. Research should focus on the strategies and pathways for upscaling successful agroecological models, bridging the gap between small-scale pilot projects and large-scale implementation.

In conclusion, advancing agroecological research and education is vital for promoting sustainable agriculture and addressing the challenges we face in the 21st century. By prioritizing biodiversity, soil health, and community resilience, agroecology offers a holistic approach to farming that can help us achieve food security, mitigate climate change, and protect ecosystems. Through innovative research, knowledge exchange, and policy support, we can transform our agricultural systems and build a greener future.

Chapter 5: Renewable Energy and Green Technologies

Chapter 5: Renewable Energy and Green Technologies

Chapter 5: Renewable Energy and Green Technologies

In this chapter, we will explore the important role that renewable energy and green technologies play in addressing the global energy crisis and mitigating the impacts of climate change. We will discuss the need for renewable energy, the different types of renewable energy technologies, and showcase some success stories in this field. Additionally, we will examine the challenges that come with the widespread adoption of renewable energy and discuss future directions in this area.

The Need for Renewable Energy

The burning of fossil fuels for energy production has been the primary cause of greenhouse gas emissions, leading to climate change and environmental degradation. To address these issues, there is an urgent need to transition from fossil fuels to cleaner and renewable energy sources.

Renewable energy refers to energy derived from natural processes that are replenished constantly, such as sunlight, wind, water, and organic materials. Unlike fossil fuels, which are finite and contribute to pollution and climate change, renewable energy sources offer a more sustainable and environmentally friendly alternative.

Renewable Energy Technologies

There are various technologies available that harness renewable energy sources. Let's explore some of the key ones:

314 *CHAPTER 5: RENEWABLE ENERGY AND GREEN TECHNOLOGIES*

1. Solar Energy and Photovoltaics:

   Solar energy is the conversion of sunlight into usable energy, typically through the use of photovoltaic (PV) cells. These cells, made of semiconducting materials, generate electricity when exposed to sunlight. Solar energy is renewable, abundant, and can be harnessed in both large-scale power plants and smaller residential systems.

2. Wind Power and Wind Turbines:

   Wind power involves harnessing the energy generated by the natural movement of the wind. Wind turbines, equipped with large blades, capture the kinetic energy of the wind and convert it into electricity. Wind power is one of the fastest-growing sources of renewable energy globally, with large-scale wind farms being established both onshore and offshore.

3. Hydroelectric Power and Tidal Energy:

   Hydroelectric power harnesses the energy of flowing or falling water to generate electricity. It typically involves building dams to store water and control its flow, which passes through turbines to generate electricity. Tidal energy, on the other hand, leverages the ebb and flow of ocean tides to generate power. Both hydroelectric and tidal energy are reliable and highly efficient sources of renewable energy.

4. Geothermal Energy and Biogas:

   Geothermal energy utilizes the heat generated by the Earth's core. This heat can be harnessed through geothermal power plants, which tap into hot water reservoirs or underground steam to generate power. Geothermal energy is abundant and highly sustainable.

   Biogas is produced by the breakdown of organic matter, such as plant waste, agricultural byproducts, or livestock manure in an oxygen-free environment. This process releases methane, which can be captured and used as a renewable energy source. Biogas can be utilized for generating electricity or heat.

5. Biomass and Bioenergy:

   Biomass refers to any organic matter derived from plants or animals. Biomass energy involves the conversion of this matter into useful forms of energy, such as heat, electricity, or biofuels. Biomass can be used directly as a fuel, or its organic content can be converted through processes like combustion, gasification, or anaerobic digestion.

CHAPTER 5: RENEWABLE ENERGY AND GREEN TECHNOLOGIES 315

Case Studies in Renewable Energy

Let's take a look at some real-life examples of successful applications of renewable energy:

1. Germany's Energy Transition:

   Germany has become a global leader in renewable energy, with a successful energy transition, also known as the "Energiewende." This initiative aims to replace fossil fuels with renewable energy sources, reduce greenhouse gas emissions, and promote energy efficiency. Through favorable policies, feed-in tariffs, and substantial investments in renewable energy infrastructure, Germany has significantly increased its share of renewable energy in the electricity mix.

2. Solar Energy in India:

   India has made impressive strides in solar energy adoption. The country has been actively investing in large-scale solar projects and implementing policies to promote rooftop solar installations. This has not only increased energy access for millions of people but also contributed to reducing carbon emissions and combating climate change.

3. Offshore Wind Farms in Europe:

   Countries like the United Kingdom, Denmark, and the Netherlands have made significant progress in harnessing the power of offshore wind. They have developed large-scale offshore wind farms that generate substantial amounts of electricity. These projects have not only reduced reliance on fossil fuels but also created job opportunities and stimulated local economies.

4. Hydropower in Brazil:

   Brazil relies heavily on hydropower, with the Itaipu Dam being one of the largest hydroelectric power plants in the world. This renewable energy source provides a significant portion of the country's electricity needs while contributing to its low-carbon energy profile.

5. Biogas Production and Utilization in Sweden:

   Sweden has successfully implemented biogas production and utilization as an alternative to fossil fuels. By capturing methane emissions from organic waste, such as agricultural residues and food waste, Sweden has been able to

316 CHAPTER 5: RENEWABLE ENERGY AND GREEN TECHNOLOGIES

produce biogas for transportation, heating, and electricity generation. This approach not only reduces greenhouse gas emissions but also helps in waste management.

Success Stories in Renewable Energy

Apart from the case studies mentioned above, there are other remarkable success stories in the field of renewable energy. Let's highlight a few:

1. Costa Rica's Renewable Energy Revolution:

   Costa Rica has established itself as a global leader in renewable energy by harnessing its abundant natural resources. The country has achieved nearly 100% renewable electricity generation, primarily through hydropower, geothermal energy, and wind power. Costa Rica's success demonstrates that small nations can transition to clean and renewable energy sources.

2. Green Building and Sustainable Architecture:

   The green building movement has gained momentum worldwide. Architects, engineers, and builders are incorporating sustainable design principles, energy-efficient technologies, and renewable energy systems into the construction of buildings. These green buildings not only reduce energy consumption and greenhouse gas emissions but also provide healthier and more comfortable living and working spaces.

3. Microgrids and Distributed Energy Systems:

   Microgrids are small-scale, localized power grids that can operate independently or in coordination with the main electrical grid. They integrate renewable energy sources, energy storage systems, and advanced control technologies to provide reliable and resilient power supply, especially in remote areas or during emergencies. Microgrids empower communities, reduce transmission losses, and contribute to a more decentralized energy landscape.

4. Renewable Energy in Remote Areas and Developing Countries:

   Renewable energy technologies have brought electricity to remote areas and developing countries with limited or no access to the main power grid. Off-grid solar systems, small wind turbines, and micro-hydro projects have improved the quality of life for millions of people, enabling them to access electricity for lighting, cooking, education, healthcare, and communication.

CHAPTER 5: RENEWABLE ENERGY AND GREEN TECHNOLOGIES 317

5. Electric Vehicles and Sustainable Transportation:

   The shift towards electric vehicles (EVs) is gaining momentum worldwide. EVs reduce dependence on fossil fuels, lower greenhouse gas emissions, and improve air quality. With advancements in battery technology and the establishment of charging infrastructure, EVs are becoming increasingly viable and are playing a crucial role in sustainable transportation.

Challenges and Future Directions in Renewable Energy

While renewable energy offers significant opportunities, there are several challenges that need to be addressed for a successful transition to a clean energy future. Some of these challenges include:

1. Energy Storage Technologies and Grid Integration:

   One of the key challenges with renewable energy is the intermittent nature of sources like solar and wind. Developing affordable and efficient energy storage technologies, such as large-scale batteries, is crucial for storing excess energy during peak generation and ensuring a steady supply to the grid when renewable sources are not available. Additionally, integrating renewable energy sources into existing power grids requires smart grid technologies and proper management systems.

2. Policy and Regulatory Support for Renewable Energy:

   An enabling policy and regulatory framework is essential for the widespread adoption of renewable energy. Governments need to provide incentives, subsidies, and long-term contracts to encourage investments in renewable energy projects. Policies supporting grid interconnection, net metering, and feed-in tariffs can also play a significant role in promoting renewable energy.

3. Addressing the Intermittency of Renewable Sources:

   Renewable energy sources like solar and wind are subject to periodic variations due to weather conditions. Developing strategies to address this intermittency and ensure a consistent power supply is crucial. This can be achieved through a combination of energy storage systems, demand response mechanisms, and a diverse portfolio of renewable energy sources.

4. Renewable Energy for Industrial Processes:

   Expanding the use of renewable energy beyond electricity generation to power industrial processes is a challenge. Industries have unique energy

318 CHAPTER 5: RENEWABLE ENERGY AND GREEN TECHNOLOGIES

requirements, and transitioning away from fossil fuels may involve significant technological and infrastructural changes. Research and innovation are needed to develop renewable energy solutions for various industrial sectors.

5. Advancements in Green Technologies and Innovations:

   Continued research and development are necessary to drive advancements in renewable energy technologies. This includes improving efficiency, reducing costs, and exploring new avenues for renewable energy generation. Emerging technologies like advanced solar cells, offshore wind turbines, and hydrogen fuel cells hold great promise for the future of renewable energy.

In conclusion, renewable energy and green technologies are vital components of a sustainable future. They offer a viable alternative to fossil fuels, reduce greenhouse gas emissions, and promote environmental preservation. Through innovative solutions, supportive policies, and ongoing research, we can accelerate the transition to a renewable energy-powered world. Let's embrace the challenges and opportunities of renewable energy and work towards a greener and more sustainable future.

The Need for Renewable Energy

Fossil Fuels and Climate Change

Introduction

Fossil fuels, including coal, oil, and natural gas, have been the backbone of global energy production for centuries. These energy sources have fueled industrialization, transportation, and technological advancements, leading to significant economic growth and improved living standards. However, the combustion of fossil fuels releases greenhouse gases (GHGs) into the atmosphere, causing climate change and posing a grave threat to our planet's future.

In this section, we will explore the link between fossil fuels and climate change, examining the mechanisms by which these fuels contribute to the greenhouse effect. We will also discuss the consequences of climate change and the urgent need to transition towards renewable energy sources.

FOSSIL FUELS AND CLIMATE CHANGE

319

The Greenhouse Effect

To understand the relationship between fossil fuels and climate change, we must first grasp the concept of the greenhouse effect. The Earth's atmosphere acts like a blanket, trapping heat and maintaining a moderate temperature suitable for life. Greenhouse gases, such as carbon dioxide (CO_2), methane (CH_4), and nitrous oxide (N_2O), play a crucial role in this natural process.

When solar radiation reaches the Earth's surface, some of it is absorbed, while the rest is reflected back into space. Greenhouse gases trap a portion of the reflected radiation, preventing it from escaping the atmosphere and thus warming the planet. This phenomenon is called the greenhouse effect.

Fossil Fuels and Greenhouse Gas Emissions

Fossil fuels are formed from the remains of ancient plants and animals that underwent geological processes over millions of years. When burned or combusted, these fuels release carbon dioxide, methane, and other GHGs into the atmosphere, augmenting the natural greenhouse effect.

Carbon Dioxide (CO_2)

Carbon dioxide is the most significant greenhouse gas emitted by human activities. It is primarily produced through the combustion of fossil fuels such as coal, oil, and natural gas. When carbon-containing fuels are burned, carbon atoms combine with oxygen from the air, forming carbon dioxide.

The burning of coal for electricity generation and industrial processes is the largest source of CO_2 emissions. Similarly, the combustion of gasoline and diesel in vehicles contributes significantly to atmospheric CO_2 levels. Deforestation and land-use changes also release large amounts of carbon dioxide, as the clearing of forests reduces the Earth's capacity to absorb CO_2 through photosynthesis.

Methane (CH_4)

Methane is another potent greenhouse gas emitted during the extraction, production, and transport of fossil fuels. It is also released by livestock, agriculture, and the decay of organic waste in landfills. Although methane concentrations are relatively lower than carbon dioxide in the atmosphere, it has a much higher warming potential over a 20-year timeframe.

Methane emissions from the fossil fuel industry occur due to leaks and venting during oil and gas extraction and transportation. These emissions can be reduced

320 CHAPTER 5: RENEWABLE ENERGY AND GREEN TECHNOLOGIES

through improved infrastructure and practices to capture and minimize methane release.

Nitrous Oxide (N2O)

Nitrous oxide is predominantly released from agricultural activities, fossil fuel combustion, and industrial processes. While its atmospheric concentration is significantly lower than CO2 and CH4, it is a highly potent greenhouse gas with a long lifespan in the atmosphere.

Fossil fuel combustion, particularly in power plants and industrial operations, contributes to nitrous oxide emissions through the oxidation of nitrogen present in the fuel. Agricultural practices, such as the use of synthetic fertilizers and the burning of crop residues, also release significant amounts of N2O into the atmosphere.

Climate Change Impacts

Climate change, driven by the increase in greenhouse gas emissions from human activities, has far-reaching consequences for the environment, ecosystems, and societies worldwide. The following are some of the key impacts of climate change:

Rising Temperature

The Earth's average surface temperature has been steadily increasing over the past century, primarily due to human-induced greenhouse gas emissions. This rise in temperature has led to heatwaves, melting polar ice caps and glaciers, and changes in weather patterns.

Warmer temperatures also amplify the intensity and frequency of extreme weather events such as hurricanes, droughts, and floods, leading to more significant economic losses and a higher risk to human lives.

Sea-Level Rise

As global temperatures increase, the melting of polar ice caps and glaciers contributes to rising sea levels. Higher sea levels pose a significant threat to coastal communities, low-lying islands, and sensitive ecosystems.

In addition to the direct risks of flooding and erosion, rising sea levels can also lead to saltwater intrusion into freshwater sources, affecting agriculture and drinking water supplies.

FOSSIL FUELS AND CLIMATE CHANGE

Changes in Precipitation Patterns

Climate change alters rainfall patterns, leading to both increased rainfall in certain regions and decreased rainfall in others. These changes can disrupt agricultural productivity, impact water availability, and exacerbate the risk of droughts and wildfires.

Ecological Disruption

Climate change poses significant risks to biodiversity and ecosystem integrity. Many species are struggling to adapt to the rapid pace of climate change, leading to habitat loss, reduced food availability, and increased vulnerability to disease and invasive species.

Disruptions in ecosystem services, such as pollination, water purification, and carbon sequestration, further compound the impacts of climate change on human societies.

Transitioning to Renewable Energy

To mitigate the adverse impacts of climate change, it is crucial to reduce our reliance on fossil fuels and transition towards renewable energy sources. Renewable energy technologies harness naturally replenishing resources, such as sunlight, wind, and water, to generate electricity without emitting greenhouse gases.

Solar Energy

Solar energy is abundant, freely available, and has enormous potential as a clean energy source. Photovoltaic (PV) systems convert sunlight directly into electricity, while solar thermal systems utilize the sun's heat for various applications. The decreasing cost of solar panels and advancements in technology have made solar energy increasingly accessible.

Wind Power

Harnessing the power of wind through wind turbines is another rapidly growing renewable energy sector. Wind power is clean, abundant, and increasingly cost-effective. Offshore wind farms and onshore wind installations are becoming more prevalent worldwide, contributing significantly to electricity generation.

Hydroelectric Power

Hydropower utilizes the energy of flowing or falling water to generate electricity. Large-scale hydropower plants, such as dams, can provide a stable and reliable source of renewable energy. However, there are concerns about their environmental impact, including habitat alteration and fish migration disruption.

Other Renewable Energy Sources

Other renewable energy sources, including geothermal energy and biomass, are also viable alternatives to fossil fuels. Geothermal energy utilizes the Earth's natural heat, while biomass energy uses organic materials for heat and electricity production.

Conclusion

The combustion of fossil fuels and the resulting greenhouse gas emissions are driving climate change, which poses severe risks to our planet and its inhabitants. Transitioning to renewable energy sources is essential to mitigate these risks and build a sustainable future.

By embracing renewable energy technologies, reducing fossil fuel consumption, and implementing energy efficiency measures, we can pave the way for a greener and more resilient future. The urgent need to address climate change requires collective action, innovative solutions, and a shared commitment to creating a healthier and more sustainable planet for generations to come.

Transitioning to Renewable Energy Sources

As we continue the fight for a greener future, one of the most crucial steps we can take is transitioning from fossil fuels to renewable energy sources. This transition is essential to mitigate the harmful effects of climate change and reduce our dependence on finite resources. In this section, we will explore the reasons why transitioning to renewable energy is necessary, the different types of renewable energy sources, and the challenges and opportunities associated with this transition.

The Need for Transition

The burning of fossil fuels, such as coal, oil, and natural gas, releases greenhouse gases into the atmosphere, contributing to global warming and climate change. To combat this, we must transition to renewable energy sources that have a lower carbon footprint and can be replenished naturally.

FOSSIL FUELS AND CLIMATE CHANGE

Renewable energy sources are derived from sustainable and naturally occurring processes. Unlike fossil fuels, which take millions of years to form, renewable energy sources can be replenished within a human lifetime. By harnessing these sources, we can reduce our greenhouse gas emissions, improve air quality, and mitigate the adverse effects of climate change.

Types of Renewable Energy Sources

There are various types of renewable energy sources that can be used to generate electricity and power our societies. In this section, we will explore some of the most prominent ones.

Solar Energy   Solar energy is derived from the radiation emitted by the sun. It can be harnessed using photovoltaic (PV) cells, which convert sunlight directly into electricity, or concentrating solar power (CSP) systems, which use mirrors or lenses to concentrate sunlight onto a receiver to generate heat and produce electricity. Solar energy is abundant, widely available, and emissions-free.

Wind Power   Wind power is generated by harnessing the kinetic energy of the wind to produce electricity. Wind turbines, consisting of rotor blades connected to a generator, convert the rotational motion caused by the wind into electrical energy. Wind power is highly scalable, and large wind farms can generate significant amounts of electricity. Additionally, offshore wind farms have the potential to harness even stronger and more consistent winds.

Hydropower   Hydropower involves harnessing the energy of moving or falling water to generate electricity. It is the most widely used form of renewable energy globally. Hydropower plants, such as dams and reservoirs, use the force of the water to turn turbines connected to generators. Hydropower is reliable, cost-effective, and can provide baseload power, making it an essential component of a renewable energy portfolio.

Geothermal Energy   Geothermal energy utilizes the heat generated by the Earth's core and crust. Geothermal power plants tap into underground reservoirs of steam or hot water to drive turbines and produce electricity. This form of energy is highly sustainable, as it relies on the Earth's natural heat. Geothermal energy can be harnessed in regions with volcanic activity or hot springs.

324 CHAPTER 5: RENEWABLE ENERGY AND GREEN TECHNOLOGIES

Biomass Biomass refers to organic matter derived from plants and animals. It can be used as a fuel to produce heat or electricity. Biomass energy is typically obtained from agricultural crops, forest residues, or dedicated energy crops. Biomass power plants combust biomass to generate steam, which drives turbines connected to generators. This process is known as bioenergy.

Challenges and Opportunities

While transitioning to renewable energy sources offers many advantages, it also presents several challenges. These challenges must be addressed to ensure a successful and sustainable transition.

Intermittency and Grid Integration One of the main challenges of renewables is their intermittent nature. Solar energy is only available during the day, wind energy relies on wind speed, and hydropower fluctuates with rainfall. Integrating these variable sources into the power grid requires the development of energy storage technologies, such as batteries, to store excess energy and release it when needed. Additionally, smart grid technologies and grid interconnections can help balance energy supply and demand.

Policy and Regulatory Support A successful transition to renewable energy requires supportive policies and regulations. Governments must incentivize the deployment of renewable energy technologies through financial incentives, feed-in tariffs, and tax credits. Long-term policy frameworks and stable regulatory environments are essential to attract investments and ensure a level playing field for renewable energy providers.

Infrastructure and Investment Transitioning to renewable energy sources requires significant infrastructure development and investment. This includes constructing renewable energy power plants, expanding transmission and distribution networks, and upgrading existing infrastructure. Investments in research and development are also necessary to enhance the efficiency and cost-effectiveness of renewable energy technologies.

Public Perception and Acceptance Public perception, awareness, and acceptance of renewable energy play a vital role in the transition. Educating the public about the benefits of renewable energy sources and addressing any concerns or misinformation is crucial. Engaging local communities in the decision-making

FOSSIL FUELS AND CLIMATE CHANGE

process and providing opportunities for participation can foster a sense of ownership and support for renewable energy projects.

Job Creation and Economic Opportunities Transitioning to renewable energy can create new job opportunities and stimulate economic growth. The renewable energy sector requires a skilled workforce in various areas, such as engineering, construction, operations, and maintenance. Investing in renewable energy can also reduce energy import dependence and create a more resilient and competitive economy.

Conclusion

Transitioning to renewable energy sources is a necessary and urgent step in our journey towards a greener future. By harnessing the power of the sun, wind, water, heat from the Earth, and biomass, we can reduce greenhouse gas emissions, mitigate climate change, improve air quality, and create a more sustainable and resilient energy system. However, this transition comes with its challenges and requires supportive policies, infrastructure development, public engagement, and investments. Despite these challenges, the opportunities offered by renewable energy are vast, including job creation, economic growth, and a cleaner and healthier environment. Let's embrace the transition and be part of the solution for a brighter and greener future.

Environmental Impacts of Energy Production

Energy production is a crucial aspect of modern society, but it also has significant environmental consequences. As we strive to transition to renewable energy sources, it is important to understand the environmental impacts associated with different forms of energy production. In this section, we will explore the main environmental issues related to traditional energy sources, such as fossil fuels, and compare them to the cleaner alternatives offered by renewable energy technologies.

Fossil Fuels and Climate Change

The combustion of fossil fuels, such as coal, oil, and natural gas, for energy production is the primary source of greenhouse gas emissions. Carbon dioxide (CO_2) is the most prominent greenhouse gas, trapping heat in the Earth's atmosphere and contributing to global climate change. The burning of fossil fuels releases large amounts of CO_2 into the atmosphere, intensifying the greenhouse effect and leading to rising temperatures.

326 CHAPTER 5: RENEWABLE ENERGY AND GREEN TECHNOLOGIES

The extraction and transportation of fossil fuels also pose environmental risks. Activities like coal mining can result in habitat destruction, soil erosion, and water pollution. The extraction and refining of oil can lead to oil spills, which have devastating effects on marine ecosystems. Additionally, the excessive use of water for hydraulic fracturing, or fracking, in natural gas production can deplete local water resources.

Transitioning to Renewable Energy Sources

Renewable energy sources, on the other hand, offer a cleaner and more sustainable alternative to fossil fuels. Let's explore the environmental impacts associated with some commonly used renewable energy technologies.

Solar Energy and Photovoltaics: Solar energy harnesses the power of the sun by converting sunlight into electricity through photovoltaic (PV) cells. One of the major advantages of solar energy is its minimal environmental impact. Solar panels produce electricity without any emissions and have a long lifespan. However, the production and disposal of solar panels can have some environmental implications, primarily due to the use of certain rare-earth metals and chemicals. Proper recycling and sustainable manufacturing practices can mitigate these concerns.

Wind Power and Wind Turbines: Wind power harnesses the kinetic energy of the wind to generate electricity through wind turbines. Wind energy is a clean and abundant resource with minimal environmental impacts once the turbines are operational. However, the installation and maintenance of wind turbines can have some environmental consequences, such as bird and bat collisions and noise pollution. Proper siting studies and careful planning help minimize these risks.

Hydroelectric Power and Tidal Energy: Hydroelectric power harnesses the energy of flowing or falling water to generate electricity. Large-scale hydroelectric dams can have significant environmental impacts, including habitat alteration, damming of rivers, and displacement of communities. Tidal energy, which utilizes the power of ocean tides, can also have impacts on marine ecosystems and fish migration patterns. However, newer technologies, such as run-of-river hydroelectric systems and tidal turbines, aim to reduce these environmental risks.

Geothermal Energy and Biogas: Geothermal energy harnesses heat from the Earth's interior to generate electricity and heat buildings. While geothermal energy is a clean and renewable source, the extraction of geothermal resources can release harmful gases and fluids from the underground. Proper monitoring and management of geothermal reservoirs help minimize these impacts. Biogas, which is produced by the decomposition of organic matter, can be used as a renewable

FOSSIL FUELS AND CLIMATE CHANGE

energy source. However, the production of biogas from agricultural waste can have air quality and odor issues if not managed properly.

Biomass and Bioenergy: Biomass energy utilizes organic matter, such as wood pellets or agricultural residues, to generate heat or electricity. While biomass can be a renewable energy source, its environmental impacts depend on the sustainability of its production and sourcing. Unsustainable biomass harvesting can lead to deforestation, habitat loss, and increased greenhouse gas emissions.

Environmental Impacts Mitigation

To mitigate the environmental impacts associated with energy production, several strategies can be employed:

Energy Efficiency: Improving energy efficiency is a crucial step towards reducing the overall environmental impact of energy production. By using energy more efficiently, we can decrease the demand for energy and reduce greenhouse gas emissions.

Emissions Control Technologies: The implementation of emissions control technologies, such as scrubbers, filters, and catalytic converters, can help reduce air pollutants released during energy production. This applies to both fossil fuel power plants and certain renewable energy technologies like biomass combustion facilities.

Environmental Impact Assessments: Conducting thorough environmental impact assessments before the construction and operation of energy projects can help identify potential risks and develop mitigation measures. These assessments should consider the entire lifecycle of the energy project, from material extraction to decommissioning.

Conservation and Land Management: Conserving and managing natural habitats and ecosystems is essential for preserving biodiversity and mitigating environmental impacts. This involves responsible land use practices, habitat restoration, and protecting sensitive areas from energy development.

Real-World Examples

A notable real-world example of the environmental impacts of energy production is the tar sands industry in Alberta, Canada. The extraction and processing of oil from tar sands result in extensive habitat destruction, water pollution, and increased greenhouse gas emissions compared to conventional oil. Efforts are underway to mitigate these impacts through improved extraction methods and reclamation strategies.

328 CHAPTER 5: RENEWABLE ENERGY AND GREEN TECHNOLOGIES

In contrast, Denmark has become a global leader in wind energy. The country has invested heavily in wind power infrastructure and has implemented strict environmental regulations to ensure minimal impacts on bird migration patterns and the marine ecosystem.

Conclusion

The environmental impacts of energy production are significant and must be taken into account when considering different energy sources. While traditional fossil fuel-based energy production has severe consequences for climate change, pollution, and biodiversity loss, renewable energy technologies offer a cleaner and more sustainable future. By adopting renewable energy sources and implementing appropriate mitigation measures, we can minimize environmental impacts and work towards a greener and more sustainable future.

Exercises

1. Research and discuss a case study of an energy project that faced significant environmental challenges. What mitigation measures were implemented, and what were the outcomes?

2. Compare the environmental impacts of hydropower dams and tidal energy projects. Analyze the advantages and disadvantages of each technology from an environmental perspective.

3. Investigate the carbon footprint of different forms of transportation, such as cars, buses, trains, and airplanes. Discuss the environmental impacts of each mode of transportation and suggest strategies to reduce emissions.

4. Research and discuss the impact of energy production on local communities, particularly those in close proximity to fossil fuel power plants or renewable energy projects. What are the social and environmental justice implications?

5. Explore the potential environmental impacts of large-scale solar energy projects in desert regions. Discuss the challenges and opportunities associated with mitigating these impacts.

Renewable Energy Technologies

Solar Energy and Photovoltaics

Solar energy is one of the most abundant and sustainable sources of renewable energy available to us. It has the potential to revolutionize our energy industry and

RENEWABLE ENERGY TECHNOLOGIES

reduce our dependence on fossil fuels. In this section, we will explore the principles of solar energy and the technology of photovoltaics that converts sunlight into electricity.

Principles of Solar Energy

The sun is a natural nuclear reactor, converting hydrogen into helium through a process called nuclear fusion. This process releases an enormous amount of energy in the form of sunlight. The Earth receives an incredible amount of solar energy, but only a fraction of it can be harnessed and utilized for our energy needs.

The amount of solar energy reaching the Earth's surface depends on various factors such as geographic location, time of day, season, and weather conditions. However, on average, the Earth receives about 1,400 watts of solar energy per square meter, known as solar irradiance. This energy can be captured and converted into usable forms, such as electricity or heat.

Photovoltaic Technology

Photovoltaic (PV) technology is the method used to convert sunlight directly into electricity. It is based on the principle of the photovoltaic effect, which was discovered by French physicist Alexandre-Edmond Becquerel in 1839. The photovoltaic effect occurs when certain materials, known as semiconductors, absorb photons (particles of light) and release electrons, creating an electric current.

The most commonly used semiconductor material in photovoltaic cells is silicon. Silicon can be doped with impurities to create a region with an excess of electrons (n-type) and a region with a deficit of electrons (p-type). When sunlight strikes the PV cell, photons with enough energy can knock loose electrons from the lattice of silicon atoms. These free electrons are then captured and directed through an external circuit, generating an electric current.

Components of a PV System

A typical photovoltaic system consists of several components that work together to convert solar energy into electricity.

1. PV Panels: These are the most crucial components of a PV system. They are made up of multiple PV cells connected in series or parallel to achieve the desired voltage and current output. The panels are usually made of silicon wafers encased in a protective glass cover and mounted in a sturdy frame.

330 CHAPTER 5: RENEWABLE ENERGY AND GREEN TECHNOLOGIES

2. Inverter: The generated direct current (DC) produced by the PV panels needs to be converted into alternating current (AC) because most electrical devices in our homes and businesses operate on AC power. The inverter is responsible for this conversion process.

3. Charge Controller (Optional): In off-grid PV systems or those that utilize batteries for energy storage, a charge controller is used to regulate the charging and discharging of the batteries. It prevents overcharging and extends the battery life.

4. Batteries (Optional): In off-grid PV systems, batteries are used to store excess energy generated during the day for use during periods of low sunlight or at night. This ensures a continuous supply of electricity.

5. Mounting and Tracking Systems: PV panels need to be mounted in a location that receives maximum sunlight. Mounting systems allow for secure installation on rooftops or other suitable structures. Tracking systems can be used to move the PV panels to follow the sun's path and maximize energy capture.

6. Electrical Wiring and Safety Devices: Proper wiring and grounding are essential for the safe and efficient operation of the PV system. Electrical safety devices, such as circuit breakers and surge protectors, protect the system from overvoltage and current surges.

Advantages and Applications of Solar Energy

Solar energy offers numerous advantages over conventional energy sources, making it an attractive choice for power generation.

1. Renewable and Clean: Solar energy is a renewable resource, meaning we will never run out of it as long as the sun is shining. It is also clean energy that does not produce harmful emissions or contribute to climate change.

2. Financial Savings: Once installed, solar panels can significantly reduce electricity bills, especially in areas with high electricity rates. Over time, the savings on energy costs can offset the initial investment of the PV system.

3. Energy Independence: Solar energy provides an opportunity to reduce dependence on fossil fuels and foreign energy sources. It allows individuals, communities, and businesses to generate their electricity on-site, promoting energy self-sufficiency.

4. Versatility: Solar energy can be harnessed in various ways, from large-scale solar farms to small-scale installations on residential rooftops. It can be used for electricity generation, heating water, or even powering vehicles.

5. Scalability: PV systems can be easily expanded or scaled down based on energy needs. Additional panels can be installed as demand increases, making solar energy a flexible option for both large and small applications.

RENEWABLE ENERGY TECHNOLOGIES

Solar energy finds applications in a wide range of sectors, including:

- Residential: Many homeowners install solar panels on their rooftops to generate electricity and lower their energy bills.

- Commercial and Industrial: Businesses and industries utilize solar energy to power their operations, reduce operating costs, and meet sustainability targets.

- Agriculture: Solar-powered irrigation systems, crop drying, and powering farm machinery are some applications in the agricultural sector.

- Transportation: Solar energy is used to power electric vehicles (EVs), charging stations, and even solar-powered boats and airplanes.

Challenges and Future Directions in Solar Energy

While solar energy has made significant progress in recent years, several challenges still need to be addressed for its widespread adoption and integration into our energy systems.

1. Intermittency: Solar energy production is dependent on sunlight, which varies throughout the day and is affected by weather conditions. Developing effective energy storage technologies and improving grid infrastructure are crucial for managing the intermittent nature of solar power.

2. Cost: Although the cost of PV panels has significantly decreased over time, the initial investment can still be high for some individuals or organizations. Continued research and development to reduce the cost of solar energy equipment will make it more accessible to a wider population.

3. Efficiency: Improving the efficiency of solar panels is essential for maximizing energy capture and reducing the overall system cost. Researchers are exploring new materials, such as perovskite, and alternative designs to make solar panels more efficient.

4. Environmental Impact: The manufacturing and disposal of PV panels can have environmental implications if not managed properly. Developing strategies for the responsible manufacturing, recycling, and disposal of solar panels is essential to ensure the long-term sustainability of solar energy.

5. Grid Integration: As solar energy becomes more prevalent, integrating it into the existing electricity grid poses several technical challenges. Upgrading and modernizing the grid infrastructure to accommodate distributed solar power generation is necessary.

The future of solar energy holds great promise. Continued advancements in technology, increased efficiency, and greater implementation of energy storage solutions will play a vital role in overcoming the challenges and making solar energy an integral part of our global energy portfolio.

332 CHAPTER 5: RENEWABLE ENERGY AND GREEN TECHNOLOGIES

Exercises

1. Calculate the amount of solar energy received by a rooftop with an area of 150 square meters on a sunny day with a solar irradiance of 1,000 watts per square meter.

2. Research and discuss a case study of a large-scale solar energy project in your region. What were the challenges faced during the project, and what were the outcomes?

3. Investigate the environmental impact of solar panel manufacturing and propose sustainable solutions for minimizing the waste generated during the production process.

4. Design a solar-powered off-grid system for a remote cabin that requires energy for lighting, small appliances, and a water pump. Consider the energy needs, available sunlight, and the components required for the system.

5. Explore the advancements in solar panel technologies beyond silicon-based PV cells. Research and discuss the potential of emerging technologies like perovskite solar cells.

Further Reading

1. O'Sullivan, F., & Pearsall, N. (2013). *Photovoltaics: A Practical Guide.* CRC Press.

2. Arbab, N., & Panwar, N. L. (2018). *Solar Photovoltaic Technology and Systems: A Manual for Technicians, Trainers, and Engineers.* Springer.

3. Twidell, J., & Weir, T. (2015). *Renewable Energy Resources.* Routledge.

4. Physics.org. (n.d.). How do Solar Photovoltaics Work? Retrieved from https://www.physics.org/article-questions.asp?id=53.

5. National Renewable Energy Laboratory (NREL). (n.d.). *Solar Photovoltaic Technology Basics.* Retrieved from https://www.nrel.gov/research/re-photovoltaics.html.

Wind Power and Wind Turbines

In this section, we will explore the principles and technologies behind wind power and wind turbines. Wind power is a renewable energy source that harnesses the kinetic energy of the wind to generate electricity. It is a clean and sustainable alternative to traditional fossil fuels, making it an important component of a greener and more sustainable future.

RENEWABLE ENERGY TECHNOLOGIES

The Power of the Wind

To understand how wind power works, we first need to grasp the basic principles behind the power of the wind. Wind is created by the uneven heating of the Earth's surface by the sun, resulting in the movement of air from areas of high pressure to areas of low pressure. This movement generates kinetic energy, which can be harnessed to perform work.

The power contained in the wind can be calculated using the equation:

$$P = \frac{1}{2}\rho A v^3 \tag{1}$$

Where:

+ P is the power

+ ρ is the air density

+ A is the area intercepted by the turbine blades

+ v is the wind speed

From this equation, we can see that the power of the wind is directly proportional to the cube of the wind speed. This means that even a small increase in wind speed can result in a significant increase in power output.

Wind Turbines: Components and Operation

Wind turbines are the devices used to convert the kinetic energy of the wind into electrical energy. They consist of several key components:

1. Tower: The tower is the tall structure that supports the wind turbine. It is typically made of steel or concrete and provides stability and height to maximize exposure to wind.

2. Blades: The blades capture the energy of the wind and convert it into rotational motion. They are usually made of lightweight and durable materials such as fiberglass or carbon fiber.

3. Nacelle: The nacelle houses the key components of the wind turbine, including the gearbox, generator, and control system. It is typically located at the top of the tower.

334 CHAPTER 5: RENEWABLE ENERGY AND GREEN TECHNOLOGIES

4. Generator: The generator is responsible for converting the rotational energy of the turbine blades into electrical energy. It uses electromagnetic induction to generate an electric current.

The operation of a wind turbine can be summarized in the following steps:

1. The wind blows, causing the turbine blades to rotate.

2. The rotation of the blades drives the rotor, which is connected to the gearbox.

3. The gearbox steps up the rotational speed and transfers the energy to the generator.

4. The generator converts the mechanical energy into electrical energy.

5. The electrical energy is then transmitted through power lines for distribution and use.

Types of Wind Turbines

There are two main types of wind turbines: horizontal-axis wind turbines (HAWT) and vertical-axis wind turbines (VAWT).

1. Horizontal-Axis Wind Turbines (HAWT): HAWTs are the most commonly used wind turbine design. They have a horizontal rotor shaft and blades that rotate parallel to the ground. HAWTs are highly efficient and can achieve higher power outputs compared to VAWTs. They are typically used in large-scale wind farms.

2. Vertical-Axis Wind Turbines (VAWT): VAWTs have a vertical rotor shaft and blades that rotate around it. VAWTs have the advantage of being able to harness wind from any direction, making them suitable for urban and decentralized applications. However, they are generally less efficient and produce lower power outputs compared to HAWTs.

Advantages and Challenges

Wind power offers numerous advantages as a renewable energy source:

+ Sustainability: Wind power is a clean and sustainable energy source, as wind is an abundant and naturally occurring resource.

RENEWABLE ENERGY TECHNOLOGIES

- Carbon-free: Wind turbines produce no greenhouse gas emissions or air pollutants, contributing to the reduction of climate change and air pollution.

- Energy independence: Wind power reduces dependence on fossil fuels and increases energy independence.

- Local economic benefits: Wind farms can bring economic benefits to local communities through job creation, increased tax revenues, and land lease payments.

However, wind power also faces several challenges:

- Intermittency: The availability of wind is not constant, which causes fluctuations in power output. This intermittency requires the integration of wind power with energy storage systems or other forms of generation to ensure a reliable electricity supply.

- Visual and noise impact: Wind turbines can have visual and noise impacts, which can be a concern for nearby residents. Proper siting and community engagement are crucial to address these issues.

- Wildlife impacts: Wind turbines can pose risks to birds and bats if not properly located and managed. Research and careful planning are necessary to minimize these impacts.

- Grid integration: The integration of wind power into existing electricity grids can pose technical challenges due to its intermittent nature and variable power output. Upgrades and advanced grid management systems are needed to accommodate large-scale wind power.

Real-World Applications

Wind power has seen significant growth and is now one of the fastest-growing renewable energy sources worldwide. Several countries have embraced wind power as part of their energy mix and have achieved remarkable success.

One notable example is Denmark, which has a long history of investing in wind power. Wind energy now provides a significant portion of Denmark's electricity demand, and the country has become a global leader in wind turbine manufacturing and technology.

Another example is Germany, which has implemented a successful energy transition program, prioritizing the development and deployment of renewable

336 CHAPTER 5: RENEWABLE ENERGY AND GREEN TECHNOLOGIES

energy sources, including wind power. Germany has made significant progress in reducing its dependence on fossil fuels and has become a pioneer in sustainable energy practices.

Conclusion

Wind power and wind turbines play a crucial role in the transition to a greener and more sustainable future. By harnessing the power of the wind, we can generate clean and renewable energy, reduce greenhouse gas emissions, and mitigate the impacts of climate change. While challenges remain, ongoing technological advancements and supportive policies are driving the continued growth and success of wind power as a viable and essential energy source.

Summary

In this section, we explored wind power and wind turbines. We learned about the power of the wind and how it can be harnessed to generate electricity. We discussed the components and operation of wind turbines, including the tower, blades, nacelle, and generator. We also examined the different types of wind turbines, namely horizontal-axis and vertical-axis turbines. Furthermore, we discussed the advantages and challenges of wind power, such as sustainability, carbon-free generation, and intermittency. Lastly, we highlighted real-world applications of wind power, including Denmark and Germany's successful adoption of wind energy. Wind power holds great promise as a renewable energy source, and its continued development and integration into our energy systems are crucial for achieving a greener and more sustainable future.

Hydroelectric Power and Tidal Energy

Hydroelectric power and tidal energy are two important sources of renewable energy that harness the natural power of water. In this section, we will explore the principles, technologies, and challenges associated with these forms of energy generation.

Principles of Hydroelectric Power

Hydroelectric power is generated by harnessing the energy of flowing or falling water. It involves the conversion of kinetic energy of moving water into electrical energy. The basic principle behind hydroelectric power is the utilization of the water cycle, where water evaporates from oceans and other bodies of water, forms

RENEWABLE ENERGY TECHNOLOGIES

clouds, precipitates as rain or snow, and then flows through rivers back to the ocean.

To generate hydroelectric power, a dam is constructed across a river, creating a reservoir. The potential energy stored in the elevated water is then converted into kinetic energy as it flows down through turbines. The rotating turbines activate generators, producing electricity. The amount of electricity generated depends on the volume of water flowing through the turbines and the height from which it falls.

Types of Hydroelectric Power Plants

There are two main types of hydroelectric power plants:

1. **Reservoir-based Hydroelectric Power Plants:** In these plants, large dams are built to store water in reservoirs. The water is released in a controlled manner to produce a steady flow that drives the turbines. Reservoir-based plants offer greater control over the amount and timing of electricity generation. They can also provide additional benefits such as flood control, irrigation, and recreational activities in the reservoirs.

2. **Run-of-the-river Hydroelectric Power Plants:** These plants do not use reservoirs and rely on the natural flow of the river to generate electricity. The water is diverted from the river into a canal or penstock, where it flows through the turbines. Run-of-the-river plants have a smaller environmental footprint compared to reservoir-based plants, as they do not require the construction of large dams.

Advantages and Challenges of Hydroelectric Power

Hydroelectric power offers several advantages as a renewable energy source:

- **Clean and Renewable:** Hydroelectric power does not produce greenhouse gas emissions, making it a clean source of energy. It relies on the continuous water cycle, making it a renewable resource.

- **Stable and Reliable:** Hydroelectric power plants can provide a steady and reliable supply of electricity. They can quickly respond to changes in demand and can be used for baseload or peakload electricity generation.

- **Multipurpose Benefits:** Reservoir-based hydroelectric power plants can provide additional benefits such as flood control, water storage for irrigation, and recreational activities in the reservoirs.

However, hydroelectric power also poses certain challenges:

- **Environmental Impact:** The construction of large dams and reservoirs can lead to habitat destruction, loss of biodiversity, and disruption of ecosystems. The alteration of natural river flow can also affect fish migration patterns.

338 CHAPTER 5: RENEWABLE ENERGY AND GREEN TECHNOLOGIES

- **Displacement of Communities:** The construction of large dams for hydroelectric power can result in the displacement of communities residing in the project area.
- **Limited Suitable Sites:** Not all locations have the topography and water resources suitable for hydroelectric power generation. Identifying suitable sites and obtaining necessary approvals can be challenging.

Tidal Energy

Tidal energy is a form of renewable energy that harnesses the power of tides. It involves the capture of kinetic energy from the rise and fall of ocean tides. Tides are caused by the gravitational forces exerted by the moon and the sun on the Earth's oceans.

Tidal energy can be harnessed through two main technologies:

1. **Tidal Barrages:** Tidal barrages are large structures built across river estuaries or bays. They work similarly to hydroelectric power plants by utilizing the potential energy of the elevated water. As the tide rises, water flows into the barrage through turbines, generating electricity. When the tide recedes, the water flows out of the barrage, generating electricity again. Tidal barrages can provide a consistent and predictable source of energy.

2. **Tidal Turbines:** Tidal turbines are devices that resemble underwater wind turbines. They are placed on the seabed in areas with strong tidal currents. As the tides flow in and out, the turbines are rotated by the kinetic energy of the moving water, generating electricity. Tidal turbines are more flexible in terms of installation locations and have a lower impact on the environment compared to tidal barrages.

Advantages and Challenges of Tidal Energy

Tidal energy offers several advantages as a renewable energy source:

- **Predictable and Reliable:** Tidal energy is highly predictable since tides are influenced by celestial motions. This predictability makes tidal energy a reliable source of electricity generation.
- **Environmentally Friendly:** Tidal energy does not produce any air pollution or greenhouse gas emissions. Tidal turbines have a minimal impact on the marine environment when properly designed and located.
- **Long Lifespan:** Tidal energy technologies have a long lifespan, typically lasting for several decades. This makes them a reliable and sustainable long-term investment.

However, tidal energy also faces certain challenges:

RENEWABLE ENERGY TECHNOLOGIES

- **High Initial Costs:** Tidal barrages and turbines require significant upfront investments for design, construction, and installation. The costs associated with tidal energy can be higher compared to other renewable energy sources.
- **Limited Site Availability:** Tidal energy is location-dependent and requires specific conditions with strong tidal currents. Identifying suitable sites for tidal energy projects can be challenging.
- **Environmental Impact:** Tidal barrages can have environmental impacts such as changes in water quality, sedimentation patterns, and disruption of marine habitats. Tidal turbines should be carefully designed to minimize their impact on marine life.

Real-World Examples

One notable example of hydroelectric power is the Three Gorges Dam in China, which is the world's largest hydroelectric power plant. It generates a massive amount of electricity using the flow of the Yangtze River.

In terms of tidal energy, the Sihwa Lake Tidal Power Station in South Korea is one of the largest tidal power plants in the world. It utilizes a tidal barrage system to generate clean electricity.

Conclusion

Hydroelectric power and tidal energy are important sources of renewable energy that utilize the power of water. While hydroelectric power offers stable and reliable electricity generation, it poses challenges in terms of environmental impact and community displacement. Tidal energy, on the other hand, provides predictable and environmentally friendly power but faces limited site availability and high initial costs. Despite these challenges, both hydroelectric power and tidal energy contribute to the transition towards a greener and more sustainable future.

Geothermal Energy and Biogas

Geothermal energy and biogas are two important renewable energy sources that have gained significant attention in recent years. They offer unique advantages in terms of energy production and environmental sustainability. In this section, we will explore the principles, applications, and challenges associated with geothermal energy and biogas.

340 *CHAPTER 5: RENEWABLE ENERGY AND GREEN TECHNOLOGIES*

Geothermal Energy

Geothermal energy refers to the heat that is stored within the Earth's crust. It is a renewable resource that can be harnessed for power generation and heating purposes. The heat is derived from the radioactive decay of minerals, primarily uranium, thorium, and potassium, present in the Earth's core.

Principles of Geothermal Energy:

The principles of geothermal energy are based on the transfer of heat from the Earth's interior to the surface. This heat transfer occurs through three main mechanisms:

1. **Conduction:** The transfer of heat through direct contact between materials.

2. **Convection:** The transfer of heat through the movement of fluids or gases.

3. **Radiation:** The transfer of heat through electromagnetic waves.

Geothermal Power Plants:

Geothermal power plants harness the heat from the Earth's interior to generate electricity. The most common type of geothermal power plant is the binary cycle plant. It uses a heat exchanger to transfer heat from the geothermal fluid to a working fluid with a lower boiling point, such as isobutane or pentane. This working fluid vaporizes and drives a turbine, which in turn generates electricity.

Geothermal Heating Systems:

Geothermal heating systems use the natural heat from the Earth to provide space heating and hot water. These systems utilize a heat pump, which extracts heat from the ground during the winter and transfers it indoors. In the summer, the process is reversed, and heat from the building is transferred back to the ground.

Advantages of Geothermal Energy:

Geothermal energy offers several advantages over traditional energy sources:

- **Renewable and Sustainable:** Geothermal energy is continuously replenished by the Earth's natural heat, making it a sustainable energy source.

- **Low Emissions:** Geothermal power plants have low emissions of greenhouse gases and other pollutants compared to fossil fuel-based power plants.

- **Reliable and Stable:** Geothermal power plants operate continuously, providing a reliable source of electricity without depending on external factors like weather conditions.

RENEWABLE ENERGY TECHNOLOGIES

341

- **Long Lifespan:** Geothermal power plants have a long lifespan, with some operating for more than 30 years, making them a viable long-term energy solution.

Challenges and Future Directions:
While geothermal energy has many advantages, there are also challenges and considerations that need to be addressed:

- **Resource Availability:** Geothermal energy is location-dependent. Not all regions have access to geothermal resources suitable for power generation or heating purposes.

- **High Upfront Cost:** Developing geothermal power plants and geothermal heating systems can be expensive due to the need for drilling and infrastructure development.

- **Environmental Impact:** Improper handling of geothermal fluids can lead to the release of harmful gases and other pollutants. Care must be taken to ensure the proper management of geothermal resources.

- **Technological Advancements:** There is ongoing research and development to improve the efficiency and cost-effectiveness of geothermal energy systems. This includes the exploration of enhanced geothermal systems and the development of advanced drilling technologies.

Biogas

Biogas is a mixture of gases produced by the breakdown of organic matter in the absence of oxygen. It is primarily composed of methane (CH_4) and carbon dioxide (CO_2), with small amounts of other gases such as hydrogen sulfide (H_2S) and nitrogen (N_2). Biogas can be produced from a variety of organic waste materials, including agricultural waste, food waste, and sewage.

Biogas Production Process:
The production of biogas involves four main stages:

1. **Feedstock Preparation:** Organic waste materials are collected and processed to remove contaminants and create a homogeneous mixture.

2. **Anaerobic Digestion:** The prepared feedstock is placed in an anaerobic digester, where it undergoes fermentation by bacteria in the absence of oxygen. This process produces biogas.

342 CHAPTER 5: RENEWABLE ENERGY AND GREEN TECHNOLOGIES

3. **Gas Storage:** The produced biogas is collected and stored in gas holders or gas storage tanks for later use.

4. **Gas Utilization:** Biogas can be used directly for heating, cooking, or electricity generation. It can also be upgraded to biomethane, which has a higher methane content and can be used as a vehicle fuel or injected into the natural gas grid.

Advantages of Biogas:
Biogas offers several advantages as a renewable energy source:

+ **Waste Management:** Biogas production helps to treat organic waste materials, reducing the amount of waste sent to landfills and decreasing methane emissions.

+ **Renewable and Clean Energy:** Biogas is a renewable energy source that can replace fossil fuels, reducing greenhouse gas emissions and air pollution.

+ **Energy Independence:** Biogas production provides a decentralized energy source, allowing communities to become more self-sufficient and less reliant on external energy supplies.

+ **Nutrient Recycling:** The byproduct of anaerobic digestion, called digestate, is a nutrient-rich fertilizer that can be used in agriculture, completing the nutrient cycle.

Challenges and Future Directions:
There are several challenges and future directions for biogas production:

+ **Feedstock Availability:** The availability and quality of organic waste materials for biogas production can vary depending on the region and season.

+ **Digester Efficiency:** The efficiency of anaerobic digestion processes can be affected by factors such as temperature, pH, and feedstock composition. Optimizing these factors is crucial for maximizing biogas production.

+ **Gas Upgrading and Utilization:** The upgrading of biogas to biomethane and its utilization in various applications require appropriate technologies and infrastructure.

+ **Public Acceptance:** Biogas production facilities can face resistance from local communities due to concerns about odor, noise, and visual impacts. Public awareness and engagement are important for successful biogas projects.

RENEWABLE ENERGY TECHNOLOGIES

Conclusion

Geothermal energy and biogas are two important components of the renewable energy landscape. Geothermal energy utilizes the Earth's natural heat for power generation and heating, while biogas production harnesses the power of organic waste to produce clean energy. Both sources offer numerous advantages in terms of sustainability, environmental impact, and energy independence. However, challenges such as resource availability, upfront costs, and technological advancements need to be addressed to fully unlock their potential. Despite these challenges, geothermal energy and biogas contribute to a greener and more sustainable future.

Biomass and Bioenergy

Biomass refers to organic materials that come from living, or recently living organisms, such as plants, animals, and microorganisms. Biomass can be used to produce bioenergy, which is energy generated from organic matter through processes like combustion, fermentation, or conversion to liquid or gas fuels. In this section, we will explore the principles, technologies, and applications of biomass and bioenergy.

Principles of Biomass and Bioenergy

The use of biomass for energy production is based on the principles of the carbon cycle and sustainable resource management. During photosynthesis, plants capture carbon dioxide from the atmosphere and convert it into organic compounds through the energy of sunlight. These organic compounds form the biomass. When biomass is used for energy, the carbon that was captured from the atmosphere is released back into the atmosphere. This results in a closed carbon cycle, where the net emissions of carbon dioxide are neutral or even negative, depending on the specific circumstances.

Sustainable resource management is crucial in biomass and bioenergy production. It ensures that biomass is harvested or grown in a way that does not deplete or degrade natural resources. Sustainable practices may involve using agricultural residues, forest residues, or purpose-grown energy crops for bioenergy production. Regenerative farming techniques and agroforestry systems can also contribute to biomass production while improving soil health and ecosystem services.

344 *CHAPTER 5: RENEWABLE ENERGY AND GREEN TECHNOLOGIES*

Biomass Conversion Technologies

There are different technologies available for converting biomass into bioenergy, depending on the characteristics of the biomass and the desired end-use. Here are some key biomass conversion technologies:

1. **Combustion:** Biomass is burned to produce heat, which can be used directly in industries or converted into electricity in power plants through steam turbines. This process is similar to the combustion of fossil fuels but with lower net carbon emissions.

2. **Anaerobic Digestion:** Microorganisms break down biomass in the absence of oxygen, producing biogas (mainly methane and carbon dioxide) as a byproduct. Biogas can be used for heat and electricity generation, as well as biofuels production.

3. **Pyrolysis:** Biomass is heated in the absence of oxygen, leading to the breakdown of complex organic molecules into bio-oil, syngas (a mixture of hydrogen, carbon monoxide, and other gases), and biochar (a form of charcoal). Bio-oil and syngas can be further processed into transportation fuels or used for power generation, while biochar can be used as a soil amendment.

4. **Fermentation:** Biomass, such as sugarcane or corn, can be fermented to produce ethanol, a biofuel that can be blended with gasoline. The process involves the breakdown of sugars by microorganisms, such as yeast, into ethanol and carbon dioxide.

Applications of Biomass and Bioenergy

Biomass and bioenergy have a wide range of applications in various sectors, including:

+ **Electricity Generation:** Biomass power plants produce electricity from the combustion of biomass. The electricity can be used for residential, commercial, or industrial purposes, providing a renewable and sustainable energy source.

+ **Heat Production:** Biomass can be burned to produce heat for space heating, water heating, or industrial processes. Biomass boilers or stoves are commonly used for this purpose, especially in rural areas and regions with abundant biomass resources.

RENEWABLE ENERGY TECHNOLOGIES

- **Transportation Fuels:** Biofuels, such as ethanol and biodiesel, can be blended with conventional fossil fuels or used as standalone fuels. They can be used in vehicles, airplanes, or ships, reducing greenhouse gas emissions and dependence on fossil fuels.

- **Bioproducts and Biochemicals:** Biomass can be converted into various bioproducts and biochemicals, such as bioplastics, bio-based chemicals, and biocomposites. These products provide sustainable alternatives to petroleum-based materials and contribute to the shift towards a bio-based circular economy.

- **Combined Heat and Power (CHP):** Biomass CHP systems generate both heat and electricity, maximizing the energy efficiency of biomass utilization. These systems are particularly suitable for district heating, where the excess heat produced can be used to meet the heating needs of buildings or industrial processes.

Challenges and Future Directions

While biomass and bioenergy offer significant potential for sustainable and renewable energy, there are several challenges that need to be addressed:

- **Feedstock Availability and Sustainability:** Ensuring a sustainable supply of biomass feedstock is a key challenge. This includes avoiding competition with food production, protecting natural ecosystems, and promoting efficient land use practices.

- **Logistics and Infrastructure:** Biomass feedstock often requires collection, transportation, and storage, which can be costly and logistically challenging, especially for remote or rural areas. Adequate infrastructure is necessary to support biomass supply chains.

- **Technological Advancements:** Continued research and development are needed to improve biomass conversion technologies, increase energy efficiency, and reduce costs. Advances in genetic engineering, biotechnology, and process innovations can contribute to the optimization of bioenergy production.

- **Environmental Impacts:** While biomass can significantly reduce greenhouse gas emissions compared to fossil fuels, certain biomass production practices and conversion technologies can have negative environmental impacts. These

346 CHAPTER 5: RENEWABLE ENERGY AND GREEN TECHNOLOGIES

include deforestation, habitat destruction, water pollution, and air emissions. Sustainable management practices and strict environmental regulations are necessary to minimize these impacts.

+ **Policy and Market Support:** Government policies, incentives, and market mechanisms play a crucial role in promoting bioenergy development. Supportive policies can include feed-in tariffs, carbon pricing, renewable energy targets, and research funding.

In conclusion, biomass and bioenergy are important components of a sustainable and low-carbon energy system. By utilizing organic waste and purpose-grown energy crops, biomass can be converted into heat, electricity, biofuels, and other valuable products. However, careful management, technological advancements, and supportive policies are necessary to address challenges and ensure the sustainable and responsible use of biomass resources for a greener future.

Case Studies in Renewable Energy

Germany's Energy Transition

Germany's energy transition, also known as the Energiewende, is a comprehensive plan aimed at transforming the country's energy system from fossil fuels to renewable sources. This ambitious undertaking is driven by the need to address climate change, reduce reliance on imported energy, and promote sustainable economic growth. The Energiewende has made significant progress since its inception in the early 2000s and serves as a model for other countries seeking to transition to a more sustainable energy future.

Background

Germany has a long history of utilizing coal for its energy needs, and this reliance on fossil fuels has contributed to significant greenhouse gas emissions. In the early 1990s, concerns about air pollution, nuclear power, and climate change prompted discussions about transitioning to cleaner and more sustainable energy sources. These discussions culminated in the passage of the Renewable Energy Sources Act (EEG) in 2000, which provided a framework for the development of renewable energy in Germany.

CASE STUDIES IN RENEWABLE ENERGY

Principles of the Energy Transition

The energy transition in Germany is guided by several key principles:

1. **Decentralization:** The energy transition aims to decentralize energy production and promote the use of locally generated renewable energy. This involves the expansion of small-scale energy systems such as solar panels on rooftops and community wind farms.

2. **Renewable Energy Expansion:** The main focus of the energy transition is to expand renewable energy capacity, particularly wind and solar power. Germany has set ambitious targets for renewable energy deployment, aiming to reach at least 80% renewable electricity by 2050.

3. **Energy Efficiency:** Alongside the expansion of renewable energy, energy efficiency measures play a crucial role in reducing overall energy consumption. Germany has implemented various programs to improve energy efficiency in buildings, industry, and transportation.

4. **Phasing Out Nuclear Power:** The Energiewende includes the phase-out of nuclear power by 2022. This decision reflects the concerns of the German public about the safety and long-term viability of nuclear energy.

Achievements and Challenges

The energy transition in Germany has achieved several notable successes:

1. **Renewable Energy Growth:** Germany has become a global leader in renewable energy deployment. The country has significantly increased its renewable energy capacity, with renewables now accounting for a major share of electricity generation.

2. **Job Creation:** The energy transition has created thousands of jobs in the renewable energy sector. This includes jobs in manufacturing, installation, and maintenance of renewable energy technologies.

3. **CO2 Emissions Reduction:** Germany has successfully reduced its greenhouse gas emissions through the increased use of renewables. This has helped the country to meet its climate change targets and contribute to global efforts to mitigate climate change.

However, the energy transition also faces significant challenges:

1. **Grid Integration:** The intermittent nature of wind and solar power poses challenges for grid integration and stability. The expansion of renewable energy requires upgrading and modernizing the grid infrastructure to accommodate the fluctuations in supply.

348 *CHAPTER 5: RENEWABLE ENERGY AND GREEN TECHNOLOGIES*

2. **Costs and Affordability:** The transition to renewable energy comes with significant costs, including subsidies for renewable energy producers. This has led to concerns about rising electricity prices, which can affect the affordability of energy for consumers and businesses.

3. **Storage and Flexibility:** The expansion of intermittent renewable energy sources necessitates the development of energy storage technologies and flexible power systems. These technologies will facilitate the integration of renewables into the grid and ensure a reliable electricity supply.

Policy and Support Mechanisms

The success of Germany's energy transition can be attributed to a range of policy and support mechanisms:

1. **Feed-in Tariffs:** The EEG introduced feed-in tariffs, which guarantee a fixed price for renewable energy producers. This mechanism has incentivized investment in renewable energy projects and facilitated their rapid expansion.

2. **Renewable Energy Sources Act (EEG):** The EEG provides the legal framework for the development of renewable energy in Germany. It sets binding targets and outlines the support mechanisms for renewable energy producers.

3. **Energy Efficiency Programs:** Germany has implemented various programs to improve energy efficiency in buildings, industry, and transportation. These programs provide incentives and financial support for energy-efficient technologies and practices.

4. **Research and Development Funding:** Germany has invested heavily in research and development of renewable energy technologies. Public funding supports innovation and the development of new technologies, contributing to the growth of the renewable energy sector.

Conclusion

Germany's energy transition represents a comprehensive and ambitious approach to shifting towards a more sustainable energy future. The country's commitment to renewable energy expansion, energy efficiency, and phasing out nuclear power has yielded significant achievements and global recognition. However, challenges remain in terms of grid integration, cost-effectiveness, and energy storage. The experiences and lessons learned from Germany's energy transition serve as valuable insights for other countries embarking on a similar path towards a greener and more sustainable energy system.

CASE STUDIES IN RENEWABLE ENERGY

Solar Energy in India

India, with its abundant sunlight throughout the year, has great potential for harnessing solar energy. Solar energy is a renewable and sustainable source of power that can be utilized to meet the country's growing energy demands while reducing greenhouse gas emissions. In this section, we will explore the development and current state of solar energy in India, including its benefits, challenges, and future prospects.

Background

India is the world's third-largest emitter of greenhouse gases, primarily due to its heavy reliance on fossil fuels for energy generation. The increasing demand for power, coupled with the need to mitigate climate change, has led to a strong push towards renewable energy sources, particularly solar energy.

Principles of Solar Energy

Solar energy is derived from the sun's radiation, which can be converted into electricity or used directly for heating or lighting purposes. The key principles underlying solar energy include:

- **Photovoltaic Effect:** Solar panels or photovoltaic (PV) cells are used to convert sunlight into electricity. When photons from sunlight strike the PV cells, they dislodge electrons from their atoms, creating a flow of electricity.

- **Solar Thermal Technology:** Solar thermal systems use the heat from the sun to generate electricity or provide hot water and space heating. Concentrated solar power (CSP) technologies focus sunlight onto receivers to produce high-temperature heat, which is then converted into electricity.

- **Net Metering:** Net metering allows solar energy system owners to feed excess electricity they generate back into the grid, offsetting their electricity costs. This enables the integration of solar energy into existing power grids.

Current Status of Solar Energy in India

India has made significant progress in the adoption of solar energy. The government has implemented various policies and initiatives to promote solar power generation, including:

350 CHAPTER 5: RENEWABLE ENERGY AND GREEN TECHNOLOGIES

- **Jawaharlal Nehru National Solar Mission (JNNSM):** Launched in 2010, JNNSM aims to promote the use of solar energy in India and increase the country's solar power capacity to 100 GW by 2022. It includes various schemes such as solar parks, rooftop solar installations, and solar off-grid applications.

- **Solar Parks:** Solar parks provide large-scale infrastructure for solar power generation. These parks facilitate the development of solar projects by providing land, transmission infrastructure, and other necessary amenities. Many states in India have established solar parks to promote solar energy adoption.

- **Rooftop Solar Program:** The rooftop solar program encourages individuals, commercial establishments, and industries to install solar panels on their rooftops. This helps reduce dependence on the grid and allows them to generate their own electricity.

- **Quality Control and Certification:** The Indian government has established quality control measures and certification standards for solar panels and other solar energy systems. This ensures the reliability and performance of solar installations in the country.

As a result of these initiatives, India's solar power capacity has increased significantly in recent years. According to the Ministry of New and Renewable Energy, as of March 2021, India has installed more than 40 GW of solar power capacity, making it one of the largest solar markets globally.

Benefits of Solar Energy in India

The adoption of solar energy in India brings several benefits, including:

- **Clean and Renewable:** Solar energy is a clean and renewable source of power, which helps reduce greenhouse gas emissions and combat climate change. By transitioning to solar energy, India can significantly reduce its carbon footprint.

- **Energy Independence:** Solar energy reduces dependence on fossil fuel imports and enhances energy security for the country. This is particularly important for a rapidly developing nation like India, which has a high energy demand.

CASE STUDIES IN RENEWABLE ENERGY

- **Job Creation:** The growth of the solar energy sector has created numerous employment opportunities in India. The installation, operation, and maintenance of solar power plants require a skilled workforce, contributing to job creation and economic development.

- **Reduced Electricity Costs:** Solar energy can help reduce electricity costs for consumers, especially in areas with high solar potential. With net metering and rooftop solar installations, consumers can generate their own electricity and lower their reliance on the grid.

Challenges and Future Prospects

While India has made significant progress in solar energy adoption, several challenges remain:

- **Storage and Grid Integration:** The intermittent nature of solar energy requires effective energy storage solutions for uninterrupted power supply. Additionally, integrating large-scale solar power into the existing grid infrastructure poses technical challenges that need to be addressed.

- **Land Availability:** Establishing large-scale solar power projects requires vast land areas. Identifying suitable land and obtaining necessary approvals can be a challenge, particularly in densely populated regions.

- **Funding and Financial Viability:** The cost of setting up and maintaining solar power plants can be high. Access to financing, favorable government policies, and incentives are crucial to ensuring the financial viability of solar projects.

Looking ahead, the future of solar energy in India appears promising. The government's continued support through policy frameworks, research and development, and financial incentives will further drive the growth of the solar energy sector. Advancements in energy storage technologies, grid integration, and the development of innovative business models will improve the efficiency and reliability of solar energy systems.

Conclusion

Solar energy holds significant potential for meeting India's growing energy needs while reducing its carbon footprint. The government's efforts in promoting solar power generation have led to substantial progress in the adoption of solar energy.

352 CHAPTER 5: RENEWABLE ENERGY AND GREEN TECHNOLOGIES

By tapping into its abundant solar resources, India can enhance its energy security, create job opportunities, and contribute to global efforts in combating climate change. Continued investments in research, technology development, and policy support will pave the way for a greener and more sustainable energy future for India.

Offshore Wind Farms in Europe

Offshore wind farms have increasingly gained attention as a promising source of renewable energy in Europe. These wind farms are located in open waters, often several kilometers away from the coastline, and consist of multiple wind turbines arranged in an array. In this section, we will explore the principles behind offshore wind farms, their benefits, challenges, and the success stories in Europe.

Principles of Offshore Wind Farms

Offshore wind farms harness the power of winds blowing over the sea to generate electricity. The basic principles governing their operation are similar to onshore wind farms. However, there are some unique considerations that must be taken into account for offshore installations.

Location Selection Choosing the right location for an offshore wind farm is crucial. It requires careful assessment of factors such as wind speed and direction, water depth, seabed conditions, and proximity to the electrical grid. Wind speed is particularly important as higher wind speeds result in greater energy generation. The water depth should be suitable for the foundation of turbines, and the seabed should have sufficient stability to support the structures.

Turbine Design Offshore wind turbines are typically larger and more robust than their onshore counterparts. They are designed to withstand harsh marine environments, including strong winds, waves, and saltwater corrosion. The turbine components are often made of corrosion-resistant materials, and the foundations are engineered to provide stability in deep water.

Electrical Infrastructure Offshore wind farms require an efficient electrical infrastructure to transmit the generated electricity to the onshore grid. Submarine cables are used to transfer the electricity from the offshore turbines to an onshore substation. These cables must be capable of withstanding the marine environment and have high transmission capacity to minimize energy loss during transmission.

CASE STUDIES IN RENEWABLE ENERGY

Environmental Impact Assessment   Before the construction of offshore wind farms, an environmental impact assessment is conducted to evaluate potential effects on marine ecosystems and wildlife.   It includes studying the migration patterns of birds and marine mammals, the impact on fish populations, and the potential disruption to the seabed and marine habitats. This assessment helps in developing appropriate mitigation measures and ensures the sustainability of the offshore wind farms.

Benefits of Offshore Wind Farms

Offshore wind farms offer numerous benefits that make them an attractive option for clean energy generation in Europe.

Abundant Wind Resources   Coastal areas in Europe are known for their strong and consistent winds. By harnessing offshore wind resources, countries can tap into a significant and reliable energy source. Offshore wind farms have the potential to produce large amounts of electricity, often exceeding the capacity of onshore wind farms.

Reduced Visual and Noise Impact   As offshore wind farms are located far from the coastline, they have a reduced visual impact compared to onshore wind farms. This makes them more socially acceptable, especially in densely populated coastal areas. Additionally, offshore installations are typically quieter than onshore turbines, minimizing noise pollution.

Higher Energy Conversion Efficiency   Offshore wind farms benefit from the unobstructed airflow over the sea, allowing turbines to operate at higher efficiency. The higher wind speeds and lower turbulence in offshore locations result in increased energy conversion.

Potential for Large-Scale Deployment   The vast expanse of open waters in Europe provides ample space for the deployment of large-scale offshore wind farms. This scalability allows for the generation of significant amounts of electricity, potentially meeting a substantial portion of the region's energy demand.

Challenges and Solutions

While offshore wind farms have many advantages, they also face several challenges that need to be addressed for successful implementation.

354 CHAPTER 5: RENEWABLE ENERGY AND GREEN TECHNOLOGIES

High Cost of Installation One of the primary challenges is the high initial cost of installing offshore wind farms. Construction and installation in marine environments are complex and require specialized equipment and techniques. However, technological advancements and economies of scale have led to a gradual reduction in the cost of offshore wind farms.

Maintenance and Accessibility Maintaining and accessing offshore wind farms pose logistical challenges. The remote location and harsh weather conditions make repairs and inspections more difficult and costly. However, advancements in remote monitoring systems and robotic maintenance solutions are helping overcome these challenges.

Environmental Concerns Offshore wind farms can have potential environmental impacts, including disturbance to marine ecosystems and wildlife. However, rigorous environmental impact assessments and proper planning can minimize these effects. Collaboration with environmental organizations and ongoing monitoring can ensure the sustainable coexistence of offshore wind farms and marine ecosystems.

Grid Integration The integration of offshore wind energy into the existing electrical grid can be challenging. Transmission infrastructure needs to be upgraded to accommodate the increased capacity and distance. Additionally, the intermittent nature of wind energy requires effective grid management and balancing with other sources of electricity. Advances in grid technologies and energy storage systems are addressing these integration challenges.

Success Stories: Offshore Wind Farms in Europe

Europe has been at the forefront of offshore wind farm development. Several countries have made significant progress in harnessing offshore wind energy to meet their clean energy targets.

Denmark Denmark is a pioneer in offshore wind farm development and has been generating electricity from offshore wind since the early 1990s. The Horns Rev offshore wind farm, located off the Danish coast in the North Sea, was one of the first large-scale projects. It has a capacity of 400 MW and provides renewable electricity to thousands of households.

CASE STUDIES IN RENEWABLE ENERGY

United Kingdom The United Kingdom has made remarkable strides in offshore wind energy. The London Array, located in the Thames Estuary, is one of the world's largest offshore wind farms, with a capacity of 630 MW. It generates clean electricity for over half a million homes. The UK government has set ambitious targets to expand its offshore wind capacity, aiming for 40 GW by 2030.

Germany Germany has a strong commitment to renewable energy, and offshore wind plays a crucial role in its energy transition. The BARD Offshore 1 wind farm in the North Sea was the first commercial-scale project in Germany, with a capacity of 400 MW. Germany has continued to invest in offshore wind farms, significantly increasing its capacity in recent years.

Netherlands The Netherlands has embraced offshore wind energy as part of its strategy to reduce greenhouse gas emissions. The Gemini Offshore Wind Park, situated in the North Sea, has a capacity of 600 MW and provides renewable energy to around 1.5 million households. The Dutch government has set ambitious targets to reach a total offshore wind capacity of 11.5 GW by 2030.

Sweden Sweden has been actively developing offshore wind farms in the Baltic Sea. The Kriegers Flak offshore wind farm, once completed, will have a capacity of 600 MW and is expected to be one of the largest in the region. Sweden aims to increase its offshore wind capacity to 30 GW by 2030, contributing to its goal of becoming climate neutral by 2045.

Conclusion

Offshore wind farms in Europe have emerged as a vital component of the region's transition to clean energy. They offer abundant wind resources, reduced visual impact, and the potential for large-scale deployment. Despite challenges, countries like Denmark, the United Kingdom, Germany, the Netherlands, and Sweden have demonstrated successful implementation of offshore wind farms. These success stories provide valuable lessons and inspiration for other nations seeking to tap into the vast potential of offshore wind energy. The continuous advancement of technologies, coupled with effective planning and collaboration, will ensure the further growth and sustainability of offshore wind farms in Europe and beyond.

356 CHAPTER 5: RENEWABLE ENERGY AND GREEN TECHNOLOGIES

Hydropower in Brazil

Hydropower is a significant renewable energy source that has been harnessed in many countries around the world. One country that stands out in terms of hydropower generation is Brazil. With its extensive river network and abundant rainfall, Brazil has leveraged hydropower to become one of the leading producers of clean and sustainable energy. In this section, we will explore the development, benefits, challenges, and future directions of hydropower in Brazil.

Development of Hydropower in Brazil

The development of hydropower in Brazil can be traced back to the early 20th century when the government recognized the potential of its rivers for electricity generation. The construction of large-scale hydropower plants started with the inauguration of the Paulo Afonso I plant in 1955, located on the São Francisco River. Since then, Brazil has continuously expanded its hydropower capacity and has become a pioneer in developing advanced technologies for this sector.

The most significant milestone in Brazil's hydropower development is the construction of the Itaipu Dam, one of the world's largest hydroelectric power plants. Located on the Paraná River, it is a joint project between Brazil and Paraguay and has a capacity of 14,000 megawatts (MW). Itaipu Dam not only generates significant amounts of electricity but also serves as a symbol of cooperation and shared resources between the two countries.

Benefits of Hydropower in Brazil

Hydropower has brought numerous benefits to Brazil, both in terms of energy production and socioeconomic impact. Some of the key benefits are:

- **Clean and Renewable Energy:** Hydropower is a clean and renewable energy source, which helps Brazil reduce its dependence on fossil fuels and decrease greenhouse gas emissions. It plays a crucial role in the country's efforts to combat climate change.

- **Large-Scale Electricity Generation:** Brazil's hydropower plants have a significant capacity for electricity generation. They supply a substantial portion of the country's total energy demand, making Brazil less reliant on imported energy sources.

- **Job Creation and Economic Growth:** The construction and operation of hydropower projects create employment opportunities and boost the local

CASE STUDIES IN RENEWABLE ENERGY

economy. It generates jobs in various sectors, such as engineering, construction, maintenance, and tourism.

+ **Flood Control and Water Management:** Hydropower reservoirs provide a water management system that helps control floods and ensures a stable water supply throughout the year. This is especially important in arid regions of Brazil, where water scarcity is a persistent issue.

Challenges and Environmental Concerns

While hydropower brings significant benefits, it also poses challenges and environmental concerns that need to be addressed. Some of the key challenges faced by hydropower development in Brazil are:

+ **Environmental Impact:** Large-scale hydropower projects can cause environmental degradation, including habitat loss, alteration of river ecosystems, and displacement of local communities. The flooding of vast areas for reservoir creation can lead to deforestation and loss of biodiversity.

+ **Social Impacts:** The construction of hydropower plants can result in the displacement of indigenous communities and local populations. Their livelihoods and cultural heritage may be severely affected. Proper mitigation measures and the involvement of affected communities are essential to address these social impacts.

+ **Reliance on Rainfall:** Hydropower generation heavily relies on rainfall patterns and water availability. Droughts and changes in precipitation patterns due to climate change can affect the reliability of hydropower production. Brazil needs to diversify its energy mix to reduce vulnerability to such changes.

Future Directions

To address the challenges and ensure the future sustainability of hydropower in Brazil, several strategies can be implemented:

+ **Environmental and Social Impact Assessment:** Before constructing hydropower projects, thorough environmental and social impact assessments should be conducted. This process should involve the local communities and stakeholders, ensuring their participation and addressing their concerns.

358 CHAPTER 5: RENEWABLE ENERGY AND GREEN TECHNOLOGIES

- **Sustainable Operations and Practices:** Hydropower plants should adopt sustainable operational practices to minimize their ecological footprint. This includes efficient water usage, implementing environmental flow regimes, and preserving the downstream ecology.

- **Investment in Research and Development:** Continued investment in research and development is essential to improve the efficiency of hydropower generation, mitigate environmental impacts, and develop innovative technologies for fish migration and sediment management.

- **Diversification of Energy Sources:** Brazil should further diversify its energy mix by promoting the development of other renewable energy sources like solar, wind, and bioenergy. This will reduce the country's dependence on hydropower and enhance energy security.

Case Study: Belo Monte Dam

One prominent example of hydropower development in Brazil is the Belo Monte Dam project. Located on the Xingu River in the state of Pará, it is one of the largest hydropower projects in the country. However, the dam has faced significant controversies due to its potential environmental and social impacts.

The Belo Monte Dam has led to the displacement of indigenous communities and the disruption of the Xingu River's ecosystem. The dam's reservoir has flooded vast areas of the Amazon rainforest, causing substantial deforestation and loss of biodiversity. The project has been a subject of legal battles and protests by environmental and indigenous rights organizations.

To mitigate the negative impacts of the Belo Monte Dam, the Brazilian government and project developers have implemented various measures. These include the relocation and compensation of affected communities, the implementation of environmental protection programs, and the establishment of fish passage systems to allow the migration of aquatic species.

While the Belo Monte Dam project highlights the challenges associated with large-scale hydropower development, it also serves as a lesson for the need to prioritize environmental and social considerations in future projects.

Conclusion

Hydropower has played a significant role in Brazil's energy landscape, contributing to its economic growth and sustainability goals. It has provided clean and renewable energy, created job opportunities, and helped manage water resources. However,

CASE STUDIES IN RENEWABLE ENERGY

environmental and social concerns associated with hydropower development should not be overlooked.

By addressing the challenges, incorporating sustainable practices, and diversifying the energy mix, Brazil can ensure the future of hydropower while minimizing its negative impacts. Hydropower will continue to be an integral part of Brazil's energy transition, working alongside other renewable energy sources to build a greener and more sustainable future for the country.

Biogas Production and Utilization in Sweden

In recent years, biogas has gained significant attention as a renewable energy source due to its environmental benefits and versatile applications. Sweden, known for its commitment to sustainability, has emerged as a global leader in biogas production and utilization. This section will delve into the process of biogas production, its various applications, and the success of Sweden in harnessing its potential.

Biogas Production

Biogas is produced through the anaerobic digestion of organic matter, such as agricultural waste, food waste, and sewage sludge. The process involves the breakdown of organic materials by microorganisms in the absence of oxygen, resulting in the release of methane-rich biogas.

The key steps in biogas production include feedstock preparation, anaerobic digestion, and biogas upgrading. Feedstock preparation involves collecting organic waste and processing it to maximize its suitability for digestion. This can include shredding, mixing, and adjusting moisture and pH levels.

Anaerobic digestion takes place in a controlled environment, such as an anaerobic digester. The organic material is introduced into the digester, where it undergoes four stages: hydrolysis, acidogenesis, acetogenesis, and methanogenesis. Each stage is facilitated by specific groups of microorganisms that break down complex organic compounds into simpler forms, ultimately producing methane and carbon dioxide.

Once produced, the raw biogas requires upgrading to remove impurities, especially carbon dioxide. This can be achieved through various techniques, such as pressure swing adsorption or membrane separation, resulting in high-quality biomethane that is suitable for injection into the natural gas grid or used as vehicle fuel.

360 CHAPTER 5: RENEWABLE ENERGY AND GREEN TECHNOLOGIES

Applications of Biogas

The versatile nature of biogas allows for its utilization in various sectors, contributing to Sweden's sustainable energy landscape. Let's explore some of the key applications:

1. **Power Generation:** Biogas can be used in combined heat and power (CHP) plants to generate electricity and heat. CHP plants utilize the energy content of biogas efficiently, making them an ideal solution for district heating systems and electricity production.

2. **Transportation Fuel:** Biomethane, derived from upgraded biogas, can be used as a sustainable alternative to fossil fuels in the transportation sector. Sweden has made significant progress in developing biogas refueling infrastructure, encouraging the adoption of biogas-powered vehicles.

3. **Industrial Processes:** Biogas finds applications in various industrial processes, such as heating and steam generation. It can replace fossil fuels in industries, contributing to significant reductions in greenhouse gas emissions.

4. **Circular Agriculture:** Biogas production provides an opportunity for farmers to manage organic waste efficiently while producing renewable energy. By using the digestate, a byproduct of the biogas process, as a nutrient-rich fertilizer, the agricultural sector can achieve greater sustainability and close the nutrient loop.

Success Stories in Sweden

Sweden has established itself as a global leader in biogas production and utilization, leveraging its strong commitment to renewable energy and sustainable development. The country has implemented several initiatives that have propelled its success in this sector:

1. **Policy Support:** Sweden's supportive regulatory framework, including financial incentives and subsidies for biogas production and infrastructure development, has been instrumental in driving growth in the biogas sector. The government's long-term vision and clear roadmap for renewable energy have provided a stable environment for investment and innovation.

2. **Strong Collaboration:** The success of biogas in Sweden can be attributed to strong collaboration between various stakeholders, including agricultural organizations, waste management companies, energy producers, and research institutions. This collaborative approach has facilitated the establishment of biogas production facilities and the development of efficient value chains.

3. **Infrastructure Development:** Sweden has made substantial investments in biogas infrastructure, including upgrading plants and refueling stations. This has

CASE STUDIES IN RENEWABLE ENERGY

created a reliable and extensive network for the production, distribution, and utilization of biogas, fostering its integration into the energy system.

4. **Public Awareness and Education:** Sweden's commitment to sustainability extends to raising public awareness about biogas and its benefits. Educational campaigns and initiatives have increased public acceptance and utilization of biogas, encouraging individuals and businesses to embrace this renewable energy source.

Challenges and Future Directions

Despite Sweden's remarkable achievements in biogas production and utilization, some challenges and opportunities lie ahead. These include:

1. **Feedstock Availability:** Ensuring a consistent and sufficient supply of organic waste for biogas production is crucial. Sweden needs to explore strategies to optimize waste collection systems and promote source separation to increase the availability of potential feedstock.

2. **Technological Innovations:** Continued research and development in biogas production and upgrading technologies are essential to improve efficiency, reduce costs, and enhance the overall sustainability of the process. Investing in innovative solutions such as advanced digestion techniques and new upgrading methods will further strengthen the biogas sector.

3. **Integration with the Energy System:** As biogas production increases, integrating it into the existing energy system becomes paramount. Optimizing the utilization of biogas for power generation, heating, and transportation will ensure a seamless transition to a more sustainable and renewable energy mix.

4. **International Cooperation:** Collaboration with other countries and knowledge-sharing platforms can accelerate the growth of the biogas sector globally. Partnering with international organizations and sharing best practices will enable Sweden to contribute to a wider shift towards a greener future.

As Sweden continues to push the boundaries of biogas production and utilization, it sets an inspiring example for other nations striving to achieve sustainability goals. By capitalizing on its success stories and addressing future challenges, Sweden paves the way for a greener, more renewable energy future.

362 CHAPTER 5: RENEWABLE ENERGY AND GREEN TECHNOLOGIES

Success Stories in Renewable Energy

Costa Rica's Renewable Energy Revolution

Costa Rica, a small country located in Central America, has become a global leader in renewable energy. This success story demonstrates the country's commitment to sustainability and its determination to reduce dependence on fossil fuels. In this section, we will explore the key factors that have contributed to Costa Rica's renewable energy revolution.

Background

Costa Rica is known for its rich biodiversity, stunning landscapes, and commitment to environmental protection. With a population of around 5 million people, the country recognized the need to find alternative sources of energy to ensure a sustainable future. The government set ambitious goals to promote the use of renewable energy, reduce greenhouse gas emissions, and protect its natural resources.

Principles of Renewable Energy

To understand Costa Rica's renewable energy revolution, we must first grasp the principles that underpin this transition. Renewable energy refers to energy sources that are naturally replenished, such as sunlight, wind, and water. The utilization of these sources has many advantages over traditional fossil fuels, including reduced carbon emissions, diversification of energy sources, and increased energy security.

One of the key principles of renewable energy is the use of solar power. Costa Rica's location close to the equator provides abundant sunlight, making solar energy an ideal choice. Photovoltaic (PV) technology, which converts sunlight into electricity using semiconductors, has seen significant progress in recent years. The country has embraced solar power by installing solar panels in both urban and rural areas, reducing dependence on traditional energy sources.

Another crucial principle is harnessing wind power. Costa Rica has harnessed its strong winds to generate electricity through wind turbines. Wind power is not only renewable but also highly efficient, with modern turbines converting wind energy into electricity. The country's windy coastal regions have become hotspots for wind farm installations, contributing to its renewable energy revolution.

Additionally, Costa Rica has tapped into its abundant water resources to generate hydroelectric power. By building dams and utilizing the kinetic energy of moving water, the country has established a robust hydropower infrastructure.

SUCCESS STORIES IN RENEWABLE ENERGY

Water used in hydroelectric plants is a constant renewable resource, making it a reliable source of energy. This emphasis on sustainable hydropower has contributed significantly to Costa Rica's renewable energy success.

Costa Rica's Renewable Energy Mix

One of the notable achievements of Costa Rica's renewable energy revolution is its increasing reliance on renewable sources for electricity generation. The country has set a goal to become carbon neutral by 2021. It has made significant progress in this regard, with renewable energy accounting for a substantial portion of its total energy mix.

Hydropower has historically been the dominant source of renewable energy in Costa Rica. The country's numerous rivers and mountainous terrain make it well-suited for hydroelectric projects. Currently, hydropower represents about 75% of Costa Rica's total renewable energy production.

In recent years, Costa Rica has also made significant strides in diversifying its energy sources. The country has heavily invested in wind power, with wind farms springing up along its coasts and in mountainous regions. Wind energy now contributes about 10% to Costa Rica's renewable energy mix.

Solar power is another growing sector in Costa Rica's renewable energy landscape. The government has implemented various incentives to encourage the adoption of solar panels in residential, commercial, and industrial buildings. Solar energy accounts for around 5% of the country's renewable energy generation.

Biogas and biomass are other sources of renewable energy being developed in the country. Biogas is produced through the anaerobic digestion of organic waste, such as agricultural residues and animal manure. Biomass refers to the use of organic matter, such as wood and agricultural waste, to generate heat and electricity. These emerging sectors play a smaller role but are gradually gaining momentum.

Benefits and Challenges

Costa Rica's renewable energy revolution has brought numerous benefits to the country and its people. Firstly, the use of renewable energy sources has significantly reduced carbon emissions, making a substantial contribution to mitigating climate change. It has also improved air quality and reduced reliance on imported fossil fuels, enhancing energy security and reducing the country's vulnerability to oil price fluctuations.

Moreover, Costa Rica's transition to renewable energy has created new employment opportunities and stimulated economic growth. The renewable

364 CHAPTER 5: RENEWABLE ENERGY AND GREEN TECHNOLOGIES

energy sector has attracted investments and provided a platform for innovation and technological advancements. The country's commitment to sustainability has also boosted its tourism industry, attracting visitors interested in eco-friendly practices and renewable energy projects.

However, like any energy transition, Costa Rica has faced certain challenges in its renewable energy journey. One of the key challenges is the variability of certain renewable energy sources. Solar and wind power generation is dependent on favorable weather conditions, which can be unpredictable. This intermittency requires the development of effective energy storage systems and a smart grid infrastructure to ensure a stable and reliable energy supply.

Another challenge is the initial investment required for renewable energy projects. While the long-term benefits are clear, the upfront costs of installing solar panels or building wind farms can be significant. The government has tackled this challenge by providing incentives, such as tax breaks and feed-in tariffs, to promote renewable energy adoption. Public-private partnerships have also played a crucial role in attracting investments and sharing the financial burden.

Lessons Learned and Future Outlook

Costa Rica's renewable energy revolution offers valuable lessons for other countries striving to reduce their reliance on fossil fuels. The key lesson is the importance of political will and long-term commitment. Costa Rica's government set clear goals and implemented supportive policies to promote renewable energy adoption. It also engaged in partnerships with private entities, civil society, and international organizations to drive the transition.

Another crucial lesson is the need for a diversified renewable energy mix. Costa Rica's success is not solely reliant on one energy source but rather a combination of hydropower, wind power, solar power, and emerging technologies like biogas and biomass. This diversified approach ensures resilience and reduces the vulnerability of the energy system to fluctuations in weather patterns and resource availability.

Looking ahead, Costa Rica aims to further expand its renewable energy capacity. The government plans to increase solar, wind, and geothermal energy installations, as well as harness the potential of emerging technologies like wave energy. Efforts are also underway to promote energy efficiency and reduce overall energy consumption.

Overall, Costa Rica's renewable energy revolution showcases the possibilities and benefits of transitioning to a sustainable energy future. With the right combination of political will, supportive policies, and public-private partnerships, countries around the world can learn from Costa Rica's example and work towards a greener and more sustainable future.

Summary

Costa Rica's renewable energy revolution has demonstrated the power of determination and a commitment to sustainability. The country's transition to a renewable energy future has relied on the principles of solar power, wind power, and hydropower. By harnessing these renewable sources, Costa Rica has made significant progress in reducing carbon emissions and diversifying its energy mix.

The country's success in renewable energy has brought numerous benefits, including improved air quality, energy security, and economic growth. However, challenges such as intermittency and upfront costs remain. Costa Rica's experience provides valuable lessons for other nations seeking to embrace renewable energy and move towards a sustainable future.

As we explore more success stories in renewable energy and green technologies, we will continue to learn from the experiences of countries like Costa Rica and endeavor to build a greener and more sustainable world.

Green Building and Sustainable Architecture

Green building and sustainable architecture play a crucial role in promoting environmental sustainability and mitigating the impacts of climate change. This section will explore the principles, techniques, and success stories in green building, with a focus on sustainable architecture.

Principles of Green Building

Green building incorporates various principles that aim to minimize the negative environmental impact of a structure throughout its entire life cycle. These principles include:

1. Energy Efficiency: Designing buildings to minimize energy consumption and maximize energy efficiency is a fundamental principle of green building. This involves using energy-efficient materials, proper insulation, and efficient HVAC systems to reduce the building's carbon footprint.

2. Water Conservation: Green buildings adopt water-saving measures, such as efficient plumbing fixtures, rainwater harvesting systems, and graywater recycling. These practices help to reduce water consumption and preserve precious water resources.

3. Site Selection and Design: Careful site selection and design can significantly influence a building's environmental impact. Green building practices

366 CHAPTER 5: RENEWABLE ENERGY AND GREEN TECHNOLOGIES

emphasize the use of existing infrastructure, preservation of natural habitats, and incorporation of open spaces and greenery.

4. Materials and Resources: Green building promotes the use of environmentally friendly and sustainable materials. This includes utilizing recycled content, selecting materials with low embodied energy, and choosing products that have a minimal ecological footprint.

5. Indoor Environmental Quality: Ensuring a healthy and comfortable living environment is crucial in green building. This involves proper ventilation, use of non-toxic materials, and adequate lighting to enhance occupants' well-being and productivity.

Sustainable Architecture

Sustainable architecture goes beyond the principles of green building and focuses on creating harmonious structures that integrate with their surrounding environment. It takes into account the social, economic, and cultural aspects in addition to the environmental aspects. Some key considerations in sustainable architecture are:

1. Passive Design: Passive design strategies aim to optimize natural lighting, ventilation, and heating/cooling in buildings. This includes orienting the building to maximize daylight, designing windows for cross ventilation, and incorporating shading devices to reduce heat gain.

2. Biomimicry: Biomimicry is a design approach that draws inspiration from nature to create innovative and sustainable solutions. It involves studying natural processes and systems to develop design strategies that mimic the efficiency and resilience found in nature.

3. Integration of Renewable Energy: Sustainable architecture promotes the integration of renewable energy systems such as solar panels, wind turbines, and geothermal systems. These systems help to generate clean energy on-site, reducing reliance on fossil fuels.

4. Adaptive Reuse: Adaptive reuse involves repurposing existing structures instead of demolishing them. By giving new life to old buildings, sustainable architecture reduces construction waste and preserves the cultural and historical value of the built environment.

SUCCESS STORIES IN RENEWABLE ENERGY

5. Biophilic Design: Biophilic design seeks to connect people with nature by incorporating natural elements and patterns into building design. This includes features such as green roofs, living walls, and indoor gardens, which provide psychological and physiological benefits to occupants.

Success Stories in Green Building and Sustainable Architecture

Numerous success stories demonstrate the transformative power of green building and sustainable architecture. Here are some notable examples:

1. The Edge, Amsterdam: The Edge is considered one of the most sustainable office buildings in the world. It features innovative energy-efficient systems, including smart LED lighting, occupancy sensors, and a sophisticated climate control system. The building generates its energy from rooftop solar panels and utilizes rainwater for irrigation and flushing toilets.

2. Nk'Mip Desert Cultural Centre, Canada: This center showcases sustainable architecture in a culturally significant context. It incorporates earth-sheltered design, using the natural thermal properties of the earth to provide insulation. The building also incorporates passive solar strategies and utilizes local and recycled materials.

3. Bosco Verticale, Milan: Bosco Verticale, or the Vertical Forest, is a pair of residential towers featuring extensive greenery on the facades. The towers host a variety of tree and plant species, providing multiple ecological benefits including air purification, noise reduction, and temperature regulation.

4. Khoo Teck Puat Hospital, Singapore: This hospital incorporates sustainable features such as natural ventilation, rainwater harvesting, and extensive greenery. The integration of biophilic design principles has created a healing environment that promotes patient well-being and reduces energy consumption.

5. Bullitt Center, Seattle: The Bullitt Center is a six-story office building that aims to be self-sustaining and carbon-neutral. It features solar panels, geothermal heating and cooling, rainwater harvesting, and composting toilets. The building consumes 80% less energy than a typical office building and has a net-zero energy goal.

368 CHAPTER 5: RENEWABLE ENERGY AND GREEN TECHNOLOGIES

Challenges and Future Directions

While green building and sustainable architecture have made significant strides, several challenges remain to be addressed:

1. Cost Implications: The upfront costs of implementing green building practices can often be higher compared to conventional construction. However, the long-term benefits in terms of energy savings and environmental impact make green building a worthwhile investment.

2. Knowledge and Skills Gap: There is a need to enhance the knowledge and skills of architects, engineers, and construction professionals in sustainable design and construction techniques. Educational programs and training initiatives can bridge this gap and promote the wider adoption of sustainable practices.

3. Codes and Regulations: Governments need to implement and enforce building codes and regulations that foster sustainable design and construction. This can create a level playing field and provide incentives for the broader adoption of green building practices.

4. Community Engagement: Sustainable architecture should prioritize community engagement and involve stakeholders throughout the design and construction process. Engaging local communities can ensure that the built environment meets their needs and aspirations while addressing environmental concerns.

5. Advancements in Technology and Innovation: Continued advancements in green technologies and innovative materials will drive the future of sustainable architecture. This includes advancements in energy storage, smart building automation, and the development of new sustainable building materials.

In conclusion, green building and sustainable architecture are crucial in creating a more sustainable and resilient future. By incorporating principles of energy efficiency, water conservation, and sustainable material selection, and integrating innovative design strategies and renewable energy systems, green buildings can minimize their environmental impact and create healthier and more comfortable living spaces. Challenges such as cost implications and knowledge gaps need to be tackled through policy support, increased education, and community engagement. With further advancements in technology and greater awareness, the future of green building and sustainable architecture looks promising.

SUCCESS STORIES IN RENEWABLE ENERGY

Microgrids and Distributed Energy Systems

Microgrids and distributed energy systems are innovative approaches to generating and distributing electricity that have gained significant attention in recent years. These systems are designed to provide reliable and sustainable power to a specific geographic area or community, independent of the larger grid infrastructure. In this section, we will explore the principles, benefits, challenges, and future directions of microgrids and distributed energy systems.

Principles of Microgrids

A microgrid is a localized energy system that consists of interconnected loads, distributed energy resources (DERs), and energy storage. The key principle of a microgrid is its ability to operate in both connected and isolated modes, which means it can draw power from the main grid or operate independently and self-sustain during grid outages. Microgrids are typically governed by a smart control system that ensures optimal operation and coordination between different energy sources.

Distributed energy resources (DERs) play a crucial role in microgrids. These resources include renewable energy sources such as solar panels, wind turbines, and small-scale hydroelectric systems, as well as non-renewable sources like diesel generators and combined heat and power (CHP) plants. Energy storage technologies, such as batteries, are also integral components of microgrids, as they enable the storage and release of excess energy for later use.

Benefits of Microgrids

Microgrids offer several benefits over traditional grid systems. Firstly, they enhance the resilience and reliability of the energy supply. As microgrids can operate independently of the main grid, they are less susceptible to power outages caused by extreme weather events or equipment failures. This feature is particularly important for critical facilities like hospitals, military bases, and remote communities.

Secondly, microgrids promote the integration of renewable energy sources. By utilizing local renewable resources, microgrids reduce dependence on fossil fuels and contribute to greenhouse gas emissions reduction. Furthermore, the distributed nature of microgrids allows for better integration of intermittent renewables, such as solar and wind, by balancing supply and demand locally.

Another significant benefit of microgrids is improved energy efficiency. With the ability to generate and store energy locally, microgrids can optimize energy usage

370 CHAPTER 5: RENEWABLE ENERGY AND GREEN TECHNOLOGIES

and minimize transmission losses. Additionally, in communities where energy costs are high, microgrids can provide economic benefits by reducing reliance on expensive imported electricity.

Challenges and Solutions

Despite their numerous advantages, microgrids face a range of challenges that need to be addressed for their widespread implementation.

One of the primary challenges is the high upfront cost of establishing microgrid infrastructure. The installation of renewable energy sources, energy storage systems, and smart control systems requires significant investment. However, the decreasing costs of renewable technologies, coupled with government incentives and private investments, are gradually making microgrids more economically viable.

Another challenge is the coordination of multiple energy sources and storage systems within a microgrid. The smart control systems used in microgrids must efficiently manage the fluctuating energy supply and demand, as well as the charging and discharging of energy storage. Advanced algorithms and predictive modeling are being developed to optimize energy flow and ensure the reliable operation of microgrids.

Furthermore, regulatory frameworks and policies often do not accommodate the unique characteristics of microgrids. Current regulations primarily focus on centralized grid systems and may not support the deployment and operation of microgrids. Policymakers need to develop flexible regulations that encourage the development of microgrid infrastructure and incentivize the integration of renewable energy sources.

Real-world Examples

Several real-world examples demonstrate the successful implementation of microgrids and distributed energy systems.

One notable example is the Brooklyn Microgrid project in New York City. The project enables neighbors to buy and sell excess solar energy generated from rooftop solar panels using blockchain technology. This peer-to-peer energy trading system not only facilitates the use of renewable energy but also creates a resilient energy network within the local community.

Another example is the Alameda County Santa Rita Jail Microgrid in California. This microgrid combines solar panels, fuel cells, and energy storage to provide reliable and independent power to the jail facility. The microgrid

SUCCESS STORIES IN RENEWABLE ENERGY

significantly reduces the facility's energy costs and enhances its resilience during grid outages.

Future Directions

The future of microgrids and distributed energy systems looks promising as advancements in technology and supportive policies continue to emerge.

One prominent direction is the incorporation of advanced monitoring and control systems. Internet of Things (IoT) sensors, machine learning algorithms, and artificial intelligence can be utilized to optimize the performance of microgrids, detect and predict faults, and enable real-time energy management.

Additionally, the integration of electric vehicles (EVs) and vehicle-to-grid (V2G) technology holds great potential for microgrids. EVs can serve as mobile energy storage units, allowing for the bidirectional flow of electricity between the grid and the EV batteries. This capability not only enhances the utilization of renewable energy but also provides grid stability and reliability.

In conclusion, microgrids and distributed energy systems represent a paradigm shift in the generation and distribution of electricity. They offer numerous benefits, including improved resilience, integration of renewable energy, and enhanced energy efficiency. However, challenges related to cost, coordination, and regulation need to be overcome for wider adoption. With the advancement of technology and supportive policies, microgrids are poised to play a significant role in the future of energy systems, paving the way towards a more sustainable and resilient energy future.

Exercise: Think about your local community. How could microgrids and distributed energy systems benefit your community? Consider the advantages, challenges, and potential renewable energy sources that could be utilized. Discuss with your peers and present your findings in a short presentation.

Additional Resources: 1. "Microgrids and Active Distribution Networks" by Antonello Monti 2. "Distributed Generation: Induction and Permanent Magnet Generators" by J.C. Wu 3. "Distributed Energy Systems: A new paradigm for sustainable energy" by R.C. Dugan

Renewable Energy in Remote Areas and Developing Countries

Access to reliable and affordable energy is essential for the socio-economic development of remote areas and developing countries. However, many of these regions currently lack access to a centralized electricity grid or rely heavily on fossil fuels, which can be expensive, polluting, and unreliable. In recent years, renewable

372 CHAPTER 5: RENEWABLE ENERGY AND GREEN TECHNOLOGIES

energy technologies have emerged as a viable and sustainable solution to address these energy challenges. This section will explore the importance of renewable energy in remote areas and developing countries, discuss the various renewable energy options available, and highlight success stories in implementing renewable energy solutions.

Importance of Renewable Energy in Remote Areas and Developing Countries

In remote areas and developing countries, access to energy is crucial for various aspects of daily life, including education, healthcare, agriculture, and economic activities. However, traditional energy sources such as kerosene lamps, diesel generators, and firewood are not only environmentally damaging but also pose health risks and limit opportunities for growth and development. Renewable energy offers an alternative, clean, and sustainable solution that can address these challenges. Here are a few key reasons why renewable energy is important in remote areas and developing countries:

1. **Energy Access:** Renewable energy technologies, such as solar, wind, and biomass, can provide decentralized and off-grid solutions for energy access in remote areas where grid connectivity is limited or non-existent. This ensures that even the most isolated communities can benefit from electricity and modern energy services.

2. **Sustainable Development:** Renewable energy promotes sustainable development by reducing reliance on fossil fuels, minimizing greenhouse gas emissions, and mitigating climate change. Additionally, renewable energy projects create job opportunities, stimulate local economies, and empower communities to become self-sufficient in meeting their energy needs.

3. **Improved Quality of Life:** Access to electricity through renewable energy sources enhances the quality of life in remote areas. It enables improved lighting for households and schools, powering medical facilities and refrigeration for vaccines, supporting communication systems, and facilitating productive activities like farming, small-scale industries, and entrepreneurship.

4. **Environmental Protection:** Traditional energy sources contribute to deforestation, air pollution, and climate change. By transitioning to renewable energy, remote areas can significantly reduce the negative

SUCCESS STORIES IN RENEWABLE ENERGY

environmental impacts associated with conventional energy generation and foster conservation efforts.

Renewable Energy Options for Remote Areas and Developing Countries

Various renewable energy technologies are suitable for implementation in remote areas and developing countries. These technologies are often scalable, cost-effective, and adaptable to local conditions. The following are some of the most commonly used renewable energy options:

1. **Solar Energy:** Solar energy is abundant in most remote areas and developing countries, making it an excellent choice for decentralized power generation. Photovoltaic (PV) systems can be installed on rooftops or in open spaces to harness the sun's energy and convert it into electricity. Solar lanterns and solar home systems are also popular for individual households and small businesses that lack access to the grid.

2. **Wind Power:** Wind turbines can be deployed in windy regions to generate electricity. Small wind turbines are suitable for individual households, while larger wind farms can supply power to communities and even feed excess electricity into the grid. Wind power provides a reliable and continuous source of energy, particularly in areas with consistent wind patterns.

3. **Hydroelectric Power:** In regions with access to water bodies like rivers or streams, micro-hydro power plants can be established to generate electricity. These systems utilize the flow of water to rotate turbines and produce clean energy. Micro-hydro power is particularly suitable for hilly or mountainous areas with strong water currents.

4. **Biomass Energy:** Biomass, such as agricultural waste, animal manure, or wood, can be used to produce biogas or generate heat and electricity. Biogas digesters capture methane emissions from organic waste and convert them into a valuable energy source. Biomass energy provides a reliable and locally available option for cooking, heating, and electricity generation.

Success Stories in Implementing Renewable Energy Solutions

Several success stories demonstrate the effectiveness of renewable energy in remote areas and developing countries. These examples showcase how renewable energy technologies have transformed communities and improved the lives of people who previously lacked access to electricity. Here are a few notable success stories:

374 CHAPTER 5: RENEWABLE ENERGY AND GREEN TECHNOLOGIES

1. **Solar Home Systems in Bangladesh:** The Grameen Shakti program in Bangladesh has successfully provided solar home systems to over 4 million households in rural areas. This initiative has empowered rural communities, enabling them to have lights, charge mobile devices, and power appliances. It has contributed to economic development, education, and women's empowerment, while reducing carbon emissions and reliance on kerosene lamps.

2. **Micro-Hydro Power in Nepal:** Nepal's Rural Energy Development Program has implemented small-scale hydroelectric power plants in remote mountainous regions. These installations have brought electricity to thousands of households, improving the living conditions and economic opportunities of rural communities. Micro-hydro power has replaced the hazardous use of kerosene lamps and diesel generators, providing cleaner and more affordable energy.

3. **Off-Grid Solar in Kenya:** The M-KOPA Solar company in Kenya has pioneered pay-as-you-go solar solutions for off-grid households. By utilizing mobile payment technology, they have made solar home systems affordable and accessible to low-income households. This innovative approach has transformed the lives of thousands of families, fostering economic growth, reducing reliance on fossil fuels, and improving energy access.

Challenges and Future Directions

While renewable energy has shown tremendous potential in remote areas and developing countries, several challenges and future directions need to be addressed for its widespread adoption:

1. **Cost and Affordability:** The initial investment cost of renewable energy technologies can be a barrier to deployment, especially in low-income regions. Governments, international organizations, and financial institutions need to come together to provide financial incentives, subsidies, and affordable financing options to make renewable energy more accessible.

2. **Technical Expertise and Capacity Building:** Building local technical expertise and capacity is essential for the successful implementation and maintenance of renewable energy projects. Training programs and knowledge-sharing initiatives should be prioritized to empower local communities and ensure the long-term sustainability of renewable energy solutions.

SUCCESS STORIES IN RENEWABLE ENERGY

3. **Policy and Regulatory Frameworks:** Governments play a crucial role in creating a supportive policy and regulatory environment for renewable energy development. Streamlined approval processes, favorable tariffs, and renewable energy targets can attract private investments and promote the growth of the renewable energy sector.

4. **Integration with Existing Infrastructure:** Introducing renewable energy technologies in remote areas may require the development of new infrastructure for transmission and distribution. Integrating renewable energy systems with existing infrastructure and designing hybrid systems can enhance reliability and maximize the utilization of clean energy resources.

5. **Community Engagement and Participation:** The success of renewable energy projects heavily relies on community engagement and participation. Local communities should be involved in the decision-making process, and their interests, concerns, and cultural aspects should be considered to ensure the acceptance and long-term success of renewable energy initiatives.

Conclusion

Renewable energy offers immense opportunities for addressing energy challenges in remote areas and developing countries. By harnessing the power of the sun, wind, water, and biomass, these regions can achieve energy access, promote sustainable development, improve living conditions, and protect the environment. The success stories and experiences shared in this section demonstrate the transformative impact of renewable energy in empowering communities and driving positive change. However, to fully unleash the potential of renewable energy, it is crucial to overcome challenges and foster collaboration among governments, organizations, and local communities. By doing so, we can create a greener and more inclusive future for all.

Electric Vehicles and Sustainable Transportation

In recent years, there has been growing concern over the environmental impact of traditional transportation systems, particularly in relation to greenhouse gas emissions and air pollution. As a result, there has been a significant shift towards sustainable transportation options, one of which is the use of electric vehicles (EVs). This section will explore the principles and benefits of electric vehicles and their role in creating a greener and more sustainable future for transportation.

376 CHAPTER 5: RENEWABLE ENERGY AND GREEN TECHNOLOGIES

Principles of Electric Vehicles

Electric vehicles, also known as EVs, are vehicles that are powered by one or more electric motors, using energy stored in rechargeable batteries. Unlike conventional vehicles that rely on internal combustion engines fueled by gasoline or diesel, EVs operate entirely on electric power. This fundamental difference in propulsion systems makes electric vehicles an attractive option in terms of reducing greenhouse gas emissions and promoting sustainable transportation.

The principle behind electric vehicles lies in the conversion of electrical energy from the battery to mechanical energy in the electric motor. The battery, typically a lithium-ion battery, stores electrical energy that is used to power the motor, which in turn drives the vehicle's wheels. The electric motor converts the electrical energy into mechanical energy, resulting in the movement of the vehicle.

Benefits of Electric Vehicles

The adoption of electric vehicles offers numerous benefits for both individuals and the environment. Here are some key advantages:

1. Environmental Benefits: The main advantage of electric vehicles is their significantly reduced carbon footprint compared to traditional vehicles. EVs produce zero tailpipe emissions, resulting in lower greenhouse gas emissions and improved air quality. This is particularly important in densely populated areas where air pollution is a major concern.

2. Energy Efficiency: Electric vehicles are highly energy-efficient compared to conventional vehicles. While internal combustion engines are notoriously inefficient, electric motors have a much higher energy conversion efficiency. This means that a higher percentage of the energy stored in the battery is used to power the vehicle, resulting in less wasted energy.

3. Reduced Dependence on Fossil Fuels: Electric vehicles can be charged using electricity generated from renewable energy sources such as solar or wind power. This reduces dependence on fossil fuels and helps to promote the use of clean and sustainable energy sources.

4. Economic Benefits: Although the initial cost of purchasing an electric vehicle may be higher than that of a conventional vehicle, EV owners can benefit from lower operating costs. Electric vehicles have lower fuel costs, as electricity is typically cheaper than gasoline or diesel. Additionally, government incentives such as tax credits and subsidies are often available to encourage the adoption of electric vehicles.

SUCCESS STORIES IN RENEWABLE ENERGY

5. Quiet and Smooth Operation: Electric vehicles operate much quieter than traditional vehicles, as they do not have internal combustion engines. This results in a more peaceful and pleasant driving experience for both the driver and surrounding communities. Additionally, electric motors provide instant torque, offering smooth acceleration and a responsive driving experience.

Challenges and Solutions

While electric vehicles offer many benefits, there are still challenges that need to be addressed in order to accelerate their adoption. Here are some key challenges and potential solutions:

1. Range Anxiety: Range anxiety refers to the concern or fear of running out of battery charge while driving. The limited range of electric vehicles compared to conventional vehicles has been a deterrent for some potential buyers. However, advancements in battery technology and the establishment of a widespread charging infrastructure are helping to alleviate this concern. High-capacity batteries with longer ranges and the availability of fast-charging stations are expanding the practicality of electric vehicles.

2. Charging Infrastructure: The availability of a reliable and convenient charging infrastructure is crucial for the widespread adoption of electric vehicles. Governments, private companies, and communities are investing in the development of public charging stations in parking lots, shopping centers, and along highways. Additionally, home-charging stations are becoming more accessible, allowing EV owners to conveniently charge their vehicles overnight.

3. Battery Life and Recycling: The lifespan of electric vehicle batteries and the potential environmental impact of battery disposal are important considerations. Battery technology is constantly improving, with longer-lasting and more durable batteries being developed. Additionally, efforts are being made to implement effective battery recycling programs to minimize waste and recover valuable materials.

4. Affordability and Accessibility: The initial cost of electric vehicles is often higher than that of conventional vehicles, presenting a barrier to entry for some consumers. However, as battery technology advances and economies of scale are achieved, the cost of electric vehicles is expected to decrease. Additionally, government incentives and subsidies are being provided to make electric vehicles more affordable and accessible to a wider range of consumers.

378 CHAPTER 5: RENEWABLE ENERGY AND GREEN TECHNOLOGIES

Real-world Example: The Rise of Electric Vehicles in Norway

One notable success story in the adoption of electric vehicles is the country of Norway. Norway has become a global leader in electric vehicle adoption, with electric vehicles accounting for a significant portion of new car sales. Several factors have contributed to this success:

1. Incentives: Norway offers a range of incentives to encourage the purchase and use of electric vehicles, including exemption from import tax and VAT, reduced annual road tax, free access to toll roads and ferries, and free municipal parking. Additionally, electric vehicle owners enjoy reduced charging costs and are exempt from certain restrictions in congested city centers.

2. Charging Infrastructure: Norway has invested heavily in the development of a comprehensive charging infrastructure, making it easy for electric vehicle owners to charge their vehicles wherever they go. Charging stations can be found in urban areas, parking lots, and along highways, providing convenience and peace of mind for EV owners.

3. Environmental Awareness: The Norwegian population has a strong environmental consciousness, and there is a societal commitment to reducing greenhouse gas emissions and air pollution. This cultural mindset has played a significant role in the widespread adoption of electric vehicles.

The success of Norway in promoting electric vehicles serves as an inspiring example of how a combination of incentives, infrastructure development, and a supportive societal attitude can drive the transition towards sustainable transportation.

Conclusion

Electric vehicles have emerged as a promising solution for creating a greener and more sustainable future for transportation. The principles underlying electric vehicles, combined with their numerous benefits, make them a compelling alternative to conventional vehicles. Overcoming challenges related to range anxiety, charging infrastructure, battery life, and affordability will be key in accelerating the adoption of electric vehicles on a global scale. Real-world success stories, such as Norway's embrace of electric vehicles, provide valuable lessons and inspiration for other countries and communities looking to transition to sustainable transportation. By embracing electric vehicles and investing in renewable energy sources, we can pave the way for a cleaner and more environmentally friendly transportation system.

Challenges and Future Directions in Renewable Energy

Energy Storage Technologies and Grid Integration

Energy storage technologies play a crucial role in overcoming one of the main challenges of renewable energy sources: their intermittent nature. As renewable energy generation heavily relies on weather conditions, energy storage systems provide a means to store excess energy during peak production periods and release it during times of low generation. This allows for a more reliable and stable energy supply, reduces waste, and enables better integration of renewable energy into the grid. In this section, we will explore various energy storage technologies and their integration into the grid.

Principles of Energy Storage

The main principle behind energy storage is the conversion of electrical energy into a form that can be stored and later converted back into electricity when needed. There are several key factors to consider when evaluating energy storage technologies:

- Energy Capacity: The total amount of energy that can be stored in a system.

- Power Capacity: The maximum power output or input of the storage system.

- Efficiency: The ratio of output energy to input energy during the storage and retrieval processes.

- Response Time: The time required for the storage system to respond to grid demands or fluctuations in renewable energy generation.

- Lifetime: The expected operating life of the storage system without significant degradation in performance.

Types of Energy Storage Technologies

There are several energy storage technologies available with varying characteristics, each suitable for different applications. Let's explore some of the most commonly used ones:

1. Batteries: Battery storage systems, such as lithium-ion batteries and flow batteries, have gained significant prominence due to their high energy density, fast response times, and scalability. They are used in various applications, from small-scale residential storage to grid-scale installations.

380 CHAPTER 5: RENEWABLE ENERGY AND GREEN TECHNOLOGIES

2. Pumped Hydroelectric Storage: This form of energy storage utilizes the potential energy of water by pumping it uphill during periods of low electricity demand and releasing it through turbines to generate electricity during peak demand. Pumped hydroelectric storage offers high energy capacity and efficiency, making it an excellent option for grid-level storage.

3. Compressed Air Energy Storage (CAES): CAES involves compressing air and storing it in underground caverns or tanks. During periods of high electricity demand, the compressed air is released and heated using natural gas or other fuels to drive turbines and generate electricity. CAES is a cost-effective energy storage solution that offers large-scale capacity.

4. Flywheels: Flywheel energy storage systems store energy by spinning a rotor at high speeds and releasing it when needed. They provide fast response times and high power capacity, making them suitable for applications requiring short-duration energy bursts, such as stabilizing grid frequency.

5. Thermal Energy Storage: This technology stores energy in the form of heat, utilizing materials with high heat capacities or phase-change materials. The stored heat can be used directly or converted back into electricity using steam turbines or thermoelectric generators. Thermal energy storage is often used in conjunction with concentrated solar power systems.

Grid Integration of Energy Storage

Effective integration of energy storage systems into the grid requires careful coordination and control mechanisms to optimize their operations. Here are some key considerations for grid integration:

- Energy Management Systems: Advanced control systems are essential for managing energy storage systems and making optimal decisions regarding energy dispatch, charging, and discharging. These systems use real-time data on electricity prices, renewable energy availability, and grid demand to ensure efficient operation.

- Grid Services: Energy storage systems can provide several valuable services to the grid, including frequency regulation, load leveling, peak shaving, and grid backup. By participating in ancillary service markets, storage operators can generate additional revenue while contributing to grid stability.

CHALLENGES AND FUTURE DIRECTIONS IN RENEWABLE ENERGY 381

- Interconnection and Sizing: Properly sizing energy storage systems and ensuring their seamless integration into the grid infrastructure is critical. It involves considering factors such as voltage and frequency compatibility, capacity planning, and grid connection requirements.

- Hybrid Energy Systems: Combining multiple energy storage technologies and renewable energy sources can enhance system flexibility and reliability. Hybrid systems, such as solar with battery storage or wind with pumped hydro, can significantly improve the overall performance and efficiency of the integrated energy system.

Challenges and Future Directions

While energy storage technologies have made significant advancements in recent years, there are still challenges and areas for improvement:

- Cost: The upfront costs of energy storage systems remain a significant barrier to their widespread adoption. Continued research and technological advancements are needed to reduce costs, improve efficiency, and increase the lifespan of storage technologies.

- Environmental Impact: Developing sustainable and environmentally friendly energy storage technologies is crucial. Efforts are being made to minimize the environmental impact of battery production, improve recycling methods, and explore alternative materials that are more abundant and less harmful.

- Grid Integration: Achieving seamless integration of energy storage systems into existing grids requires addressing technical and regulatory challenges. Standardization of communication protocols, grid codes, and market mechanisms are essential for efficient integration and optimal utilization of energy storage assets.

- Research and Development: Continued research and development are necessary to explore new energy storage technologies and enhance the performance and capabilities of existing ones. This includes advancements in materials science, electrochemistry, control systems, and grid management techniques.

Overall, energy storage technologies and their integration into the grid are vital components in transitioning to a more sustainable and renewable energy future. Their continued development and deployment will play a crucial role in achieving a

382 CHAPTER 5: RENEWABLE ENERGY AND GREEN TECHNOLOGIES

reliable, resilient, and decarbonized energy system. By effectively managing energy storage systems and addressing existing challenges, we can unlock the full potential of renewable energy sources and pave the way for a greener and more sustainable world.

CHALLENGES AND FUTURE DIRECTIONS IN RENEWABLE ENERGY 383

Exercises

1. Research and compare the different energy storage technologies mentioned in this section. Discuss their advantages, limitations, and suitable applications.

2. Calculate the efficiency of a battery storage system that stores 500 kWh of energy and provides 400 kWh of usable energy. Determine the percentage efficiency of the system.

3. Investigate a specific energy storage installation or project in your region or country. Discuss its objectives, challenges faced, and the impact it has had on the local grid and renewable energy integration.

4. Design a hypothetical hybrid energy system that combines two renewable energy sources with different storage technologies. Provide a detailed explanation of how the system operates and the benefits it offers compared to standalone systems.

5. Discuss the current policies and regulations in place in your country or state regarding grid integration and the utilization of energy storage systems. Analyze how these policies impact the adoption and deployment of energy storage technologies.

6. Explore emerging energy storage technologies that are currently being researched or developed. Discuss their potential applications, advantages, and challenges in comparison to existing storage solutions.

Additional Resources

+ Energy Storage Association (ESA): `https://energystorage.org/`

+ U.S. Department of Energy - Energy Storage: `https://www.energy.gov/energy-storage`

+ International Renewable Energy Agency (IRENA) - Energy Storage: `https://www.irena.org/topics/energy-storage`

+ World Energy Council - Energy Storage: `https://www.worldenergy.org/technology/energy-storage/`

+ The Electricity Storage Network: `https://www.electricitystorage.co.uk/`

384 CHAPTER 5: RENEWABLE ENERGY AND GREEN TECHNOLOGIES

By engaging with the exercises and exploring the additional resources, you will gain a deeper understanding of energy storage technologies and their important role in the integration of renewable energy into the grid.

Policy and Regulatory Support for Renewable Energy

Policy and regulatory frameworks play a crucial role in driving the adoption and growth of renewable energy sources. These frameworks provide the necessary guidelines, incentives, and regulations for the development, deployment, and integration of renewable energy technologies into existing energy systems. In this section, we will explore the importance of policy and regulatory support for renewable energy and discuss various strategies that can be implemented to accelerate the transition to a greener and more sustainable energy future.

The Need for Policy and Regulatory Support

Renewable energy sources, such as solar, wind, hydro, geothermal, and biomass, have the potential to significantly reduce greenhouse gas emissions, improve energy security, and create new economic opportunities. However, the widespread adoption of renewable energy technologies faces several challenges, including high upfront costs, technological barriers, and a lack of supportive infrastructure. Policy and regulatory support can address these challenges and provide the necessary incentives and frameworks to overcome barriers to renewable energy deployment.

One key aspect of policy and regulatory support is the establishment of renewable energy targets. These targets set clear objectives for the share of renewable energy in the overall energy mix and provide guidance for policymakers, industry stakeholders, and investors. By setting ambitious and enforceable targets, governments can signal their commitment to renewable energy development and provide a framework for long-term planning and investment.

Another important aspect of policy support is the implementation of financial incentives, such as feed-in tariffs, tax credits, grants, and subsidies. These incentives help to reduce the high initial costs associated with renewable energy projects and encourage private investment. Feed-in tariffs, for example, guarantee a fixed price for electricity generated from renewable sources, providing a stable and predictable revenue stream for project developers and investors.

Regulatory support is equally important in creating a favorable environment for renewable energy. Streamlining the permitting and licensing processes, reducing bureaucratic barriers, and ensuring grid access for renewable energy projects are key regulatory measures that can accelerate the development and

CHALLENGES AND FUTURE DIRECTIONS IN RENEWABLE ENERGY 385

deployment of renewable energy technologies. Additionally, the establishment of clear and consistent regulatory frameworks can provide certainty for investors and reduce investment risks.

International Examples of Policy and Regulatory Support

Countries around the world have implemented various policy and regulatory measures to support the growth of renewable energy. Let's look at some examples:

Germany: Germany has been at the forefront of renewable energy development, thanks in large part to its comprehensive policy framework. The Renewable Energy Sources Act (EEG) guarantees fixed feed-in tariffs for renewable energy projects and provides long-term support for the industry. The EEG has been highly successful in promoting the rapid expansion of renewable energy capacity in Germany, particularly in the solar and wind sectors.

China: China has become the world's largest producer and consumer of renewable energy, largely due to its ambitious renewable energy targets and supportive policies. The country has implemented various financial incentives, including feed-in tariffs, tax incentives, and subsidies, to promote the development and deployment of renewable energy technologies. China's commitment to renewable energy has also resulted in significant investments in research and development, driving technological advancements in the sector.

Denmark: Denmark has set ambitious targets to transition to a 100% renewable energy system by 2050. The country has implemented a combination of policy and regulatory measures to achieve this goal, including feed-in tariffs, tax exemptions, green certificates, and extensive research and development programs. Denmark's success in renewable energy can be attributed to its long-term planning, strong political commitment, and collaboration between industry stakeholders and government agencies.

Challenges and Future Directions

While policy and regulatory support have played a critical role in advancing renewable energy deployment, several challenges and future directions need to be addressed for a successful transition to a sustainable energy future.

Grid Integration: The intermittent nature of renewable energy sources presents challenges for grid integration. As more renewable energy capacity comes online, it is crucial to develop smart grid technologies, energy storage solutions, and demand management strategies to ensure a stable and reliable electricity supply.

386 CHAPTER 5: RENEWABLE ENERGY AND GREEN TECHNOLOGIES

Market Design: Existing electricity markets may not always reflect the true value of renewable energy, leading to market barriers and underinvestment. Policy and regulatory frameworks should be designed to accurately account for the environmental and societal benefits of renewable energy, ensuring a level playing field and fair competition.

International Collaboration: Renewable energy deployment requires international collaboration and cooperation. Countries can learn from each other's experiences and share best practices to accelerate the transition to renewable energy. International agreements and partnerships can facilitate technology transfer, capacity building, and the exchange of knowledge.

Emerging Technologies: Continued policy and regulatory support is needed to drive the adoption of emerging renewable energy technologies, such as ocean energy, hydrogen, and advanced biofuels. These technologies have the potential to further enhance the sustainability and efficiency of the energy system.

In conclusion, policy and regulatory support are essential in driving the adoption and growth of renewable energy. Governments and regulatory bodies play a key role in creating a favorable environment for investment, streamlining the deployment process, and ensuring the integration of renewable energy into existing energy systems. By implementing comprehensive and supportive policy frameworks, we can accelerate the transition to a greener and more sustainable energy future.

Addressing the Intermittency of Renewable Sources

Reliance on renewable energy sources is a key strategy for mitigating climate change and reducing our dependence on fossil fuels. However, one of the challenges associated with renewable energy is its intermittency. Unlike traditional power sources, such as coal and gas, renewable sources like solar and wind are dependent on external factors like weather conditions and time of day, leading to fluctuations in energy production. In this section, we will explore the methods and technologies used to address the intermittency of renewable sources.

Understanding the Intermittency Issue

Renewable energy sources, particularly solar and wind, are variable in nature. Solar energy production is dependent on the availability of sunlight, which varies throughout the day and is influenced by factors such as cloud cover and seasonal changes. Similarly, wind energy production is influenced by the speed and consistency of wind patterns, which can change with weather conditions and location.

This intermittency poses challenges for grid operators and energy consumers since the supply of renewable energy may not always align with the demand. There are two main aspects of the intermittency issue:

1. **Variability:** Fluctuations in the availability of renewable energy due to natural factors.

2. **Predictability:** Difficulty in accurately forecasting the energy production from renewable sources.

To address the intermittency issue, various strategies and technologies have been developed, which can be broadly categorized into three approaches: grid integration, energy storage, and demand-side management.

Grid Integration

Grid integration refers to the incorporation of renewable energy sources into the existing electricity grid system. It involves optimizing the operation and management of the grid to accommodate intermittent energy generation.

One approach to grid integration is the use of **smart grids**, which employ advanced sensing, metering, and communication technologies to monitor and control the flow of electricity. By providing real-time data on energy production and demand, smart grids enable grid operators to make informed decisions and effectively manage the variability of renewable energy sources.

Another approach is the **expansion of transmission networks.** By connecting geographically dispersed renewable energy generation sites, the transmission capacity is increased, reducing the impact of local fluctuations. This allows surplus energy from regions with high production to be transported to areas with high demand, improving the overall reliability of the grid.

Energy Storage

Energy storage technologies play a crucial role in addressing the intermittency of renewable sources. They allow excess energy generated during periods of high production to be stored and used when production is low. Here are some commonly used energy storage solutions:

1. **Batteries:** Battery storage systems, such as lithium-ion batteries, are widely used to store energy at both utility and individual levels. They can be charged

388 CHAPTER 5: RENEWABLE ENERGY AND GREEN TECHNOLOGIES

during periods of excess renewable energy production and discharged when needed.

2. **Pumped Hydro Storage:** This method utilizes the potential energy of water by pumping it to an elevated reservoir during periods of excess energy production. The water is then released to flow downhill, spinning turbines and generating electricity during periods of peak demand.

3. **Compressed Air Energy Storage (CAES):** CAES systems store excess energy by compressing air and storing it in underground caverns. During periods of high demand, the compressed air is released, driving turbines to generate electricity.

4. **Hydrogen Storage:** Hydrogen can be produced by electrolyzing water using excess electricity and stored for later use in fuel cells or other applications.

5. **Thermal Energy Storage (TES):** TES systems store excess energy in the form of heat or cold and release it when required. They can be used in conjunction with solar thermal or geothermal power plants to store thermal energy for electricity generation.

Energy storage technologies are continually evolving, and ongoing research aims to develop more efficient, cost-effective, and environmentally friendly solutions.

Demand-Side Management

Demand-side management focuses on influencing energy consumption patterns to better align with renewable energy production. By shifting the timing of energy-intensive activities to periods of high renewable energy generation, demand-side management helps to maximize the utilization of renewable sources.

Time-of-use (TOU) pricing is commonly used to incentivize consumers to shift their energy usage to off-peak hours when renewable energy generation is high. By adjusting electricity rates based on the time of day, consumers are encouraged to reduce their energy consumption during peak demand periods and utilize more energy during periods of high renewable energy availability.

Demand response programs involve a direct response from consumers to adjust their electricity usage during periods of high demand or low renewable energy availability. This can be achieved through automated controls or by providing incentives for consumers to reduce their energy consumption.

Additionally, advances in **energy management systems** and **smart appliances** allow for more efficient scheduling and control of electricity usage. For example,

CHALLENGES AND FUTURE DIRECTIONS IN RENEWABLE ENERGY 389

appliances can be programmed to operate during periods of high renewable energy availability or during off-peak hours.

Case Study: Virtual Power Plants

Virtual Power Plants (VPPs) are emerging as a promising solution to address intermittency in renewable energy sources. A VPP is a network of decentralized, grid-connected power sources, energy storage systems, and flexible demand units that are aggregated and managed as a single entity.

The VPP acts as a virtual power plant operator, optimizing the generation, storage, and consumption of energy resources within the network. By intelligently managing the flow of electricity, VPPs can balance supply and demand, ensuring that renewable energy sources are utilized effectively.

For example, a VPP could coordinate the charging and discharging of electric vehicle batteries, the operation of distributed solar panels, and the use of energy storage systems in residential buildings. By controlling these distributed energy resources, a VPP can respond to fluctuations in energy availability and demand, enhancing grid stability and reducing reliance on traditional power plants.

Conclusion

Addressing the intermittency of renewable sources is essential for achieving a reliable and sustainable energy system. Grid integration, energy storage, and demand-side management strategies provide effective means to mitigate the challenges posed by the variable nature of renewable energy sources.

As technology continues to advance and costs decrease, renewable energy generation is becoming more competitive with traditional sources. However, further research and development are still needed to optimize and scale up the solutions discussed in this section. By overcoming the intermittency issue, we can unlock the full potential of renewable energy and accelerate the transition to a greener future.

Renewable Energy for Industrial Processes

In the quest for a greener future, one of the key areas that requires significant attention is the industrial sector, which accounts for a large portion of global energy consumption and greenhouse gas emissions. To combat climate change and reduce environmental impacts, integrating renewable energy sources into industrial processes is crucial. In this section, we will explore the challenges, opportunities, and success stories of utilizing renewable energy for industrial purposes.

390 CHAPTER 5: RENEWABLE ENERGY AND GREEN TECHNOLOGIES

Importance of Renewable Energy in Industry

The industrial sector plays a vital role in global economic development, but it also generates a considerable amount of carbon emissions through its energy-intensive operations. By integrating renewable energy sources into industrial processes, we can achieve multiple benefits:

- **Reduced Carbon Footprint:** Industrial processes powered by fossil fuels contribute significantly to greenhouse gas emissions. Shifting to renewable energy sources such as solar, wind, and biomass can help factories, refineries, and other industrial facilities reduce their carbon footprint and mitigate climate change.

- **Energy Cost Savings:** Traditional energy sources, such as coal and natural gas, are subject to price fluctuations, making energy costs unpredictable for industries. Adopting renewable energy can provide stability in energy prices and even lead to long-term cost savings.

- **Enhanced Energy Security:** Relying on fossil fuels for industrial operations often entails dependence on imports and geopolitical vulnerabilities. Investing in renewable energy sources can strengthen a country's energy security by reducing reliance on external energy sources.

- **Job Creation and Economic Growth:** The renewable energy sector has the potential to generate employment opportunities and stimulate economic growth. Shifting towards renewable energy in industry can create jobs in manufacturing, installation, and maintenance of renewable energy infrastructure.

Challenges in Implementing Renewable Energy for Industry

Despite the advantages of renewable energy, there are several challenges to its widespread integration into industrial processes. These challenges must be addressed to ensure a successful transition:

1. **Intermittency and Variability:** Renewable energy sources such as solar and wind are intermittent by nature, and their availability may not align with industrial energy demand, which often requires a continuous and reliable power supply. Developing energy storage technologies and implementing demand-response strategies is essential to mitigate the variability of renewables and ensure a constant energy supply.

CHALLENGES AND FUTURE DIRECTIONS IN RENEWABLE ENERGY 391

2. **High Initial Costs:** The initial capital investment required for renewable energy infrastructure, such as solar panels and wind turbines, can be substantial. Industries may hesitate to make the transition due to the perceived high costs. However, with advancements in technology and economies of scale, the costs of renewable energy systems have been steadily decreasing, making them more financially viable in the long run.

3. **Technological Limitations:** Certain industrial processes require high-temperature heat or specific types of energy that may not be easily provided by current renewable energy technologies. Research and development efforts are needed to identify and develop efficient and cost-effective renewable energy solutions for these niche industrial applications.

4. **Grid Integration and Infrastructure Upgrades:** Connecting industrial facilities to the grid and ensuring their seamless integration with renewable energy sources require proper planning and infrastructure upgrades. Upgrading electrical grids to handle intermittent power generation and incorporating advanced grid management technologies are essential steps towards achieving a reliable and resilient renewable energy system for industrial use.

5. **Policy and Regulatory Frameworks:** The absence of supportive policies and incentives can hinder industrial adoption of renewable energy. Governments need to implement favorable policies, such as feed-in tariffs, tax credits, and carbon pricing mechanisms, to encourage industries to invest in renewable energy systems.

Industrial Applications of Renewable Energy

Renewable energy technologies can be integrated into various industrial processes to reduce carbon emissions and enhance sustainability. Let's explore some of the key applications:

- **Solar Thermal Systems:** Solar collectors can be used to generate high-temperature heat for industrial processes like water heating, steam production, and space heating. Concentrated solar power (CSP) plants can also provide industrial-scale electricity generation.

- **Bioenergy and Biogas:** Biomass, such as agricultural residues, wood waste, and energy crops, can be converted into biogas or biofuels for industrial heat

392 CHAPTER 5: RENEWABLE ENERGY AND GREEN TECHNOLOGIES

and power generation. Anaerobic digestion and biomass combustion technologies are commonly used for this purpose.

+ **Wind Power:** Industrial facilities located in windy regions can harness wind power through on-site wind turbines or through off-site power purchase agreements. Wind energy can directly contribute to electricity demand or be converted into other forms of energy, such as hydrogen or compressed air.

+ **Hydroelectric Power:** Industries located near rivers or water bodies can utilize hydroelectric power for their energy needs. Small-scale hydropower systems can be installed on-site or integrated with existing water infrastructure.

Case Studies in Industrial Renewable Energy

To illustrate the successful implementation of renewable energy in industrial processes, let's explore a few notable case studies:

1. **IKEA:** The multinational furniture retailer has committed to produce more renewable energy than it consumes by 2020. IKEA has installed solar panels on its stores and warehouses, invested in wind farms, and has integrated geothermal systems in its buildings to achieve this goal.

2. **Tata Steel:** Tackling the challenge of decarbonizing the steel industry, Tata Steel in the Netherlands has implemented a project called HIsarna, which uses hydrogen produced from renewable energy sources instead of coal in the steelmaking process. This innovative approach significantly reduces carbon emissions in steel production.

3. **Suzlon Energy:** Suzlon Energy, an Indian wind turbine manufacturer, has installed wind energy projects across various industrial facilities to power their operations. The company not only reduces its own carbon footprint but also generates surplus electricity, contributing to the grid.

These case studies demonstrate that renewable energy integration in industry is both feasible and profitable, setting an example for other companies and sectors to follow.

CHALLENGES AND FUTURE DIRECTIONS IN RENEWABLE ENERGY 393

Future Directions and Opportunities

To overcome the challenges and further promote the use of renewable energy in industrial processes, several strategies and opportunities should be pursued:

- **Technological Advancements:** Continued research and development in renewable energy technologies and energy storage systems will enable more efficient and cost-effective solutions for various industrial applications.

- **Collaboration and Partnerships:** Collaboration between governments, industries, research institutions, and renewable energy companies is crucial for knowledge sharing, resource pooling, and establishing favorable conditions for renewable energy adoption in the industrial sector.

- **Incentives and Subsidies:** Governments should provide financial incentives, subsidies, and tax breaks to encourage industries to transition to renewable energy. These incentives can offset the initial investment costs and promote the sustainable growth of renewable energy in the industrial sector.

- **Smart Grid Technologies:** Developing smart grids that can efficiently manage and integrate renewable energy sources with industrial processes will be essential to ensure a reliable and stable power supply. Smart grid technologies enable the monitoring and control of energy consumption, facilitate demand response, and optimize energy distribution networks.

In conclusion, integrating renewable energy sources into industrial processes is vital for achieving a greener and more sustainable future. The challenges faced, including intermittency, high initial costs, technological limitations, grid integration, and policy frameworks, can be overcome through innovation, collaboration, and supportive policies. Through case studies and successful examples, we have seen the tremendous potential of renewable energy in the industrial sector. With continued advancements and collective efforts, we can pave the way for a cleaner and more sustainable industrial landscape.

Advancements in Green Technologies and Innovations

As the world grapples with the challenges of climate change and environmental degradation, there has been an increasing focus on the development of green technologies and innovations. These advancements aim to provide sustainable solutions to pressing environmental issues, reduce our dependence on fossil fuels, and promote a cleaner and healthier future. In this section, we will explore some of

394 CHAPTER 5: RENEWABLE ENERGY AND GREEN TECHNOLOGIES

the key advancements in green technologies and the potential they hold in shaping a more sustainable world.

Harnessing Renewable Energy

Renewable energy sources, such as solar, wind, hydroelectric, geothermal, and biomass, have gained tremendous momentum in recent years. Technological advancements have made these energy sources more efficient, reliable, and cost-effective, opening up new possibilities for large-scale deployment. For example, in the solar energy sector, the development of high-efficiency photovoltaic cells and innovations in concentrated solar power systems have significantly increased the energy output and reduced the manufacturing costs. Similarly, in wind power, the development of larger and more efficient wind turbines has significantly improved the energy generation capacity.

One key advancement in renewable energy technology is the development of energy storage systems. These systems allow for the capture and storage of excess energy produced during periods of high generation, which can be utilized during times of low generation. Advancements in battery technologies, such as lithium-ion batteries, have made energy storage more feasible and economically viable. This innovation is crucial in addressing the intermittent nature of renewable energy sources and ensuring a stable and reliable power supply.

Smart Grid and Energy Management Systems

Another significant advancement in green technologies is the development of smart grid and energy management systems. A smart grid is an intelligent electricity distribution network that utilizes advanced communication and information technologies to optimize the generation, distribution, and consumption of electricity. Smart grid systems enable the integration of renewable energy sources, facilitate real-time monitoring and control of energy flow, and allow for demand response programs. By adopting smart grid technologies, energy utilities can efficiently balance the supply and demand of electricity, reduce transmission losses, and optimize energy consumption patterns.

Energy management systems, which are an integral part of smart grid infrastructure, enable consumers to monitor and control their energy usage. These systems provide real-time data on energy consumption, allowing users to make informed decisions and adopt energy-saving practices. Additionally, energy management systems can integrate with smart home devices, allowing for automated control and optimization of energy usage. Such advancements

CHALLENGES AND FUTURE DIRECTIONS IN RENEWABLE ENERGY

empower individuals and businesses to actively participate in energy conservation and reduce their carbon footprint.

Green Building and Sustainable Architecture

The construction and operation of buildings account for a significant portion of global energy consumption and greenhouse gas emissions. As a result, there has been a growing emphasis on green building practices and sustainable architecture. Green buildings are designed to minimize the use of energy, water, and materials, while also creating a healthy and comfortable indoor environment.

Advancements in green building technologies have led to the development of energy-efficient building materials, such as high-performance insulation, smart windows, and energy-efficient lighting systems. These technologies reduce energy demand and improve thermal comfort within buildings. Additionally, innovations in building automation and control systems enable optimized energy management, allowing for real-time monitoring of energy usage and the implementation of energy-saving measures.

Sustainable architecture focuses on integrating sustainable design principles into the planning and construction of buildings. This includes considerations such as passive solar design, natural ventilation systems, and the use of recycled and environmentally friendly building materials. By incorporating these principles, sustainable architecture aims to minimize the negative environmental impacts of buildings while creating healthy and sustainable living spaces.

Circular Economy and Waste Management

Traditional linear economic models, which involve the extraction of raw materials, production, consumption, and disposal, contribute to resource depletion and waste generation. In contrast, the concept of a circular economy aims to minimize waste and maximize resource efficiency by promoting the reuse, recycling, and recovery of materials.

Advancements in waste management technologies play a crucial role in achieving a circular economy. Innovations such as advanced recycling techniques, waste-to-energy conversion systems, and anaerobic digestion have improved the efficiency of waste management processes. These technologies enable the extraction of valuable resources from waste streams, reduce the reliance on landfills, and generate renewable energy from organic waste.

Furthermore, the development of sustainable packaging materials and the promotion of product life extension strategies, such as repair, refurbishment, and

396 CHAPTER 5: RENEWABLE ENERGY AND GREEN TECHNOLOGIES

remanufacturing, contribute to reducing waste and promoting the principles of a circular economy.

Nature-Based Solutions

Nature-based solutions (NBS) refer to the strategic use of ecosystems and natural processes to address various environmental challenges. NBS capitalize on the benefits provided by nature, such as carbon sequestration, water purification, flood regulation, and biodiversity conservation. These solutions are cost-effective, sustainable, and often provide co-benefits for human well-being.

Advancements in NBS include the restoration and conservation of natural habitats, such as forests, wetlands, and coastal ecosystems. These activities not only enhance biodiversity but also provide essential ecosystem services, such as climate regulation, water purification, and storm protection.

Additionally, green infrastructure, which incorporates natural elements into urban environments, has gained prominence. Examples of green infrastructure include green roofs, urban parks, and vertical gardens. These features enhance urban resilience, mitigate the heat island effect, improve air quality, and promote overall well-being.

Challenges and Future Directions

While advancements in green technologies and innovations show promise, several challenges must be addressed to fully unleash their potential. These challenges include:

- **Cost-effectiveness:** Many green technologies and innovations are still relatively expensive, limiting their adoption, especially in developing countries. Continued research and development efforts are needed to reduce the costs and improve the affordability of these technologies.

- **Infrastructure and scalability:** The integration of renewable energy sources, such as solar and wind, into existing energy grids requires significant infrastructural modifications. The scalability of green technologies is crucial to ensure their widespread implementation and their ability to meet growing energy demands.

- **Policy and regulatory support:** Governments play a crucial role in creating an enabling environment for the deployment of green technologies. Policies that incentivize the adoption of renewable energy sources, promote sustainable

CHALLENGES AND FUTURE DIRECTIONS IN RENEWABLE ENERGY 397

practices, and provide support for research and development are essential for the success of these technologies.

- **Public awareness and education:** Increasing public awareness and understanding of the benefits of green technologies is vital. Education programs and campaigns can help foster a culture of sustainability and encourage individuals and businesses to embrace green practices.

Looking ahead, further advancements in green technologies and innovations are expected. Emerging fields such as artificial intelligence, nanotechnology, and biotechnology hold immense potential in driving the development of more efficient and sustainable solutions. Collaboration between governments, industries, and research institutions will be crucial in overcoming challenges and harnessing the full potential of green technologies in building a greener future.

Summary

The advancements in green technologies and innovations discussed in this section represent significant progress in our journey towards a more sustainable future. Harnessing renewable energy, implementing smart grid systems, adopting green building practices, promoting a circular economy, and embracing nature-based solutions all contribute to reducing our environmental footprint and mitigating climate change. However, challenges such as cost-effectiveness, scalability, policy support, and public awareness need to be addressed to unlock the full potential of these technologies. By continuing to invest in research and development, fostering collaboration, and promoting sustainable practices, we can drive further advancements in green technologies and create a resilient and sustainable world for future generations.

Index

-effectiveness, 324, 348, 397

a, 1–5, 7–11, 14–37, 39, 42, 43, 45, 47, 48, 50–53, 55–62, 64–68, 70, 71, 73–75, 80–82, 84, 85, 88–93, 97–102, 104, 106, 107, 109–118, 120, 122–126, 129, 131, 133–142, 146–153, 157, 158, 161, 163–166, 168, 170, 171, 173, 174, 176, 177, 179, 181, 183, 184, 186–188, 190, 192–194, 196–202, 204–206, 208, 209, 211, 212, 215–229, 231, 232, 234–240, 243–245, 247, 248, 251–253, 255–261, 263–268, 270, 272, 274–278, 280, 281, 284–287, 289–291, 293, 296–298, 300–302, 304–307, 311, 313–320, 322–326, 328–340, 342–344, 346, 348–350, 352–356, 358–360, 362–365, 368–373, 375–379, 381, 382, 384–387, 389, 390, 392–397

ability, 8, 16, 25, 34, 45, 60, 88, 91, 225, 235, 289, 305, 307, 369

absence, 140, 200, 209, 258, 359

abundance, 33, 57, 112, 171, 174, 190, 225

acceleration, 377

acceptance, 324

access, 36, 74, 89, 90, 93, 176, 187, 198, 222, 225, 237, 238, 264, 278, 287–290, 293, 294, 296, 297, 301, 307, 315, 316, 371–373, 375, 378, 384

accessibility, 255, 294

accident, 82

account, 23, 91, 152, 174, 186, 287, 328, 352, 366, 395

accountability, 24, 90, 290

accounting, 363, 378

accumulation, 112, 147

accuracy, 19, 42, 45, 149

acetogenesis, 359

achievement, 199

acidification, 56, 152, 164, 198

acidity, 117, 225

acidogenesis, 359
acquisition, 32
act, 50, 109, 116, 119, 163, 168,
      193, 222
action, 19, 25, 29, 36, 50, 53, 55, 59,
      62, 66, 73, 89, 106, 190,
      228, 287, 290, 322
activism, 21, 26
activity, 35, 248, 323
adaptability, 168
adaptation, 6, 10, 15, 25, 37, 47, 49,
      50, 52, 54–56, 84, 150,
      152, 155, 164, 199,
      226–229
addition, 33, 161, 163, 214, 217,
      236, 281, 320, 366
address, 3, 6, 9, 11, 14, 17, 19, 22,
      24, 28, 29, 32, 34, 36, 43,
      53, 55, 61, 66, 72–74, 89,
      91–93, 97, 100, 115, 118,
      119, 125, 129, 133–135,
      139, 140, 142, 143, 146,
      153, 161, 167, 177, 186,
      190, 192, 198, 201, 211,
      220, 224, 226, 229, 232,
      233, 250, 255, 263, 265,
      268, 275, 285, 287, 293,
      295, 296, 313, 317, 322,
      346, 357, 372, 384, 386,
      387
addressing, 2, 4, 6, 10, 15, 17, 20, 37,
      50, 56, 59, 74, 80, 82, 90,
      97, 100, 102, 106, 109,
      110, 114, 122, 125, 129,
      144, 155, 167, 168, 171,
      186, 196, 200–202, 208,
      215, 218, 221, 227–229,
      238, 239, 251, 263, 278,
      281, 284, 297, 307–309,

      311, 313, 324, 354, 359,
      375, 382, 387, 394
adherence, 20
adjustment, 145
adoption, 8, 50, 72, 81, 89, 91, 92,
      99, 100, 238, 251, 254,
      255, 259, 270, 271, 274,
      277, 285–287, 297, 307,
      313, 315, 317, 331, 336,
      349–351, 364, 371, 374,
      376–378, 384, 386
adsorption, 359
adult, 191
advance, 48, 100, 128, 228, 250,
      310, 389
advancement, 19, 278, 355, 371, 394
advantage, 376
advertising, 71
advocacy, 19, 20, 28
advocate, 19, 21, 24, 36, 152
affordability, 100, 378
afforestation, 104, 118, 119, 121,
      122, 134
Africa, 232
agreement, 29
agriculture, 9, 34, 50, 57, 60, 61, 63,
      85, 92, 93, 98, 104, 111,
      117, 119, 122, 124, 140,
      176, 184, 193, 198, 199,
      216, 220, 224, 235–240,
      242, 243, 247, 251, 255,
      257–261, 263, 265, 268,
      269, 275, 277, 278, 281,
      284, 286, 287, 289, 290,
      296, 297, 299, 300,
      304–308, 311, 319, 320,
      372
agrobiodiversity, 276, 278
agroecology, 243, 246, 247,

Index 401

281–284, 287, 296,
308–311
agroforestry, 149, 237, 243,
245–247, 251, 273–275,
343
aid, 185, 224
aim, 6, 8, 25, 44, 53–55, 58, 66, 84,
97, 98, 104, 105, 115, 116,
118, 124, 140, 152, 185,
224, 226, 243, 247, 251,
264, 266, 287, 291, 305,
365, 393
air, 1, 5, 7, 9, 10, 16–18, 28, 32, 36,
43, 57, 58, 65, 71–75,
80–82, 85, 88, 98, 99, 101,
126, 161, 319, 323, 325,
333, 363, 365, 376, 378,
396
aircraft, 46
airflow, 353
Alameda County, 370
Alberta, 327
Aldo Leopold, 23
Aldo Leopold - This, 23
Alecia M. Spooner, 15
algae, 117, 164
alteration, 104, 151, 322
alternative, 72, 98, 184, 185, 187,
199, 218, 220, 313, 315,
318, 326, 331, 332, 362,
372, 378
altitude, 225
America, 23, 202
ammunition, 190
amount, 39, 53, 98, 238, 329, 332,
337, 339, 390
amphibian, 224
analysis, 5, 10, 18, 32, 39, 42–45, 48,
158, 171, 173, 201

Andaman, 231
Andrew Stutzman, 15
animal, 36, 108, 112, 119, 122,
135–138, 147, 173, 176,
201, 217, 219, 363
annoyance, 71, 86
anthropocentrism, 35
Antonello Monti, 371
anxiety, 377, 378
application, 30, 31, 36, 158, 161,
243, 263
appreciation, 187
approach, 1, 9–11, 14, 31, 32, 56,
59, 61, 84, 86, 89, 91, 99,
102, 106, 110, 113–115,
122, 136, 139, 150, 151,
153, 167, 174, 187, 194,
196, 200, 205, 217, 218,
220, 221, 223, 224, 229,
235, 238, 240, 251, 263,
268, 269, 275, 278, 281,
284–286, 288, 297, 302,
311, 316, 348, 364
approval, 143
aquaculture, 108, 290, 292, 293
architecture, 60, 365–368, 395
Arctic, 227
area, 17, 18, 26, 81, 110, 124, 133,
139, 232, 236, 287, 313,
332, 369
army, 185
array, 116, 136, 137, 352
arrival, 225
art, 187
aspect, 10, 58, 98, 149, 164, 190,
201, 325, 384
assessment, 115, 136, 171–174, 352,
353
assistance, 185, 288, 289

association, 287
assurance, 290
atmosphere, 15, 48, 53, 71, 75, 109, 116, 119, 266, 319, 320, 322, 343
attachment, 117
attention, 3, 70, 138, 178, 183, 239, 257, 259, 339, 352, 359, 369, 389
attitude, 378
Australia, 7, 55, 104, 146, 150, 152
automation, 257, 395
availability, 31, 57, 58, 91, 99, 103, 112, 134, 136, 138, 186, 201, 220, 225, 226, 237, 276, 291, 293, 294, 305, 306, 321, 339, 343, 364, 377, 386, 389
awareness, 2, 18, 19, 23, 24, 26, 59, 62, 66, 73, 74, 91, 92, 97, 140, 163, 185, 186, 198, 199, 211, 219–221, 224, 255, 259, 265, 275, 304, 324, 368, 397
awe, 20
axis, 336

balance, 3, 7, 10, 16, 23, 25, 26, 36, 56, 57, 81, 91, 93, 117, 119, 137, 141, 149, 150, 165, 181, 184, 185, 187, 190, 193, 197, 198, 201, 202, 219, 251, 275, 291, 324, 389, 394
bamboo, 181
ban, 28, 187
bank, 111
bargaining, 288
barrage, 339

barrier, 142, 285, 377
base, 265
baseline, 136
baseload, 323
basic, 30, 116, 333, 336, 352
basis, 2, 5, 29, 60
battery, 376–378, 394
bear, 227
beauty, 20, 21, 60, 61, 106, 193, 200
beginning, 3, 51
behavior, 1, 32, 39, 71, 73, 81, 86, 91, 150, 185, 186, 190, 201, 202, 220, 256
being, 1, 5, 7, 9, 25, 31, 35, 37, 40, 50, 55, 56, 58–60, 62, 66, 70, 77, 82, 83, 85, 86, 91, 93, 97, 115, 129, 145, 151, 161, 164, 222, 225, 229, 267, 272, 277, 290, 291, 304, 314, 315, 363, 370, 377, 396
benefit, 35, 129, 140, 150, 205, 251, 263, 353, 369, 371, 376
beta, 263
biocentrism, 35
biodiversity, 1, 4, 6, 7, 9, 15, 17, 18, 21, 22, 26, 31, 34, 47, 48, 56–67, 70, 71, 85, 101, 104, 106, 108, 110, 112, 115, 117, 119, 124–126, 131, 133–135, 138–141, 147, 148, 150, 151, 157, 159, 161, 163–165, 167, 168, 177, 178, 181, 193, 197–202, 205–209, 219–222, 224, 225, 231, 232, 235, 237, 242, 243, 247, 265, 266, 268, 275, 276, 291, 293, 306, 308,

Index

311, 321, 328, 358, 362, 396
bioenergy, 324, 343–346
biofuels, 98, 314, 346
biogas, 315, 316, 339, 341–343, 359–361, 364
biography, 23
biologist, 24
biology, 1, 9–11, 15, 36, 157–161, 164, 168, 171, 174, 183, 202, 229, 260, 263
biomass, 46, 97, 111, 112, 119, 147, 322, 324, 325, 343–346, 363, 364, 375, 384, 394
biomethane, 359
biotechnology, 115, 260, 262, 263, 397
bird, 112, 137, 138, 167, 176, 209, 211, 224, 328
birth, 28, 33
blast, 117
bleaching, 56, 116, 117, 146, 152, 164, 226
blindness, 263
blockchain, 370
blood, 72
bloodstream, 72
boiling, 340
book, 3, 20, 23, 24, 27, 28
Borneo, 16, 193, 196
branch, 35
Brazil, 296, 297, 315, 356–359
break, 359
breakdown, 314, 359
breakthrough, 100
breeding, 58, 83, 153, 158, 190, 226, 261, 263, 276, 305, 306
brink, 182, 188, 190, 199
Brooklyn Microgrid, 370

Bruce Walsh, 259
buffer, 61, 109
building, 6, 49, 52, 84, 98, 100, 140, 141, 152, 153, 155, 185, 237, 238, 266, 272, 284, 290, 297, 304, 306, 314, 316, 340, 362, 364–368, 395, 397
bullet, 188
burden, 35, 364
burning, 6, 36, 48, 53, 71, 73, 149, 150, 212, 313, 319, 320, 322
business, 351
buying, 219
bycatch, 291

cabin, 332
calcium, 116
California, 58, 188, 370
call, 24, 26
camera, 173, 174
can, 1, 2, 4, 7, 10, 11, 14–16, 18, 29–35, 37, 39, 43, 45–53, 55–62, 64–67, 70–76, 80–86, 88–93, 97, 98, 101, 103, 105, 106, 110–119, 122, 123, 125, 129–131, 133, 135, 136, 140–145, 148–153, 161–166, 168, 171, 173, 174, 178, 181, 183, 185–187, 193, 196, 200, 202, 205, 206, 211, 215–230, 234, 235, 238, 243, 247, 249, 251, 255, 257–260, 263, 265, 266, 268, 272, 275, 276, 278, 280, 281, 284, 285, 288–291, 293, 294, 297,

300, 301, 304–308, 311, 314, 316–331, 333, 334, 336, 338, 340, 343, 346, 348, 349, 352–354, 356, 357, 359, 364, 368–372, 375, 376, 378, 379, 382, 384–387, 389–391, 393, 394, 397

Canada, 201, 327

cancer, 60, 74, 83

canopy, 147, 193

capacity, 33, 84, 113, 140, 141, 155, 185, 236, 248, 290, 307, 319, 350, 352–356, 364, 377, 394

capital, 56, 288, 289

capture, 46, 173, 314, 320, 330, 331, 338, 343, 394

car, 378

carbon, 15, 31, 37, 55, 57, 60, 71, 98, 104, 108, 109, 116, 119, 161, 206, 215, 237, 238, 248, 266, 306, 308, 315, 319, 321, 322, 328, 336, 343, 346, 351, 359, 362, 363, 365, 376, 390, 391, 395

carbonate, 116

Caribbean, 117

carotene, 263

carrying, 33, 83

Carson, 24, 28

cascade, 57

case, 14, 17, 26, 42, 55, 73, 80, 81, 88, 104, 105, 112, 119, 121, 122, 124, 129, 131, 136, 146, 158, 170, 174, 176, 177, 229, 231, 238, 243, 246, 263, 269, 286,

316, 328, 332, 392, 393

catalyst, 24

cattle, 136, 224

cause, 35, 39, 57, 74, 85, 117, 151, 193, 263, 305, 313

cell, 260

cement, 117

center, 35

Central America, 362

century, 2, 3, 21, 23, 25, 26, 135, 308, 311, 320, 356

certainty, 385

certification, 285, 289

chain, 197, 238, 290, 301

challenge, 6, 14, 17, 31, 32, 59, 90, 118, 134, 137, 150, 196, 197, 201, 225, 228, 238, 265, 285, 289, 297, 317, 364, 370

chance, 201

change, 1, 2, 6, 9, 10, 14, 15, 17–19, 22, 25, 26, 29, 31, 34, 36, 37, 46–57, 59, 61, 64, 68, 70, 80, 83–85, 89, 96–98, 104, 106, 109, 112, 116–119, 122, 125, 138, 146, 150–153, 155, 157, 164, 165, 168, 178, 183, 186, 197–200, 208, 211, 212, 215, 219, 225–229, 237, 238, 247, 251, 265, 276, 286, 287, 289, 290, 297, 304–308, 311, 313, 315, 318, 320–323, 325, 328, 330, 336, 346, 349, 352, 363, 365, 375, 386, 387, 389, 393, 397

channel, 131

channelization, 66, 104

Index 405

chapter, 101, 112, 157, 212, 235, 313
charge, 377, 378
charging, 100, 370, 377, 378, 389
check, 124
chemical, 5, 7, 9, 24, 26, 43, 45, 57, 237, 251, 258, 285, 286, 307
chemistry, 1, 9–11, 15
Chernobyl, 24
childhood, 263
China, 7, 16, 104, 124, 133, 135, 181, 183, 339
choice, 39, 40, 42, 43, 114, 151, 330
cholera, 83
circuit, 330
circulation, 108
cite, 82
citizen, 168, 229
city, 74, 99, 272, 378
clash, 143, 216
classification, 47
cleaner, 72, 81, 84, 91, 99, 216, 313, 325, 326, 328, 378, 393
clearing, 319
climate, 1, 2, 6, 7, 9, 10, 14, 15, 17–19, 22, 25, 26, 29, 31, 32, 34, 36, 37, 46–56, 58–61, 63, 64, 70, 80, 83–85, 89, 98, 104, 106, 109, 116–119, 122, 125, 133, 134, 138, 146, 150–153, 155, 157, 161, 164, 165, 168, 178, 183, 186, 197, 200, 208, 211, 212, 215, 219, 225–229, 237, 238, 247, 251, 257, 276, 278, 286, 287, 289, 290, 297, 305–308, 311,

313, 315, 318, 320–323, 325, 328, 330, 336, 346, 349, 352, 355, 363, 365, 386, 389, 393, 396, 397
climax, 102
cloud, 386
coal, 212, 319, 322, 326, 386
coast, 55, 146, 197, 354
coastline, 352, 353
coexistence, 217, 218, 354
coffee, 289
collaboration, 17, 19, 20, 32, 55, 59, 73, 74, 89, 92, 97, 103, 106, 138, 140, 141, 144, 149, 152, 153, 168, 186, 200, 202, 205, 220, 221, 231, 304, 307, 355, 375, 393, 397
collaborative, 115, 116, 161, 190
collapse, 281
collateral, 289
collection, 5, 23, 39, 40, 42, 43, 167, 171
Colombia, 289
colonization, 102, 110, 112, 117
color, 116
combat, 98, 184, 220, 221, 322, 389
combination, 50, 55, 58, 109, 124, 133, 141, 146, 147, 163, 173, 190, 219, 226, 273, 285, 304, 307, 317, 364, 378
combustion, 314, 319, 320, 322, 343, 376, 377
comfort, 395
commitment, 59, 135, 138, 141, 149, 183, 200, 213, 228, 265, 322, 348, 355, 359, 360, 362, 364, 365, 378, 384

communication, 19, 20, 202, 264, 265, 316, 394

community, 17, 19, 26, 28, 33, 35, 53, 58, 60, 91, 102, 103, 110, 112, 115, 122, 135, 139–141, 149, 163, 164, 181–187, 199, 200, 205, 211, 218, 220, 221, 236, 237, 264, 265, 272, 289, 290, 308, 311, 339, 368–371

compaction, 251

compensation, 134, 358

competition, 33, 111, 216, 225, 285

complexity, 4, 10, 30–32, 106, 206

compliance, 73, 81

component, 4, 8, 56, 97, 133, 158, 251, 304, 323, 332, 355

composition, 1, 5, 7, 33, 43, 102, 110, 112, 151

composting, 98, 242, 251, 276

compound, 321

comprehensiveness, 149

concentration, 260, 320

concept, 3, 4, 25, 26, 30, 32, 33, 88, 90, 91, 157, 161, 164, 165, 235, 236, 248, 257, 287, 290, 395

concern, 28, 56, 59, 70, 83, 293, 376, 377

conclusion, 34, 50, 56, 59, 70, 93, 106, 118, 125, 129, 133, 138, 148, 155, 161, 178, 186, 196, 200, 202, 208, 221, 224, 238, 247, 272, 275, 278, 284, 286, 300, 311, 318, 346, 368, 371, 386, 393

concrete, 117, 131, 193

condition, 123, 294

condor, 58, 188, 190

conflict, 218

connection, 20, 23, 61, 202, 287

connectivity, 118, 133, 136, 224

consciousness, 24, 378

consensus, 228

consequence, 198

conservation, 1–3, 6–9, 11, 15, 16, 18–23, 31, 32, 36, 47, 48, 58, 59, 61, 62, 66, 69, 70, 89–93, 104, 106, 110, 112, 125, 134, 135, 138–141, 143, 147, 157, 158, 160, 161, 163–165, 167–171, 174, 178, 182–187, 190, 192–200, 202–206, 208, 209, 211, 216, 218, 220–235, 242, 243, 247, 251, 276, 368, 395, 396

conservationist, 21, 23

conserving, 33, 92, 106, 115, 125, 153, 177, 178, 181, 207, 208, 210, 232, 251, 258, 276, 301

consideration, 35, 36, 91, 152, 172, 201

consistency, 387

construction, 18, 71, 73, 86, 124, 135, 316, 325, 353, 356, 395

consultation, 201

consumer, 185, 186, 220, 264

consumption, 28, 73, 92, 98–100, 236, 264, 291, 301, 316, 322, 364, 388, 389, 394, 395

contact, 216

Index 407

contamination, 7, 17, 71, 76, 77, 79, 85
content, 248, 314
context, 2, 22, 118, 136, 229, 278, 285
continent, 186
continuation, 277
contour, 124
contrary, 35
contrast, 266, 328, 395
contribute, 6, 9, 11, 16, 19, 36, 48, 53, 56, 58–60, 66, 71, 73, 74, 83, 100, 110, 122, 125, 129, 150, 156, 165, 167, 174, 178, 184, 186, 193, 198, 206, 211, 224, 235, 237, 238, 243, 251, 255, 258, 259, 263, 265, 266, 268, 272, 275, 276, 281, 291, 300, 301, 304, 313, 316, 318, 330, 339, 343, 352, 369, 395–397
contribution, 307, 363
contributor, 57, 98, 212
control, 2, 7, 8, 11, 16, 19, 59, 70, 72–74, 80, 81, 85, 87, 88, 122, 124, 131, 134, 139, 150, 176, 199, 200, 211, 242, 257, 264, 276, 285, 286, 289, 306, 307, 314, 316, 370, 380, 394, 395
controversy, 138
convenience, 265, 378
conversion, 57, 63, 98, 104, 134, 176, 216, 314, 336, 343, 344, 353, 376, 379, 395
cooking, 316
cooperation, 28, 29, 50, 58, 92, 134, 155, 163, 184, 186, 220, 221, 228
cooperative, 29, 287, 289, 290
coordination, 55, 316, 370, 371, 380
coral, 56, 57, 64, 104, 116–118, 146, 147, 152, 153, 164, 198, 206, 226, 227, 231, 291
core, 1, 9, 43, 314, 323, 340
corrosion, 352
corruption, 90, 219
cost, 100, 116, 255, 258, 321, 323, 324, 331, 348, 354, 368, 370, 371, 373, 376, 377, 388, 394, 397
Costa Rica, 213, 362, 364, 365
Costa Rica's, 214, 362–365
cotton, 281
count, 174
country, 22, 31, 99, 138, 237, 281, 315, 328, 335, 346, 348, 349, 356, 358–360, 362–365, 378
couple, 179
course, 138, 190
cover, 3, 46–48, 117–120, 134, 146, 147, 235, 251, 265, 285, 306, 308, 329, 386
craft, 140
creation, 21, 115, 206, 222, 232, 239, 325
credibility, 19
credit, 289
crime, 184
criminal, 186, 219, 220
crisis, 28, 116, 237, 281, 313
crop, 57, 58, 86, 98, 235, 236, 238, 242, 251, 252, 256–258, 260, 263, 276, 278, 285–287, 289, 305, 306, 308, 320, 331

cropping, 251, 306

crust, 323, 340

Cuba, 237, 281–284

cube, 333

cultivation, 236, 276, 285

culture, 60, 115, 260

current, 14, 17, 23, 70, 92, 93, 136,
171, 251, 308, 329, 330,
349

curricula, 278

curve, 33

customer, 265

cutting, 115

cyanide, 117

cycle, 111, 116, 119, 191, 264, 336,
340, 343, 365

cycling, 7, 57, 60, 112, 222, 236,
248, 308

dam, 105, 131, 337, 358

damage, 25, 83, 101, 106, 117, 118,
151, 216, 285, 305

Daniel B. Botkin, 15

darkness, 86

data, 1, 2, 5, 10, 17, 18, 31, 32,
39–43, 45–48, 81, 137,
145, 148, 149, 167, 171,
198, 236, 251, 255, 394

date, 74, 187

day, 324, 329, 331, 332, 386

death, 33, 83, 117, 164, 226

debate, 23, 24, 36, 138, 201

decay, 319, 340

decision, 3–5, 8, 11, 25, 26, 30, 34,
36, 37, 45, 59, 89, 90, 92,
93, 115, 145, 149, 198,
228, 236, 289, 324

decline, 24, 56, 57, 64, 117, 118,
152, 153, 168, 181, 184,

190, 194, 200, 209, 217,
220, 224, 281, 291

decrease, 111, 216, 266, 301, 377,
389

dedication, 190

deer, 137

deficiency, 263

deforestation, 2, 3, 6, 9, 22, 31, 36,
46, 53, 57, 59, 63, 108,
119, 122, 133, 147, 153,
209, 224, 358

degradation, 2, 6, 7, 9, 17, 24, 61, 66,
79, 82, 91, 93, 101,
103–105, 110, 113, 117,
119, 123, 124, 129,
133–135, 147, 153, 175,
197, 198, 230, 247, 258,
297, 313, 393

dehydration, 83

delay, 141

Delhi, 73, 74

Delhi, 73

delivery, 258, 265

demand, 57, 65, 98, 184–186, 216,
219–221, 258, 281,
291–293, 317, 324, 330,
335, 349, 353, 369, 370,
387–389, 394, 395

dengue, 83

Denmark, 99, 100, 315, 328, 335,
336, 354, 355

Denmark, 99

density, 104, 126, 138, 171, 173, 258

dependence, 4, 98, 212, 237, 308,
322, 325, 329, 330, 336,
362, 369, 376, 386, 393

depletion, 2, 9, 26, 57, 93, 200, 395

deployment, 104, 263, 285, 324,
335, 353, 355, 370, 381,

Index

384–386, 394

deposition, 108

depth, 17, 352

desert, 328

design, 4, 47, 60, 115, 206, 240, 281,
316, 368, 395

destruction, 2, 6, 15, 22, 34, 56, 57,
61, 63, 66, 101, 106, 119,
138, 197, 198, 218, 219,
326, 327

detection, 43, 45, 174

deterioration, 71

determination, 362, 365

deterrent, 377

development, 2, 3, 7–9, 11, 15–20,
22, 23, 25, 34, 36, 44, 45,
47, 50, 57, 58, 60, 66, 72,
88–93, 96, 99, 100, 102,
104, 108, 110, 112, 113,
117, 119, 122, 124,
138–141, 163, 199, 205,
209, 213, 216, 220–224,
228, 238, 261, 263, 277,
286, 287, 289–291, 297,
300, 305, 318, 324, 325,
331, 335, 336, 349, 351,
352, 354, 356–360, 364,
370–372, 375, 377, 378,
381, 384, 389, 390,
393–395, 397

dialogue, 92

Dickson Despommier, 259

diesel, 319, 372, 376

diet, 161, 181, 276

difference, 376

difficulty, 31

digester, 359

digestion, 99, 314, 359, 363, 395

dioxide, 15, 55, 71, 80, 109, 116,
119, 306, 319, 343, 359

direction, 278, 352

disappearance, 165

disaster, 24, 26

discarding, 301

discharge, 71

discharging, 370, 389

discipline, 14, 243

discussion, 23

disease, 33, 60, 83, 84, 164, 248,
306, 321

dispersal, 57, 114

displacement, 53, 57, 66, 339, 358

disposal, 7, 71, 73, 85, 331, 377, 395

disruption, 53, 57, 83, 113, 198,
217, 322, 353, 358

dissemination, 286

distance, 48, 354

distribution, 36, 39, 47, 57, 64, 90,
92, 150, 151, 171, 201,
225, 265, 293, 301, 324,
371, 394

disturbance, 71, 102, 110–112, 176,
306, 354

diversification, 236, 289, 290, 362

diversity, 33, 56, 59–61, 63, 106,
111, 112, 117, 118, 133,
157, 158, 161, 162, 164,
168, 169, 171, 173, 176,
177, 185, 193, 201, 222,
224, 225, 251, 264, 287,
308

dominance, 61

donor, 117, 118, 178

Douglas Brinkley, 23

downgrading, 183

drainage, 122, 124

draining, 104, 136

drift, 169
drinking, 320
drip, 236, 242, 306, 307
driver, 216, 222, 377
driving, 143, 222, 322, 336, 375,
        377, 384, 386, 397
drought, 238, 305
drug, 60
drying, 265, 331
dump, 26
duration, 71
dwelling, 225
dynamic, 4, 112, 136
dynamism, 150
dysentery, 83
dysfunction, 72

eagle, 7
Earth, 1, 3, 4, 11, 15, 35, 37, 46–48,
        56, 59, 63, 67, 129, 157,
        161, 164, 165, 223, 314,
        319, 320, 322, 323, 325,
        329, 333, 338, 340, 343
earth, 75
East Africa, 138
ebb, 314
eco, 91, 184, 364
ecocentrism, 35
ecology, 9–11, 32, 33, 35, 112, 137,
        157, 168, 308
economic, 3, 8, 16, 17, 20, 25, 26,
        34, 36, 47, 53, 60, 88–93,
        97, 102, 115, 134,
        139–142, 184, 186, 187,
        200, 215–218, 220, 221,
        228, 235, 237, 245, 251,
        269, 275, 281, 287, 290,
        291, 294, 297, 298, 300,
        301, 307, 320, 325, 346,

358, 363, 365, 366,
        370–372, 384, 390, 395
economy, 31, 50, 100, 325, 395–397
ecosystem, 4, 7, 10, 31–34, 39, 40,
        53, 55–58, 60–64, 70, 71,
        101, 102, 104–117, 119,
        123, 124, 126, 131,
        135–139, 141–144,
        146–153, 155–157, 161,
        162, 164, 167, 178, 193,
        197, 199–202, 206, 219,
        224–226, 230, 231, 235,
        248, 268, 285, 306, 308,
        321, 328, 343, 358, 396
ecotourism, 138, 140, 141, 198
Ecuador, 197
edge, 115, 226
education, 59, 61, 66, 91, 92,
        140–142, 185, 198, 200,
        218, 220, 238, 239, 296,
        308, 310, 311, 316, 368,
        372
Edward A. Keller, 15
effect, 48, 119, 318, 319, 396
effectiveness, 28, 81, 94, 104, 116,
        142, 145, 147, 148, 150,
        151, 156, 183, 187, 199,
        206, 207, 224, 303, 324,
        348, 373, 397
efficacy, 277
efficiency, 6, 49, 89, 91, 93, 97, 98,
        100, 227, 257, 315, 318,
        322, 324, 331, 348, 351,
        353, 364, 368, 369, 371,
        376, 394, 395
effort, 56, 118, 202, 212, 296, 304
electricity, 98, 99, 314–317, 319,
        321–324, 329–332,
        335–337, 339, 340, 346,

Index 411

349, 352–356, 362, 363, 369–371, 373, 376, 379, 384, 387, 389, 394
elephant, 184, 186, 187, 216, 218, 220
Elizabeth Kolbert, 229
elk, 200, 201
embrace, 239, 264, 318, 325, 365, 378
emigration, 33
emission, 6, 73, 81, 99
emitter, 349
empathy, 190
emphasis, 235, 237, 281, 363, 395
employment, 134, 363
empowerment, 139, 149, 278, 284
encounter, 70
encroachment, 139, 141
end, 235, 344
endangerment, 157, 165–168
endeavor, 193, 201, 202, 211, 365
endocrine, 72
energy, 4, 6, 8, 9, 16, 24, 25, 32–34, 37, 48–50, 52, 53, 55, 72, 73, 84, 85, 89, 93, 97–100, 140, 212–216, 227, 235, 258, 313–318, 321–343, 345–356, 358–360, 362–365, 368–376, 378–382, 384–395, 397
enforcement, 73, 139, 184, 186, 219–221
engagement, 50, 90, 102, 139, 141, 149, 156, 163, 164, 181, 182, 186, 190, 211, 220, 221, 224, 228, 231, 233, 236, 272, 325, 368
engineering, 60, 100, 115, 124, 131, 260, 263, 325

England, 21
enhance, 45, 49, 54, 56, 60, 91, 92, 101, 106, 109–111, 114–116, 118, 124, 131, 136, 140, 146, 148, 150, 178, 233, 236, 237, 243, 251, 259, 260, 263, 265, 276, 278, 289, 290, 305–307, 324, 352, 369, 391, 396
enjoyment, 21
enterprise, 219, 287
entertainment, 36
entry, 285, 377
environment, 1–5, 7–11, 15, 16, 18–20, 22–29, 31–33, 35–37, 39, 43, 45, 47, 50, 56, 57, 65, 66, 70, 73–75, 77–82, 85, 86, 88, 90, 92, 97, 101, 109–111, 119, 138, 141, 148, 149, 161, 200, 201, 227, 229, 238, 241, 243, 247, 253, 257–259, 265, 266, 274–276, 285, 308, 314, 320, 325, 352, 359, 366, 375, 376, 384, 386, 395
environmentalism, 23–26
environmentalist, 23
equality, 89, 289
equation, 333
equipment, 142, 265, 331, 354, 369
equity, 4, 8, 16, 25, 35, 88, 90–93, 97, 282, 297, 300, 308
eradication, 90, 198–200
erosion, 66, 104, 109, 124, 133–135, 176, 236, 247, 248, 251, 278, 305, 306, 320, 326
escalation, 3

essay, 22
essence, 61
establishment, 3, 7, 8, 21, 22, 28, 29, 36, 58, 61, 110–112, 120, 134, 146, 153, 164, 199, 200, 232, 290, 358, 377, 384, 385
estimate, 147, 171, 173, 174
ethic, 23
Europe, 21, 227, 352–355
evaluation, 36, 145–149
evaporation, 242, 258, 305
event, 24, 99, 112, 165, 281
Everglades, 104, 124
evidence, 11, 19, 25, 198, 263
evolution, 22
example, 1, 10, 16, 31–33, 47, 60, 80, 89, 92, 104, 111, 112, 114, 115, 135, 141, 150–152, 161, 173, 183, 190, 197, 216, 218–220, 225–227, 242, 289, 296, 327, 335, 339, 358, 364, 370, 378, 384, 389, 392, 394
exception, 260
excess, 317, 324, 370, 379, 387, 394
exchange, 136, 205, 250, 288, 311
exchanger, 340
exclusion, 33
exemption, 378
exhaust, 71
expanse, 353
expansion, 108, 119, 147, 193, 222, 223, 348
expense, 90
experience, 20, 140, 365, 377
expert, 308
expertise, 2, 11, 19, 32, 92, 109, 259

exploitation, 219
explosion, 24
exposure, 8, 71, 83
extension, 288, 395
extent, 103, 153
extinction, 6, 57, 63, 65, 119, 157, 165–168, 178, 182, 188, 190, 199, 209, 219, 226
extraction, 66, 91, 100, 319, 326, 327, 395
extreme, 48, 49, 53, 54, 57, 64, 83, 84, 109, 151, 198, 304, 305, 307, 320, 369

fabric, 276
face, 2, 15, 20, 25, 29, 54, 56, 63, 104, 109, 117, 129, 132, 147, 152, 153, 164, 178, 190, 191, 193, 197, 203, 208, 222, 225, 228–230, 251, 272, 276–278, 287, 289, 295, 306, 308, 311, 353, 370
facilitation, 102, 111
facility, 370, 371
factor, 59
failure, 306
fairness, 89
fall, 338
farm, 242, 264, 265, 272, 331, 352, 354, 355, 362
farmer, 242, 264
farming, 61, 92, 99, 100, 133, 149, 235–243, 251, 255, 257–259, 261, 263–272, 275–281, 285, 287–289, 292, 297–300, 306, 308, 311, 343
farmland, 134, 135

Index 413

fashion, 219
fate, 7
fauna, 124, 149, 197, 199
fear, 217, 377
feasibility, 55
feature, 369
feed, 33, 89, 206, 307, 315, 317, 324, 364, 384
feedback, 4, 30, 31, 145, 272
feeding, 32, 238
feedstock, 359
fencing, 176
fermentation, 343
ferret, 58
fertility, 57, 86, 111, 157, 168, 236, 237, 243, 247, 266, 275, 306, 308
fever, 83
field, 1, 3–5, 9, 11, 14–16, 18–20, 23, 29, 36, 43, 45, 106, 115, 118, 147, 157, 158, 165, 168, 212, 251, 260, 263, 308, 310, 313, 316, 324
fieldwork, 10
fight, 11, 97, 322
filtration, 122, 124
finance, 29
financing, 118, 290
finding, 1, 2, 25, 63
fire, 111, 150
firearm, 184
firewood, 372
fish, 57, 105, 108, 109, 116, 131, 197, 199–201, 208, 291, 292, 322, 353, 358
fishing, 57, 61, 65, 104, 117, 164, 197, 199, 206, 208, 220, 227, 291

flavor, 264
Flevoland, 135
flexibility, 150, 152
flightlessness, 209
flood, 104, 115, 122, 124, 131
flooding, 47, 131, 135, 320
flora, 124, 149, 197, 199
Floreana Island's, 199
Florida, 104
flow, 4, 33, 105, 108, 124, 129, 169, 222, 314, 339, 370, 389, 394
flowering, 225
flowing, 314, 322, 336, 337
fluid, 340
focus, 6, 25, 84, 97, 98, 100, 106, 124, 152, 161, 163, 177, 178, 227, 228, 266, 365, 370, 393
folklore, 187
following, 42, 82, 97, 102, 108, 110, 112, 119, 175, 192, 201, 235, 242, 259, 269, 272, 281, 287, 305, 320, 334, 373
food, 10, 33, 36, 48, 57, 58, 60, 63, 65, 71, 82, 92, 93, 99, 101, 134, 137, 138, 161, 164, 197, 200, 216, 220, 222, 225, 226, 235–239, 242, 247, 251, 255–259, 263–265, 268–270, 272, 275, 276, 281, 282, 284, 287, 291, 293–297, 301–305, 308, 311, 315, 321, 359
footprint, 37, 39, 258, 322, 328, 351, 376, 395, 397
forage, 134

force, 26, 323
forefront, 11, 24, 354
forest, 31, 101, 104, 108–111, 118–120, 134, 147, 184, 193, 224, 324, 343
forestry, 8, 58, 60
form, 5, 32, 116, 276, 323, 329, 338, 343, 379
formation, 28, 60, 92, 112, 133
formula, 260
formulation, 26, 73
fossil, 6, 36, 48, 53, 71, 89, 98, 99, 212, 235, 313, 315, 318–323, 325, 326, 328–330, 332, 336, 346, 349, 362–364, 369, 371, 376, 386, 393
foster, 59, 140, 187, 200, 205, 215, 264, 289, 325, 375
foundation, 4, 20, 32, 352
fracking, 326
fraction, 258, 329
fracturing, 326
fragility, 163
fragment, 224
fragmentation, 63, 70, 176, 183, 198, 209, 217, 222–225
frame, 329
framework, 4, 30–32, 34, 35, 37, 89, 93, 107, 317, 384
frequency, 56, 71, 109, 198, 305, 320
freshwater, 129, 133, 320
fuel, 150, 314, 318–320, 322, 324, 328, 359, 370, 376
function, 34, 112–114, 248
functionality, 98, 106, 109, 110, 126, 133, 167, 175
functioning, 4, 7, 57, 59–61, 63, 101, 129, 131, 157, 163,

224, 306
fund, 139
funding, 118, 142, 211, 228
fur, 181
fusion, 329
future, 2, 3, 8–11, 16, 17, 20–26, 29, 32, 34–37, 48, 50, 52, 53, 55, 56, 59, 62, 66, 69, 70, 73, 74, 80, 82, 85, 88, 89, 91–93, 97, 99, 100, 105, 106, 111, 118, 122, 125, 128, 132, 133, 135, 137, 138, 141, 145, 152, 153, 161, 164, 168, 170, 171, 177, 181, 183, 186, 187, 190, 193, 195, 197, 205, 209, 211, 212, 215, 216, 218, 221, 228, 229, 233–235, 238, 239, 243, 247, 251, 255, 259, 263, 265, 268, 272, 276, 278, 283, 293, 297, 299, 300, 304, 311, 313, 317, 318, 322, 325, 328, 331, 332, 336, 339, 342, 343, 346, 348, 349, 351, 352, 356–359, 362, 364, 365, 368, 369, 371, 374, 375, 378, 381, 384–386, 389, 393, 397

gain, 31, 32, 235, 384
Gansu, 181
gardening, 104, 227
gas, 6, 15, 25, 29, 36, 49, 52, 53, 56, 59, 66, 84, 98, 99, 153, 212, 226, 227, 301, 307, 313, 315, 316, 318–320, 322, 323, 325–327, 336,

Index 415

343, 349, 355, 359, 362,
369, 376, 378, 384, 386,
389, 395
gasification, 314
gasoline, 319, 376
gel, 256
gender, 289
gene, 168, 169, 222
generation, 8, 16, 34, 88, 91, 98, 149,
235, 278, 297, 316–319,
321, 330, 331, 336, 339,
340, 343, 349, 351–353,
356, 363, 364, 371, 379,
387–389, 394, 395
generator, 323, 336
genetic, 33, 56, 59, 63, 115, 117,
118, 136, 158, 162, 164,
168–171, 173, 185, 222,
224, 225, 260, 263, 276,
278, 306
geology, 1, 9–11, 15
geothermal, 314, 322, 339–341,
343, 364, 384, 394
Germany, 99, 100, 315, 335, 336,
346–348, 355
Germany, 99
GIS, 47
glass, 329
globalization, 278
globe, 204, 293
goal, 2, 30, 106, 136, 238, 301, 355,
363
Golden Rice, 263
good, 42, 186, 248, 304
gorilla, 139, 140
governance, 58, 90, 141, 186, 221,
289, 290, 297
government, 2, 18, 21, 56, 74, 99,
115, 134, 135, 138, 139,

200, 281, 297, 300, 349,
351, 355, 356, 358, 362,
364, 370, 376, 377
grass, 134
grassland, 114, 134, 177
grazing, 57, 114, 137, 139, 202, 216,
237, 307
green, 9, 17, 34, 50, 97–100, 104,
126, 212, 215, 313, 316,
318, 365–368, 393–397
greenhouse, 6, 15, 25, 29, 36, 48, 49,
51–53, 56, 59, 84, 98, 99,
119, 153, 212, 226, 227,
266, 285, 301, 307, 313,
315, 316, 318–323, 325,
327, 336, 349, 355, 362,
369, 376, 378, 384, 389,
395
grid, 98, 316, 317, 324, 331, 332,
348, 351, 352, 354, 359,
364, 369–371, 379–381,
384, 387, 389, 393, 394,
397
ground, 221, 340
groundbreaking, 3, 23, 27
grounding, 330
groundwater, 26, 76
groundwork, 22
group, 272
growth, 3, 17, 22, 33, 36, 46, 71, 88,
90, 91, 97, 98, 104, 108,
110, 111, 137, 146, 147,
150, 151, 153, 176, 181,
199, 215, 237, 247, 252,
257, 258, 280, 285, 305,
308, 325, 335, 336, 346,
351, 355, 358, 363, 365,
372, 384–386
guidance, 11, 18, 384

guide, 3, 4, 8, 35, 36, 113, 119, 136, 239, 269, 279

habitat, 2, 6, 7, 15, 17, 18, 22, 33, 34, 53, 56–58, 61, 63, 70, 104, 106, 108, 112, 115, 122, 126, 129, 131, 133, 134, 138, 139, 141, 147, 157, 163–165, 168, 174–178, 181, 183, 184, 186–188, 193, 197, 201, 209, 211, 216, 219, 222–226, 232, 321, 322, 326, 327

half, 355

hand, 6, 84, 102, 110, 120, 165, 178, 205, 285, 301, 314, 326, 339

handling, 184, 301

harm, 5, 18, 25, 35, 39, 57, 84, 86, 92, 276

harmony, 20, 93, 148, 149, 275

harness, 32, 114, 212, 313, 321, 323, 336, 340, 352, 364

harvest, 238, 264, 265, 301

harvesting, 57, 65, 236, 306, 307

haven, 135

health, 2, 7–9, 16, 25, 26, 31, 35, 45, 46, 48, 53, 56–58, 60–62, 65, 70–74, 77, 81–86, 88, 92, 104, 106, 109–111, 116–118, 126, 129, 139, 141, 147, 158, 163, 168, 173, 178, 190, 193, 201, 206, 227, 236, 237, 242, 247, 248, 250, 251, 263–266, 268, 278, 281, 285, 291, 306, 308, 311, 343, 372

healthcare, 140, 142, 185, 289, 296, 316, 372

hearing, 71

heart, 72

heat, 33, 48, 83, 84, 99, 126, 153, 305, 314, 322–325, 329, 340, 343, 346, 363, 396

heating, 316, 330, 333, 340, 343, 349

heatstroke, 83

height, 337

helium, 329

help, 4, 9, 16, 32, 36, 56, 88, 90, 94, 98, 99, 112, 118, 145, 152, 164, 173, 181, 186, 190, 217, 220, 221, 223, 224, 226, 258, 259, 265, 278, 288–290, 305–307, 311, 324, 384

Henry David Thoreau, 20, 22, 23

herbivore, 137

heritage, 18, 90, 140, 141, 149, 198, 200, 218, 276, 278, 287

history, 1, 21, 22, 24, 27, 131, 135, 165, 335

hole, 3

home, 116, 119, 137–139, 197, 209, 231, 377, 394

homestay, 140

honey, 264

hope, 116, 138, 181

hotspot, 231

household, 301

Howard M. Resh, 259

human, 1, 2, 4–11, 15, 16, 18, 20, 22, 23, 25, 26, 31, 34–36, 48, 50, 53, 56–63, 66, 70, 71, 74, 75, 77, 79, 81–86, 88, 93, 97, 101, 104, 106,

Index

108, 110, 111, 113, 116, 119, 122, 126, 129, 133, 141, 166, 169, 176, 184, 187, 193, 197, 198, 201, 206, 208, 216–218, 222, 232, 260, 301, 308, 319–321, 323
humanity, 35, 56
humidity, 257
humility, 150, 151
hunger, 238, 293, 295–297, 301
hunting, 57, 65, 193, 194, 216
hurdle, 59
hybrid, 98, 114, 115
hydraulic, 326
hydro, 316, 384
hydrogen, 98, 318, 329
hydrology, 32
hydrolysis, 359
hydroponic, 258
hydropower, 315, 322, 324, 328, 356–359, 362–365

ice, 15, 48, 53, 64, 225, 227, 320
identification, 115, 136
identity, 62, 161
illness, 83
imagery, 5, 47, 98, 115, 146, 224
immigration, 33
impact, 1, 2, 9–11, 15, 16, 18, 20, 27, 28, 31, 39, 47, 58, 64, 65, 73, 85, 88, 92, 99, 131, 152, 155, 166, 169, 193, 201, 225, 229, 236, 237, 242, 258, 260, 264, 265, 281, 285, 305, 306, 321, 322, 328, 332, 339, 343, 353–356, 365, 368, 375, 377

implementation, 3, 8, 28, 55, 73, 88, 100, 106, 113, 115, 121, 134, 140, 189, 200, 251, 278, 282, 285–287, 307, 331, 353, 355, 358, 370, 373, 384, 392, 395
import, 325, 378
importance, 2, 4, 10, 11, 15, 17, 18, 20–22, 26, 28, 29, 35, 39, 42, 50, 60–62, 66, 70, 80, 82, 89, 91, 92, 114, 115, 118, 122, 135, 141, 145, 148–150, 161, 163, 164, 176, 177, 187, 190, 193, 197, 201, 202, 218, 220, 224, 234–236, 238, 275, 276, 278, 290–292, 297, 307, 364, 372, 384
improvement, 125, 136, 145, 208, 276, 280, 283, 381
inauguration, 356
inbreeding, 158, 168, 169, 171, 222
inception, 346
incident, 24, 82
incineration, 99
inclusivity, 90
income, 134, 135, 139, 185, 220, 275, 291
incorporation, 155, 224, 387
increase, 34, 48, 49, 51, 53, 100, 112, 116, 118, 139, 181, 185, 224, 227, 236, 237, 255, 288–290, 305–307, 320, 333, 355, 364
independence, 343
India, 73, 218, 231, 315, 349–352
indigenous, 53, 59, 61, 102, 115, 147–152, 155, 202, 203, 205, 275–278, 281, 296,

358
individual, 39, 50, 53, 59, 173, 194, 228, 288
industrialization, 3, 9, 23
industry, 201, 215, 251, 281, 287, 291, 293, 319, 327, 328, 364, 384, 392
inequality, 238, 296
infection, 83
infertility, 72
infiltration, 308
inflammation, 72
influence, 1, 11, 28, 112, 185, 307
information, 2, 10, 18, 19, 39, 43, 46, 74, 136, 168, 171, 187, 289, 290, 307, 394
infrastructure, 6, 49, 50, 57, 66, 89, 100, 104, 119, 126, 186, 216, 222, 224, 265, 289, 301, 305, 315, 320, 324, 325, 328, 331, 352, 354, 362, 364, 369, 370, 377, 378, 384, 394, 396
inhibition, 102, 111
initiative, 134, 139, 155, 187, 202, 296, 297, 315
injection, 359
inlet, 135
innovation, 56, 100, 141, 153, 183, 250, 259, 263, 318, 364, 393, 394
input, 276
insect, 285
inspiration, 60, 135, 141, 161, 202, 355, 378
instability, 36, 186
installation, 330, 354, 370
instance, 16, 31, 47, 90, 92, 112, 140, 198

insulation, 98, 395
insurance, 289
integration, 8, 32, 88, 98, 116, 138, 151, 174, 205, 255, 263, 281, 284, 331, 336, 348, 351, 354, 369–371, 379–381, 384, 386, 387, 389, 390, 392–394
integrity, 19, 66, 91, 93, 116, 133, 147, 181, 321
intelligence, 185–187, 221, 397
intensity, 150, 305, 320
interconnectedness, 4, 26, 30, 32, 35, 37, 89, 149, 234
interconnection, 317
intercropping, 276
interdependence, 35, 37
interest, 43, 90, 152
interference, 206
interior, 340
intermittency, 317, 336, 364, 365, 386, 387, 389, 393
interplay, 1
intersection, 260
intervention, 28, 113, 116, 126
introduction, 57, 70, 74, 75, 85, 165, 197
intrusion, 320
investment, 89, 99, 100, 118, 142, 221, 265, 324, 330, 331, 364, 370, 384–386
involvement, 50, 58, 59, 74, 89, 103, 114, 118, 122, 135, 141, 167, 183, 199, 200, 220, 229, 232
ion, 376, 394
irradiance, 329, 332
irrigation, 236, 242, 305–307, 331
island, 84, 126, 199, 396

Index 419

isobutane, 340

isolation, 224

issue, 14, 17, 22, 23, 28, 31, 36, 56, 61, 73, 74, 80, 219, 221, 224, 228, 263, 293, 302, 387, 389

ivory, 184–187, 216, 220

J.C. Wu, 371

jail, 370

Japan, 256

Jason Green, 259

job, 315, 325, 352, 358

John Muir, 2, 20–22

John Muir - Muir's, 23

journey, 141, 183, 281, 325, 364, 397

justice, 4, 8, 36, 37, 92, 220, 328

Kakapo, 209

Kenya, 232

kerosene, 372

keystone, 190, 202, 219

kidney, 83

killing, 217

knowledge, 1, 2, 8, 10, 11, 14–17, 19, 53, 59, 60, 73, 92, 109, 115, 135, 138, 147–152, 155, 161, 163, 187, 202, 205, 230, 234, 250, 251, 275, 276, 278, 281, 286–290, 307, 308, 311, 368

labor, 142, 265

laboratory, 10, 16, 42–45, 137

lack, 59, 90, 134, 142, 219, 228, 278, 287, 307, 371, 384

land, 2, 3, 6, 16, 21, 23, 31, 46–48, 50, 52, 53, 57, 66, 84, 93, 99, 104, 113, 133–136, 148–150, 163, 176, 202, 209, 216, 218, 223, 224, 237, 257, 258, 264, 276, 277, 287, 288, 307, 319

landfill, 99

landmark, 3

landscape, 98, 99, 103, 131, 133, 136–138, 150, 224, 316, 343, 358, 360, 393

larvae, 117, 206

larval, 117

launch, 188

lava, 110

law, 33, 36, 139, 184, 186, 219–221

layer, 28

lead, 14, 33, 48, 57, 61, 64, 65, 71, 73, 83, 93, 117, 150, 151, 161, 188, 190, 216, 217, 220, 222, 225, 226, 229, 305, 320, 326

leader, 99, 141, 315, 328, 335, 359, 360, 362, 378

leadership, 290

leakage, 71

learning, 29, 129, 136, 151, 183, 211

lease, 287

Lee Hannah, 229

legacy, 23

legality, 187

legislation, 3, 36, 88, 185, 186, 220, 221

lesson, 135, 358, 364

level, 6, 32, 33, 47, 50, 54, 71, 83, 109, 164, 165, 301, 324

licensing, 384

LiDAR, 115, 147

lie, 361

life, 4, 15, 56, 59, 61, 63, 104, 110,
116, 157, 191, 227, 247,
260, 294, 315, 316, 365,
372, 378, 395
lifespan, 320, 377
lifetime, 323
light, 27, 46, 47, 71, 74, 78, 81, 85
lighting, 257, 258, 316, 332, 349,
395
limit, 29, 37, 49, 226, 372
line, 91
link, 8, 15, 318
liquid, 343
liter, 260
literacy, 2, 289
livelihood, 140, 184, 199, 218, 230,
291
livestock, 57, 60, 134, 140, 176, 222,
273, 275, 278, 305–308,
314, 319
living, 4, 15, 20, 23, 32, 33, 35, 37,
70, 75, 85, 101, 110, 117,
134, 139, 149, 161, 231,
248, 316, 343, 368, 375,
395
location, 329, 330, 352, 354, 387
Loess Plateau, 7
logging, 57, 66, 111, 119, 147, 222
logistic, 33
London, 99
London, 99
longevity, 111
look, 179, 315, 385
loop, 258
loss, 6, 7, 9, 15, 17, 18, 26, 31, 48,
53, 56–61, 63–66, 71, 85,
104, 110, 119, 126, 133,
134, 139, 157, 158, 161,
163–165, 168, 178, 181,

188, 193, 198, 209, 216,
219, 220, 222–227, 230,
232, 238, 242, 247, 258,
264, 281, 301–304, 321,
328, 352, 358
lung, 71
luxury, 65

machinery, 71, 281, 288, 331
magnitude, 226
mainstream, 277, 290
maintenance, 206, 325, 354
makeup, 169
making, 1, 3–5, 8, 11, 25, 26, 30,
34–37, 45, 59, 71, 89–93,
100, 111, 115, 116, 118,
142, 145, 149, 151, 168,
171, 181, 194, 198, 219,
228, 236, 258, 289, 323,
324, 330–332, 350, 363,
370, 378
malaria, 83
malnutrition, 293, 297
mammal, 224
man, 119
management, 1, 3–5, 7, 8, 11, 16,
18–23, 28, 30–32, 36, 42,
47–49, 56, 58, 59, 61, 65,
73, 74, 84, 90, 92–98, 100,
109, 110, 113, 115, 124,
134–139, 148–150, 157,
158, 161, 163, 168, 170,
171, 174, 176–178, 182,
183, 185, 197, 199–202,
206, 209, 232, 238, 242,
247–251, 255, 258, 285,
290–293, 305–307, 316,
317, 343, 346, 354,
387–389, 394, 395

Index 421

Mangrove, 108
mangrove, 108, 109, 206
manner, 19, 198, 265
Manuel C. Molles Jr., 15
manufacturing, 331, 332, 335, 394
manure, 314, 363
map, 46, 47
mapping, 47, 115, 224
marine, 57, 104, 116, 164, 190, 193,
    197–200, 206–209, 219,
    225, 226, 231, 232, 291,
    293, 326, 328, 352–354
mark, 173, 174
market, 219, 287, 289, 290, 301
marketing, 288, 289
marking, 24, 140
Mary Ann Cunningham, 14, 17
mass, 117, 165, 285
material, 168, 260, 359, 368
matter, 4, 32, 71, 72, 80, 236, 237,
    251, 266, 308, 314, 324,
    343, 359, 363
mean, 161
means, 48, 91, 151, 165, 281, 333,
    376, 379, 389
measure, 18, 43, 45, 90, 260
measurement, 43
meat, 264
mechanism, 111
medicine, 65, 101, 161, 184, 219
melting, 53, 320
membership, 265
membrane, 359
meter, 39, 329, 332
metering, 317
methane, 314, 315, 319, 320, 359
methanogenesis, 359
method, 43, 113, 117, 257
microbiology, 118

microfinance, 289
microgrid, 370
microscopy, 5
Midwest, 176
migration, 186, 217, 229, 322, 328,
    353, 358
milestone, 21, 24, 29, 140
militia, 184
million, 134, 165, 355, 362
mind, 152, 378
mindset, 92, 378
mining, 71, 206, 222, 326
misinformation, 324
misuse, 93
mitigation, 6, 9, 15, 49, 50, 53, 55,
    56, 84, 104, 125, 152, 153,
    164, 218, 223, 226–228,
    237, 328, 353
mix, 109, 315, 335, 359, 363–365,
    384
mobility, 98
mode, 328
model, 33, 47, 48, 265, 346
modeling, 15, 32, 228, 370
moisture, 98, 151, 359
momentum, 3, 21, 278, 316, 363,
    394
monitoring, 42, 45–47, 56, 73, 74,
    98, 113–115, 122, 135,
    137, 145–149, 164, 167,
    181, 184, 190, 198–200,
    202, 221, 224, 229, 285,
    354, 394, 395
monoculture, 57
monoxide, 71
monsoon, 133
moon, 338
mortality, 219
mosaic, 137

motion, 55, 323
motor, 376
mountain, 138, 139, 225
movement, 2, 3, 20–24, 26, 28, 136, 173, 174, 185, 222, 224, 237, 278–281, 314, 316, 333, 376
moving, 97, 178, 323, 336, 362
Muir, 21
mulching, 285
multitude, 126, 161

nacelle, 336
Namibia, 220
nanotechnology, 397
nation, 141
naturalist, 20, 21
nature, 1, 4, 10, 11, 14, 20–23, 43, 60, 104, 105, 129, 148, 149, 161, 163, 173, 181, 185, 221, 234, 264, 275, 317, 324, 331, 354, 360, 369, 379, 386, 389, 394, 397
need, 3, 5, 20–25, 28, 29, 32, 35, 36, 50, 59, 74, 91–93, 95, 99, 100, 122, 125, 147, 149–153, 168, 186, 202, 203, 212, 215, 225, 238, 247, 250, 254, 257, 258, 262, 263, 265, 271, 274, 286, 287, 290, 305, 313, 317, 318, 322, 330, 331, 333, 341, 343, 345, 346, 349, 353, 357, 358, 362, 364, 368, 370, 371, 374, 377, 385, 397
negotiation, 290
neighborhood, 26

Neil Stutzman, 15
nest, 190
net, 61, 317, 343
Netherlands, 135, 138, 315, 355
network, 219, 356, 370, 389, 394
New York, 26
New York City, 104, 370
New Zealand, 209–211
New Zealand's, 211
Niagara Falls, 26
niche, 33
Nicobar Islands, 231
night, 86
nitrogen, 71, 80, 260, 320
noise, 71, 74, 77, 85, 86, 353
non, 2, 5, 7, 9, 18, 36, 64, 73, 97, 117, 120, 235
Norway, 378
Norway, 378
number, 139, 165, 168, 173, 198
nursery, 109
nutrient, 7, 32, 57, 98, 112, 117, 222, 236, 242, 248, 251, 257–260, 264, 308

objective, 42, 101, 119
ocean, 15, 56, 117, 152, 164, 198, 206, 225, 227, 314, 337, 338
off, 55, 99, 146, 176, 197, 332, 354
offer, 117, 237, 241, 245, 247, 253, 255, 258, 264, 265, 267, 268, 271, 275, 276, 278, 288, 289, 313, 318, 326, 328, 339, 343, 345, 353, 355, 369, 371, 377
oil, 66, 193, 212, 319, 322, 326, 327, 363

Index

423

one, 4, 6, 22, 30, 31, 34, 50, 55, 57,
83, 116, 131, 146, 149,
151, 165, 181, 199, 202,
219, 222, 223, 232, 278,
289, 304, 314, 315, 322,
328, 335, 339, 350,
354–356, 358, 364, 376,
379, 386, 389
onset, 84
operating, 331, 376
operation, 257, 330, 334, 336, 352,
370, 387, 389, 395
operator, 389
opinion, 28
opportunity, 105, 129, 136, 190,
264, 265, 330
opposition, 143, 263
optimization, 394
option, 330, 353, 376
orangutan, 194–197
order, 93, 95, 151, 165, 291, 305,
377
organism, 39
organization, 112, 265
other, 6, 9, 11, 19, 22, 26, 28, 30, 31,
33, 34, 61, 80, 83, 84, 88,
91, 102, 103, 108, 110,
111, 114, 116, 117, 120,
135, 139, 141, 142, 150,
161, 165, 176–178, 181,
183, 186, 187, 190, 194,
199–201, 205, 214, 219,
222, 225, 258, 263–265,
278, 281, 284, 288, 290,
291, 301, 305, 314, 316,
319, 326, 330, 336, 339,
346, 348, 354, 355, 359,
363–365, 378, 392
out, 3, 28, 137, 179, 198, 225, 288,

330, 348, 356, 377
output, 287, 329, 333, 394
outrage, 28
outreach, 19, 201
overexploitation, 7, 56, 61, 93, 157,
164, 165, 168
overfishing, 55, 57, 116, 117, 164,
197, 199, 291, 293
overgrazing, 133, 201
overgrowth, 117
overhunting, 178
overvoltage, 330
ownership, 115, 149, 168, 198, 325
oxidation, 320
oxide, 320
oxygen, 45, 314, 319, 359
oyster, 115
ozone, 3, 28

pace, 6, 56, 221, 321
Pacific, 60
packaging, 301, 395
packing, 265
palm, 193
panda, 16, 181–183
panel, 23, 332
paradigm, 371
parallel, 329
park, 21, 22, 139, 140, 185, 186,
197, 200, 201, 231, 232
parking, 377, 378
part, 14, 30, 31, 111, 135, 276, 325,
331, 335, 355, 359, 394
participation, 89, 92, 115, 134, 135,
184, 289, 296, 325
particulate, 71, 72, 80
partner, 185, 281
partnership, 139
Pará, 358

passage, 3, 24, 358
past, 29, 134, 320
path, 97, 330, 348
pathway, 234
peace, 378
peak, 264, 317, 379
Peak District National Park, 21
peer, 370
pentane, 340
people, 24, 74, 89, 134, 137, 190, 293, 315, 316, 362, 363, 373
percentage, 376
perception, 24, 142, 324
performance, 90, 98, 395
period, 173, 225
permaculture, 92, 239–242
permitting, 384
perovskite, 331, 332
perspective, 1, 14, 35, 37, 328
pest, 242, 276, 285, 286, 306, 307
pesticide, 3, 285, 286, 305
pet, 193, 194
petroleum, 71
pharmaceutical, 60
phase, 28
phasing, 3, 348
phenomenon, 48, 117, 226, 319
pheromone, 285
philosophy, 23, 35
photography, 147
photosynthesis, 32, 119, 319, 343
photovoltaic, 329, 394
physics, 1, 9–11, 15
piece, 23
pioneer, 99, 102, 110–112, 137, 336, 354, 356
place, 56, 119, 137, 153, 198, 199, 227, 229, 359

plan, 272, 346
planet, 1, 9, 10, 18, 20, 26, 29, 50–52, 56, 58, 60, 63, 66, 75, 85, 116, 151, 161, 163, 164, 168, 193, 225, 297, 319, 322
planning, 16, 47, 49, 84, 100, 102, 107, 114, 115, 136, 138, 149, 163, 181, 200, 218, 223, 224, 228, 265, 305, 354, 355, 384, 395
plant, 60, 71, 108, 111, 112, 114, 115, 119, 122, 135–137, 147, 150–152, 161, 176, 177, 225, 247, 248, 257, 258, 260, 261, 263, 276, 308, 314, 339, 340, 356, 389
planting, 56, 109, 118, 124, 139, 176
plastic, 73
plateau, 133
platform, 364
play, 2, 5, 7, 10, 11, 16, 18–20, 43, 56–59, 61, 66, 73, 81, 85, 97, 100, 111, 116, 118, 119, 137, 147, 148, 161, 163, 176, 181, 184, 187, 190, 193, 202, 206, 208, 224, 232, 248, 258, 272, 275, 276, 287, 291, 293, 297, 298, 300, 313, 317, 324, 331, 336, 363, 365, 371, 379, 381, 384, 386, 387, 395
playing, 167, 324
poaching, 65, 139, 181, 184–187, 194, 216, 219, 220
point, 22, 24, 28, 301, 340
poisoning, 188

Index 425

policy, 10, 11, 19, 20, 22, 24, 28, 36, 45, 59, 61, 90, 100, 122, 152, 163, 186, 215, 297, 300, 302, 307, 311, 317, 324, 348, 351, 352, 368, 384–386, 393, 397

pollen, 256

pollination, 7, 57, 60, 157, 225, 306, 321

pollinator, 225, 256

pollutant, 45, 73

pollution, 1, 2, 6–11, 15–19, 22, 24, 28, 34, 43, 47, 55–58, 61, 65, 70–88, 97–99, 104, 106, 113, 116, 118, 126, 129, 133, 146, 153, 157, 164, 165, 168, 215, 227, 236, 266, 313, 326–328, 353, 376, 378

pond, 101

pool, 168, 288

popularity, 269, 279

population, 3, 5, 7, 22, 33, 39, 40, 65, 74, 90, 104, 112, 126, 138, 139, 165, 168, 169, 171–174, 181, 183, 199, 201, 220, 222, 224, 238, 258, 291, 305, 331, 362, 378

portfolio, 317, 323, 331

portion, 39, 99, 140, 157, 287, 315, 319, 335, 353, 363, 378, 389, 395

potassium, 340

potential, 7, 10, 16, 18, 19, 22, 24, 25, 32, 35, 47, 81, 92, 95–97, 99, 105, 113–116, 118, 142, 152, 161, 164, 168, 179, 199, 201, 202,

229, 236, 238, 239, 246, 255, 257, 259, 262–264, 268, 272, 275, 284, 286, 319, 323, 328, 332, 337, 343, 345, 349, 351, 353–356, 358, 359, 364, 371, 374, 375, 377, 382, 384, 389, 393, 394, 396, 397

poverty, 92, 139, 185, 186, 219–221, 238, 296, 307

power, 22, 24, 55, 88, 98–100, 131, 135, 138, 141, 212, 260, 287–289, 314–317, 320, 321, 323–325, 328, 330–340, 343, 348–352, 362, 364, 365, 367, 369, 370, 375, 376, 386, 389, 394

ppm, 260

practicality, 377

practice, 20, 107, 194, 269

prairie, 176

precipitation, 54, 64, 83, 119, 138, 151, 225, 305

precision, 42, 236, 251, 255, 306

precursor, 263

predation, 33, 201, 216

predator, 33, 200, 211

pregnancy, 72

preparation, 43, 45, 114, 359

presence, 5, 64, 75, 200, 201

present, 8, 16, 19, 23, 25, 34, 43, 88, 91, 106, 111, 221, 235, 297, 320, 340, 371

presentation, 14, 371

preservation, 2, 3, 16, 20, 21, 23, 28, 35, 36, 50, 75, 93, 140, 141, 152, 181, 183, 200,

202, 206, 211, 235, 265, 276, 278, 287, 318
presidency, 21
pressure, 72, 201, 202, 333, 359
preventing, 73, 163, 165, 201, 222, 259, 319
prevention, 7, 8, 10, 16, 72–74, 85–88
prey, 33, 200, 201, 225
price, 289, 363, 384
principle, 4, 25, 26, 32, 33, 35, 91, 92, 251, 336, 362, 376, 379
priority, 134, 163
problem, 10, 14, 65, 74, 118, 119, 212, 221
process, 101, 102, 106, 110–113, 115, 116, 119, 123, 126, 136, 137, 145, 149, 168, 178, 201, 222, 314, 324, 325, 329, 332, 340, 359, 386
processing, 288–290, 301, 327, 359
produce, 99, 235, 237, 255, 257, 263–265, 270, 278, 288, 316, 323, 324, 330, 343, 353, 376
product, 290, 395
production, 10, 28, 57, 58, 63, 91, 99, 134, 161, 212, 224, 236, 256–258, 264, 269, 276, 278, 281, 282, 285, 287, 289, 291, 301, 308, 313, 315, 319, 322, 325–328, 331, 332, 339, 341–343, 356, 359–361, 379, 386–388, 395
productivity, 53, 61, 86, 92, 133, 134, 161, 235, 236, 238, 243, 248, 250, 251, 255,

257, 258, 260, 263, 270, 286, 289, 291, 292, 297, 305–307, 321
profile, 23, 315
profiling, 173
profit, 2, 7, 18, 117
program, 134, 135, 139, 140, 176, 189, 190, 199, 281, 335
progress, 2, 50, 74, 90, 99, 115, 136, 145, 148, 160, 186, 192, 195, 211, 228, 299, 303, 315, 331, 336, 346, 349, 351, 354, 363, 365, 397
project, 14, 17, 18, 101, 103, 108, 109, 114, 130, 131, 135–137, 145, 148, 201, 202, 229, 328, 332, 355, 358, 370, 384
prominence, 396
promise, 9, 100, 105, 171, 221, 250, 318, 331, 336, 396
promotion, 8, 21, 199, 235, 395
proof, 35
propagation, 117
property, 150, 216, 263
proposal, 17, 272
propulsion, 376
protecting, 3, 7, 21, 22, 28, 61, 66, 70, 81, 88, 109, 140, 159, 161, 165, 181, 193, 205, 206, 208, 209, 211, 224, 232, 248, 251, 293
protection, 2, 16, 21, 23, 25, 28, 36, 58, 61, 90, 91, 93, 108, 115–117, 184, 186, 190, 202, 206, 224, 228, 296, 358, 362, 396
protest, 24
province, 135

Index

provision, 58, 104, 106, 110, 135

proximity, 328, 352

prudence, 35

public, 2, 8, 19, 21–24, 26, 28, 50, 59, 73, 74, 81, 84, 85, 118, 156, 167, 183, 190, 199, 201, 224, 259, 265, 324, 325, 364, 377, 397

publication, 24, 27, 29

pump, 332, 340

purchase, 265, 378

purification, 60, 63, 104, 161, 321, 396

purpose, 343, 346

pursuit, 138, 141

push, 226, 349

PV, 332

pyramid, 33

quality, 18, 41–43, 55, 57, 71, 74, 79, 81, 82, 88, 91, 99, 104, 105, 125, 131, 147, 153, 174, 176, 178, 236, 242, 248, 289, 290, 301, 316, 323, 325, 359, 363, 365, 376, 396

quantity, 301

quest, 389

question, 36

R.C. Dugan, 371

Rachel Carson - A, 23

Rachel Carson's, 3, 24, 26, 27

radiation, 46, 47, 319, 349

radio, 201

railway, 104

rain, 88, 242, 337

rainfall, 48, 57, 83, 133, 151, 305, 321, 324, 356

rainforest, 147, 193, 223, 358

rainwater, 242, 306, 307

ranching, 224

range, 1, 5, 9–11, 17, 32, 51, 60, 63, 67, 74, 82, 83, 116, 126, 135–137, 139, 157, 161, 163, 181, 187, 197, 206, 209, 219, 226, 245, 252, 261, 264, 276, 277, 331, 344, 348, 370, 377, 378

rate, 33, 49, 53, 134, 168, 181

re, 124

reach, 265, 288, 355

reactor, 329

reading, 259

realm, 36, 292

rearing, 117

recapture, 173, 174

reclamation, 135, 327

recognition, 3, 20–22, 25, 26, 28, 29, 148, 150, 237, 348

record, 46

recovery, 7, 28, 56, 58, 99, 101, 102, 105, 106, 108, 112–114, 116, 117, 131, 146, 147, 151, 163, 164, 170, 178, 181, 183, 188, 190, 199, 201, 206, 395

recycling, 9, 73, 84, 100, 235, 331, 377, 395

reduction, 49, 61, 84, 85, 88, 139, 186, 190, 220, 221, 354, 369

reed, 136

reef, 55, 56, 104, 116–118, 146, 153, 164

reestablishment, 102, 111, 113

refining, 137, 326

reforestation, 46, 58, 104, 119, 121, 122, 134
refuge, 137, 206
refurbishment, 395
regard, 363
regeneration, 113, 114, 150, 184, 193, 237
region, 22, 117, 124, 133, 134, 147, 176, 187, 206, 332, 353, 355
regrowth, 114
regulation, 7, 58, 60, 63, 80, 81, 157, 161, 257, 371, 396
rehabilitation, 139
reintroduction, 58, 139, 140, 178–181, 190, 199–202
relatedness, 173
relation, 4
relationship, 26, 264
release, 8, 24, 51, 71, 73, 85, 111, 201, 266, 319, 320, 324, 359, 379
relevance, 15, 35
reliability, 41, 98, 351, 369
reliance, 9, 20, 98, 99, 114, 220, 258, 286, 307, 308, 315, 321, 346, 349, 363, 364, 370, 389, 395
relocation, 227, 358
remain, 2, 22, 118, 140, 183, 192, 195, 199, 299, 300, 303, 336, 348, 351, 365, 368
remanufacturing, 396
remediation, 81
reminder, 22
remote, 3, 5, 32, 46, 48, 115, 147, 224, 316, 332, 354, 369, 371–375
removal, 56, 105, 119, 131

renewable, 6, 8, 9, 16, 25, 34, 37, 49, 50, 52, 53, 55, 73, 85, 89, 91, 93, 97–100, 212–215, 227, 235, 258, 313–318, 321–326, 328, 330, 332, 334–340, 342, 343, 345, 346, 348, 349, 352, 354–356, 358–360, 362–365, 368–376, 378, 379, 381, 382, 384–390, 392–395, 397
repair, 106, 136, 395
replacement, 102
report, 22, 81
reporting, 187
reproduction, 60, 181, 222
research, 2, 3, 10, 11, 14–19, 21, 36, 39, 40, 42, 50, 56, 89, 100, 117, 118, 136, 137, 163–165, 167, 190, 198–201, 205, 215, 221, 238, 239, 250, 278, 286, 288, 297, 300, 307, 308, 310, 311, 318, 324, 331, 351, 352, 388, 389, 397
resentment, 217
reserve, 22, 135–138, 173, 174
reservoir, 337, 358
resident, 137
resilience, 4, 15, 31, 34, 49, 54, 56, 57, 59–62, 84, 93, 101, 105, 109, 110, 113, 115, 116, 118, 131, 138, 139, 141, 146, 150, 152, 161, 164, 227, 228, 235, 236, 238, 247, 251, 263, 268, 276, 278, 290, 305–308, 311, 364, 369, 371, 396
resistance, 143, 260

Index 429

resolution, 47

resource, 2, 3, 8, 9, 11, 16, 18–23, 26, 31, 59, 61, 89–100, 112, 118, 137, 157, 163, 216, 242, 247, 251, 255, 257, 258, 287, 290, 330, 340, 343, 363, 364, 395

respect, 149, 151, 187

response, 28, 33, 98, 104, 138, 152, 178, 184, 199, 201, 217, 281, 317, 394

responsibility, 2, 3, 19, 22, 35–37, 50, 59, 62, 164, 168, 198, 251, 293

rest, 319

restoration, 7, 58, 61, 101–110, 112–118, 122–126, 128–153, 155, 156, 163, 164, 174–178, 199, 202, 211, 224, 306, 396

result, 39, 63, 66, 81, 83, 116, 133, 153, 176, 199, 216, 217, 222, 225, 232, 258, 305, 326, 327, 333, 350, 352, 353, 395

resurgence, 198, 237

retaliation, 217

retention, 111, 124, 266

return, 106, 142, 177

reuse, 9, 395

revenue, 139, 140, 184, 384

reverence, 20

revolution, 214, 362–365

reward, 264

rhino, 219

rhinoceros, 220

Rice, 285

rice, 263, 285

richness, 161, 197

right, 272, 352, 364

ripeness, 264

rise, 6, 15, 29, 47, 49, 54, 83, 109, 151, 225, 304, 320, 338

risk, 34, 35, 46, 47, 61, 66, 71, 72, 157, 165, 178, 229, 258, 264, 277, 305, 306, 320, 321

river, 42, 105, 131–133, 337, 356

road, 378

robot, 256

rock, 110

role, 2, 5, 7–11, 14–22, 24, 29, 33, 35, 37, 43, 47, 56–61, 66, 73, 74, 81, 85, 92, 97, 100, 102, 111, 116, 118, 120, 125, 137, 147, 161, 163, 167, 170, 171, 174, 176, 181, 184, 187, 190, 193, 200, 202, 206, 208, 220, 222, 224, 229, 232, 247, 248, 258, 265, 272, 275, 276, 284, 287, 289, 291, 293, 297, 298, 300, 313, 317, 324, 331, 336, 355, 358, 363–365, 371, 378, 379, 381, 384–387, 390, 395

rooftop, 237, 269–272, 315, 332, 370

root, 55, 153, 220, 221, 238, 242, 257, 296

rotation, 242, 251, 276, 285, 308

rotor, 323

round, 236, 257, 265

run, 330

runoff, 7, 55, 65, 71, 81, 85, 117, 164, 248

Rwanda, 138–141

safety, 24, 61, 71, 216, 217, 259, 330
salmon, 105, 131, 176
salt, 109
saltwater, 135, 320, 352
sample, 5, 43, 45
sampling, 5, 39, 40, 42, 43
sand, 110
sanitation, 285
Santa Cruz Island, 199
satellite, 5, 47, 98, 187, 224
savannahs, 138, 139
saving, 197, 276, 394, 395
scalability, 353, 397
scale, 32, 57, 101, 118, 134, 135,
    142, 153, 155, 156, 206,
    236, 237, 258, 281, 290,
    296, 308, 314–317, 322,
    328, 330, 332, 353–356,
    358, 377, 378, 389, 394
scaling, 32, 100, 155, 228, 251
scarcity, 48, 53, 216, 305, 306
scat, 173, 174
science, 1–11, 14–17, 20, 27–30,
    39, 40, 42, 43, 46, 48, 70,
    75, 91, 97, 100, 106, 118,
    133, 157, 161, 164, 167,
    168, 202, 205, 229
scope, 1, 9
Scotland, 21
scrutiny, 24
sea, 6, 16, 47, 48, 53, 54, 56, 83, 109,
    117, 164, 190, 192, 193,
    198, 227, 320, 352, 353
seabed, 352, 353
seafood, 291–293
seagrass, 206
search, 216
season, 133, 264, 329

section, 15, 17, 18, 20, 23, 27, 30,
    32, 35, 39, 43, 50, 53, 60,
    63, 66, 70, 75, 82, 97, 113,
    118, 126, 138, 142, 145,
    161, 174, 190, 193, 202,
    209, 216, 229, 239, 243,
    247, 251, 257, 260, 269,
    275, 278, 287, 290, 293,
    297, 318, 322, 323, 325,
    329, 332, 336, 339, 343,
    349, 352, 356, 359, 362,
    365, 369, 372, 375, 379,
    384, 386, 389, 393, 397
sector, 268, 281, 288, 300, 304, 307,
    321, 325, 331, 351, 356,
    360, 364, 389, 390, 393,
    394
security, 48, 57, 60, 89, 93, 99, 161,
    164, 185, 235, 237, 238,
    247, 251, 259, 272, 276,
    281, 284, 287, 293–297,
    304, 305, 311, 352, 362,
    363, 365, 384
sediment, 164
sedimentation, 133
seed, 57, 111, 114, 184, 193, 276
selecting, 39, 40, 102, 112, 122, 152,
    178
selection, 39, 103, 118, 145, 149,
    152, 168, 201, 305, 368
self, 20, 113, 136, 239, 240, 243,
    281, 330
selling, 219
sense, 20, 36, 59, 115, 149, 168, 190,
    198, 200, 264, 265, 325
sensing, 3, 5, 32, 46, 48, 115, 147,
    224
sensitivity, 150, 151
sensor, 45

Index

431

separation, 359
sequence, 102, 110
sequencing, 5
sequester, 308
sequestering, 109, 116, 266, 306
sequestration, 57, 60, 104, 108, 161,
206, 237, 238, 248, 306,
321
series, 112, 329
set, 3, 22, 49, 55, 73, 74, 80, 174,
190, 264, 266, 355,
362–364, 384
setting, 384, 392
settlement, 140
setup, 258
severity, 52, 71, 226
sewage, 71, 85, 359
Shaanxi, 181
shade, 111, 176
share, 19, 32, 181, 315, 384
sharing, 73, 92, 139, 140, 150, 185,
186, 221, 264, 287, 289,
290, 307, 364
shelter, 137, 216
shift, 83, 225, 237, 371
shopping, 377
Sichuan, 181
side, 131, 387–389
significance, 61, 119, 161, 164, 187,
200
signing, 29
silicon, 329, 332
simplicity, 20
Singapore, 99, 100
sink, 109
site, 114, 123, 136, 145, 197, 251,
330, 339
situ, 45
situation, 187

size, 33, 173, 206
skeleton, 116
skepticism, 259
sky, 86
sleep, 71, 86
sludge, 359
smallholder, 287–290
smog, 88
smuggling, 185, 219
sniffer, 185
snow, 337
society, 1, 2, 18, 59, 60, 62, 92, 101,
119, 241, 244, 253, 290,
325, 364
socio, 47, 134, 155, 184, 187, 220,
371
sociology, 11, 32
soil, 1, 5, 7, 9, 16, 18, 26, 31, 32, 43,
57, 60, 66, 71, 74, 77, 81,
83, 85, 86, 92, 98, 102,
104, 110–112, 124,
133–135, 151, 157,
236–238, 242, 243,
247–251, 257, 265, 266,
268, 275, 276, 305, 306,
308, 311, 326, 343
solar, 55, 89, 98–100, 315–319,
328–332, 349–352, 364,
365, 369, 370, 376, 384,
386, 389, 394, 395
solidarity, 289
solution, 255, 257, 259, 260, 286,
325, 372, 378
solving, 10, 11, 14, 28
source, 32, 80, 88, 161, 200, 220,
263, 291, 314, 315, 319,
322, 332, 334, 336–338,
342, 349, 352, 353, 356,
359, 363, 364

South Korea, 339
Southeast Asia, 104
sovereignty, 264, 308
soybean, 224
space, 104, 113, 216, 257, 258, 270, 272, 319, 340, 353
span, 32, 165
spawning, 117
specialty, 289
species, 4, 6, 7, 15, 16, 18, 21, 33, 34, 36, 47, 56–61, 63–66, 101, 102, 105, 108–116, 118, 119, 122, 124, 126, 129, 134–139, 141, 147, 150–152, 157, 158, 161–171, 174, 176–178, 180–183, 186, 188, 190, 193, 197–202, 206, 208, 209, 211, 217, 219, 222, 224–231, 321, 358
spectroscopy, 5
speed, 112, 324, 333, 352, 386
spillover, 206
spoilage, 301
sprawl, 3
spread, 33, 57, 117, 150, 176
stability, 15, 60, 61, 63, 111, 161, 164, 184, 206, 238, 264, 294, 352, 389
staff, 290
stage, 3, 22, 359
stakeholder, 90, 92, 114, 228
standardization, 42
state, 34, 106, 131, 152, 165, 349, 358
station, 199
status, 183, 184
steam, 314, 323, 324
steel, 117

stem, 216
step, 150, 236, 257, 325
steward, 3
stewardship, 2, 26, 88, 115, 141, 149, 200, 237, 297, 300
stock, 260
storage, 48, 89, 100, 238, 288, 301, 316, 317, 324, 331, 348, 351, 354, 364, 370, 379–382, 384, 387–389, 394
store, 242, 314, 324, 369, 379
storing, 55, 119, 317
storm, 109, 396
story, 117, 140, 199, 200, 202, 220, 362, 378
storytelling, 187
strain, 83
strategy, 115, 229, 355, 386
stream, 66, 129–133, 176, 384
strengthening, 85, 184, 186, 187, 290
stress, 55, 71, 72, 83, 305
stressor, 226
stripe, 173
structure, 4, 32–34, 101, 102, 106, 110, 112–114, 116, 117, 147, 151, 168, 169, 194, 237, 248, 308, 365
struggle, 26, 226
study, 1, 3, 6, 7, 9–11, 14–18, 32, 39, 42, 46, 55, 60, 80, 81, 88, 165, 168, 169, 176, 177, 198, 201, 229, 328, 332
subject, 317, 358
subsistence, 197
substation, 352
substrate, 110, 117

Index 433

success, 7, 22, 46, 58, 66, 68–70, 97, 99, 102, 104, 105, 114–118, 121, 125, 126, 134–141, 145, 148, 150, 151, 159, 163, 176, 178, 181, 187, 190, 192, 195, 199, 200, 202, 204, 209, 213, 214, 220, 221, 225, 229, 237, 238, 259, 265, 270, 275, 277, 280, 281, 284, 291, 292, 297, 303, 309, 313, 316, 335, 336, 348, 352, 355, 359, 360, 362–365, 367, 372, 373, 375, 378, 389

succession, 102, 106, 110–113, 136, 137

sufficiency, 240, 281, 330

suitability, 201, 359

sulfur, 71, 80

Sumatra, 16, 193, 196

summary, 9, 11, 116, 171

summer, 340

sun, 48, 325, 329, 330, 333, 338, 349, 375

sunlight, 32, 97, 111, 305, 313, 321, 329–332, 343, 349, 362, 386

supply, 124, 238, 301, 316, 317, 324, 364, 369, 370, 379, 387, 389, 394

support, 22, 24, 29, 37, 56, 63, 64, 84, 99, 101, 104, 109, 116, 118, 122, 126, 135–137, 139, 143, 151–153, 163, 174, 176, 178, 184, 185, 190, 202, 222, 229, 238, 239, 248, 250, 251, 264, 265, 268, 272, 275, 278,

288, 290, 291, 293, 297, 298, 300, 307, 308, 311, 325, 348, 351, 352, 368, 370, 384–386, 397

suppress, 285

suppression, 248

surface, 46–48, 117, 319, 320, 329, 333, 340

surge, 330

surrounding, 101, 124, 206, 222, 285, 366, 377

surveillance, 184, 185

survey, 134

survival, 56, 60, 64, 65, 70, 93, 111, 118, 134, 139, 151, 164, 165, 168, 181, 183, 185, 187, 188, 193, 196, 197, 199, 206, 211, 219, 222, 225, 227, 229

susceptibility, 71, 168

sustain, 115, 143, 202, 247

sustainability, 20, 22, 26, 34, 61, 75, 92, 100, 115, 125, 138, 141, 148, 149, 156, 161, 184, 193, 199, 202, 208, 237, 238, 240, 243, 251, 259, 265, 268, 272, 275, 278, 282, 284, 285, 292, 331, 336, 339, 343, 353, 355, 357–359, 362, 364, 365, 391

Sweden, 315, 355, 359–361

swing, 359

symbol, 184, 188

sync, 225

system, 15, 21, 30, 31, 33, 131, 248, 255, 264, 268, 275, 281, 290, 297, 301, 304, 325, 329–332, 339, 346, 348,

364, 370, 378, 382, 387, 389

tailpipe, 376
Takahe, 209
take, 5, 19, 50, 66, 74, 91, 102, 119, 137, 138, 186, 229, 315, 322, 323
tar, 327
target, 112, 117, 174, 219
task, 148, 221
tax, 89, 324, 364, 376, 378, 384
team, 14
technique, 5, 39, 40, 117, 236
technology, 3, 23, 29, 32, 45, 48, 98, 161, 184, 187, 221, 255, 260, 328, 329, 331, 335, 352, 368, 370, 371, 377, 389, 394
telemetry, 201
temperature, 15, 29, 45, 49, 54, 79, 83, 225, 257, 276, 305, 320
tenure, 277, 307
term, 8, 20, 22, 25, 34, 56, 58, 59, 61, 69–71, 74, 82, 89, 91–93, 105, 114–116, 118, 122, 125, 134, 135, 138, 139, 142, 143, 148, 149, 152, 161, 163, 164, 168, 185, 187, 199, 200, 202, 206, 208, 211, 221, 229, 237, 251, 264, 265, 275, 278, 290–292, 317, 324, 331, 364, 384
testament, 138
textbook, 17
textile, 281
Thames River, 131
the American West, 188

the Andaman Sea, 104
the Baltic Sea, 355
the Florida Keys, 104
The Galapagos Islands, 197
the Galapagos Islands, 197–200
the Galapagos Islands National Park, 58
the Hulun Lake, 104
The Loess Plateau, 124, 133
the Loess Plateau, 133–135
the Louisiana, 115
The Maasai Mara, 232
the Maasai Mara, 232
the Mesoamerican Reef, 117
the Mississippi River Delta, 115
the North Sea, 354, 355
the Pacific Northwest, 176
the Pacific Ocean, 197
the Sihwa Lake Tidal Power Station, 339
the Soviet Union, 281
the São Francisco River, 356
the Thames Estuary, 355
The United Kingdom, 355
the United Kingdom, 21, 315, 355
the United States, 3, 7, 21, 28, 80, 88, 104, 176, 200
the Virunga Massif, 139
the Xingu River, 358
the Xingu River's, 358
the Yangtze River, 339
Theodore Roosevelt, 2, 21–23
Theodore Roosevelt's, 23
theory, 4, 30–32, 35
thinking, 4, 31, 89
third, 349
Thomas E. Lovejoy, 229
Thoreau, 20, 23
thorium, 340

Index 435

threat, 29, 57, 64–66, 104, 116, 117,
140, 176, 184, 197–199,
219, 221, 320
thrive, 58, 64, 151, 152, 168, 178,
186, 218, 224, 305
tiger, 173, 174
tillage, 236, 251, 266
timber, 57
time, 15, 24, 31, 33–35, 39, 45, 47,
50, 71, 74, 83, 91, 98, 102,
110–113, 117, 165, 177,
187, 201, 264, 291,
329–331, 386, 394, 395
timeframe, 102, 113, 319
timing, 150, 225, 388
tissue, 115
today, 2, 10, 11, 15, 17, 20, 22, 35,
56, 222, 287, 293, 304
tolerance, 109, 152
toll, 378
tool, 106, 290
top, 26
topic, 23, 180
torque, 377
tourism, 90, 91, 116, 138–141, 184,
198–201, 232, 364
tower, 336
track, 90, 112, 136, 146, 147, 201
tracking, 184, 187, 221, 265
traction, 21
trade, 16, 36, 57, 61, 65, 70, 90, 92,
184–187, 193, 194, 216,
219–221, 289
trading, 281, 370
traffic, 72
trafficking, 184, 220
training, 200
trajectory, 102, 183
trampling, 137

tranquility, 60
transfer, 115, 256, 340, 352
transform, 126, 129, 255, 268, 272,
311
transformation, 136, 238, 284
transition, 37, 50, 100, 212, 213,
293, 313, 315, 317, 318,
321, 322, 324, 325, 335,
336, 339, 346–348, 355,
359, 362–365, 378,
384–386, 389, 390
translocation, 178–181
transmission, 83, 100, 278, 316, 324,
352, 370, 394
transparency, 19, 90, 264, 290
transplantation, 117
transport, 7, 57, 319
transportation, 9, 48, 71, 73, 84–86,
98, 100, 219, 237, 238,
264, 288, 301, 316, 319,
326, 328, 376, 378
trap, 48, 319
trapping, 285
travel, 90, 193
tree, 60, 111, 134, 139, 147
trekking, 139
triumph, 202
trust, 149, 264
trustworthiness, 19
turbine, 99, 334, 335, 340, 352
turbulence, 353
turn, 151, 164, 200, 220, 323, 340,
376
turning, 22, 24, 28
turtle, 192
type, 71, 101, 103, 340

UK, 99, 355
uncertainty, 25, 142

underground, 219, 314, 323
understanding, 1–4, 6–11, 14, 15,
17, 18, 20, 22, 25, 27,
29–34, 38, 40, 45, 60, 74,
88, 106, 110, 112, 141,
148, 149, 152, 153, 161,
163, 164, 167, 168, 171,
174, 181, 190, 198, 202,
205, 218, 224, 229, 234,
235, 275, 276, 384
undertaking, 35, 118, 346
ungulate, 201
uniqueness, 264
unit, 236
unrest, 138, 220
up, 24, 32, 74, 100, 112, 153, 155,
174, 187, 190, 228, 239,
251, 258, 329, 389, 394
upgrading, 324, 359
uptake, 257
uranium, 340
urbanization, 9, 57, 63, 176, 198,
216, 222, 278
USA, 104, 105
usage, 174, 258, 369, 394, 395
use, 3, 6, 8, 16, 18, 25, 31, 36, 47, 50,
52, 53, 66, 71, 73, 82,
90–93, 97, 98, 104, 112,
113, 116, 133, 150, 158,
174, 184, 190, 218, 220,
221, 223, 224, 235, 236,
239, 242, 243, 257, 258,
265, 266, 269, 276, 278,
281, 285, 286, 305–308,
317, 319, 320, 323, 326,
340, 343, 344, 346, 362,
363, 370, 376, 378, 389,
393, 395
usefulness, 35

utilization, 61, 100, 255, 257, 294,
315, 336, 359–362, 388

validation, 42
value, 4, 7, 10, 18, 21, 22, 35, 59, 60,
62, 131, 141, 163, 219,
264, 265, 276, 288–290
variability, 305, 308, 364
variable, 324, 386, 389
variation, 169, 171, 174
variety, 5, 10, 15, 33, 56, 59, 63, 102,
122, 126, 136, 152, 157,
161, 164, 165, 168, 176,
177, 206, 240, 263, 265,
276, 306
vegetable, 242
vegetation, 46, 48, 102, 112, 114,
124, 125, 137, 147, 150,
176, 197, 198, 200, 201,
222
vehicle, 359, 376–378, 389
ventilation, 395
venting, 319
verge, 209
viability, 24, 25, 58, 61, 224, 251,
290, 291, 297, 300
video, 190
visibility, 71
visit, 137, 265
visitor, 137
visualization, 48
vitamin, 263
voltage, 329
volume, 337
vulnerability, 47, 49, 54, 57, 58, 61,
63, 134, 161, 209, 287,
289, 321, 363, 364

Wales, 21

Index 437

warming, 36, 49, 119, 225, 227, 319, 322

warning, 6, 84

Washington State, 105

wastage, 251

waste, 7, 9, 11, 26, 28, 57, 65, 71, 73, 74, 83–85, 89, 91, 93, 97–100, 236, 238, 252, 301–304, 314–316, 319, 332, 343, 346, 359, 363, 377, 379, 395, 396

water, 1, 5, 7, 9, 10, 16–18, 22, 28, 31, 32, 36, 42, 43, 48, 49, 53, 55, 57, 58, 60, 63, 65, 71, 73, 74, 76, 79, 81–85, 92, 93, 97, 98, 101, 104, 105, 108, 111, 119, 122, 124–126, 129, 131, 133, 134, 136, 151, 161, 176, 202, 216, 220, 226, 235, 236, 238, 242, 248, 252, 257, 258, 260, 266, 276, 289, 304–308, 313, 314, 320–323, 325–327, 330, 332, 336, 337, 339, 340, 352, 358, 362, 368, 375, 395, 396

waterborne, 8, 71, 83

waterway, 131

wave, 364

wavelength, 46

way, 8, 11, 20–22, 27, 28, 39, 75, 85, 91, 106, 110, 137, 141, 144, 222, 227, 235, 255, 257, 270, 278, 322, 343, 352, 371, 378, 382, 393

wealth, 138, 148

weather, 6, 48, 49, 53, 54, 57, 64, 83, 84, 109, 119, 151, 198,

258, 264, 276, 304, 305, 307, 317, 320, 329, 331, 354, 364, 369, 379, 386, 387

web, 4, 9, 61, 83, 200, 222, 225, 291

webs, 225

website, 17

weed, 285

weeding, 285

welfare, 36, 138, 219

well, 1, 5, 7, 9, 10, 25, 29, 31, 32, 35, 37, 49, 50, 56, 58–60, 62, 66, 70–72, 77, 82, 83, 85, 86, 91, 93, 97, 108, 118, 129, 142, 161, 164, 165, 168, 201, 209, 216, 218, 235, 243, 251, 257, 259, 265, 269, 272, 281, 290, 291, 305, 364, 370, 396

wetland, 101, 104, 123–125, 136, 138, 139

whole, 60, 244, 253

wild, 158, 188

wilderness, 21

wildfire, 150

wildlife, 2, 7, 21, 24, 27, 33, 61, 65, 70, 86, 108, 134, 138, 140, 141, 157, 158, 161, 171–174, 176, 183–186, 216–221, 224, 229, 232, 235, 353, 354

will, 15, 18, 20, 23, 27, 30, 32, 34, 35, 39, 43, 45, 48, 50, 59, 60, 63, 66, 70, 75, 82, 90, 97, 100, 108, 109, 112, 113, 118, 125, 126, 132, 142, 143, 145, 156, 157, 161, 172, 174, 181, 185, 190, 193, 200, 202, 209,

212, 216, 224, 228, 229,
235, 239, 243, 247, 251,
257, 258, 260, 269, 275,
278, 287, 290, 291, 293,
297, 313, 318, 322, 323,
325, 329–332, 336, 339,
343, 349, 351, 352, 355,
356, 359, 362, 364, 365,
369, 372, 378, 379, 381,
384, 386, 389, 393, 397

William P. Cunningham, 14, 17

wind, 55, 89, 97–100, 313–318,
321, 323–325, 328,
332–336, 352–355, 362,
364, 365, 369, 375, 376,
384, 386, 387, 394

wingspan, 188

winter, 73, 340

wiring, 330

wisdom, 205, 278

Wolf, 202

wolf, 200–202

wood, 363

woodland, 119

work, 1, 6, 8, 11, 18–20, 23, 26, 27,
30, 32, 37, 50, 55, 56, 59,
82, 85, 88, 93, 133, 152,
153, 161, 164, 168, 186,
199, 205, 218, 224, 234,
284, 293, 318, 328, 329,

333, 364

workforce, 325

working, 11, 14, 19, 37, 56, 70, 168,
192, 221, 224, 236, 266,
289, 316, 340, 359

world, 4, 7, 10, 11, 15, 17, 18,
20–24, 26, 27, 35, 37, 55,
56, 73–75, 96, 104, 106,
116, 117, 131, 137, 138,
141, 146, 151, 159, 161,
178, 181, 187, 193, 197,
198, 227, 229, 237, 239,
251, 254, 267, 270, 274,
275, 280, 289, 291, 293,
297, 300, 303, 304, 309,
315, 318, 327, 336, 339,
349, 355, 356, 364, 365,
370, 378, 382, 385, 393,
394, 397

worldview, 202

worth, 35, 219

writer, 20, 24

year, 20, 133, 140, 236, 257, 258,
265, 319, 349

Yellowstone, 200–202

yew, 60

yield, 251, 256

zone, 109, 133

Zuiderzee, 135

Milton Keynes UK
Ingram Content Group UK Ltd.
UKHW050649310824
447605UK00014B/241